Mathematics in Nature

Mathematics in Nature

Modeling Patterns in the Natural World

JOHN A. ADAM

PRINCETON UNIVERSITY PRESS

PRINCETON AND OXFORD

Published by Princeton University Press, 41 William Street, Princeton,
New Jersey 08540
In the United Kingdom: Princeton University Press, 3 Market Place,
Woodstock, Oxfordshire OX20 1SY

LIBRARY OF CONGRESS CATALOGING-IN-PUBLICATION DATA
Adam, John A.
    Mathematics in nature : modeling patterns in the natural world / by John A. Adam.
       p. cm.
    Includes bibliographical references and index.
    ISBN: 0-691-11429-3 (acid-free paper)
    1. Mathematical models. I. Title.
  QA401.A27 2003
  511′.8–dc22   2003055616

British Library Cataloging-in-Publication Data is available

This book has been composed in Sabon and Swiss 721

Printed on acid-free paper.∞

www.pupress.princeton.edu

Printed in the United States of America

10 9 8 7 6 5 4 3 2 1

Dedicated to

MY MOTHER JOAN

and to the memory of

MY FATHER ALBERT,

*in gratitude for the many sacrifices
they made for my education.*

CONTENTS

formation by activator and inhibitor mechanisms; seashells; mechanisms of activation and inhibition; reaction-diffusion equations—a linear model; butterfly wing spots: a simplistic but informative mathematical model. Other applications of diffusion models: the size of plankton blooms; earth(l)y applications of historical interest: the diurnal and annual temperature variations below the surface; the "age" of the earth. APPENDIX: *the analogy with the normal modes of rectangular and circular membranes.*

PREFACE

Mathematics in nature: this book grew out of a course of the same name, with the rather long subtitle "the beauty of nature as revealed by mathematics and the beauty of mathematics as revealed in nature." That course in turn grew out of an awakened awareness of both such facets of nature, even in a suburban environment, enhanced by occasional trips to national and state parks armed with binoculars and camera. I decided that it might be fun to develop a course that included *some* of the mathematics that lies behind *some* of the phenomena we encounter in the natural world around us. Such a course would have to be developed from "scratch," so to speak, because to my limited knowledge there was no such course in existence at that time, and apart from some fascinating books by Ian Stewart (one of which, *Nature's Numbers*, was a valuable supplement to the course) there was nothing that I felt was suitable as a rather free-ranging and perhaps "offbeat" textbook for such a course, at least in the way I wanted to teach it. Two books written by Pat Murphy for the Exploratorium (one written with Paul Doherty) contained some magnificent photographs of the kinds of things I wanted to describe in mathematical terms, so these books were recommended for the students also. The course was offered in two successive years, first at the junior level and finally at the senior/first-year graduate level, which was therefore presented at a more sophisticated mathematical level. I have appreciated the enthusiasm of the students in these classes—we all learned much and had a great deal of fun in the process, despite the class being held during "prime sleeping time" according to one of the participants! I was fortunate to have students in this class who ranged in ability from "bright" to "really, really bright," and I thank all of them for their active participation, questions, comments (and yes, Brandon, corrections). There was a strong temptation on my part (resisted) to fail them all so that I could see them in class the next time around. Bonnie Burke "caught the vision" for the course as I started to explain my early ideas about it; thank you for the continued encouragement, Bonnie.

I wanted to limit the topics covered to those objects that could be seen with the naked eye by anyone who takes his or her eyes outside. There are many books written on the mathematical principles behind phenomena that take place at the microscopic and submicroscopic levels, and also from planetary to galactic scales. But leaves, trees, spider webs, bubbles, waves, clouds, rainbows, . . . these are elements of the stuff we can easily see. The length scales extend roughly from 0.1 mm (the thickness of a human hair or the size of some ice crystals and diatoms) to almost 1000 km (large storm systems), a factor of about 10^{10}; the timescales of the phenomena we seek to describe range from a fraction of a second (the period of some ripples on a puddle) through the order of a day (tidal periods) to the time a tree takes to mature (perhaps thirty years in some cases), corresponding to a factor of about 10^8 or 10^9. Of course, in all likelihood we would not be around to see a sequoia tree reach maturity (!), so I have drawn the line somewhat arbitrarily at thirty years, but not in order to suggest that we stand around and "watch trees grow."

Many patterns are readily identified in nature. A visit to the zoo reminds us, no doubt unnecessarily, that tigers and zebras have striped patterns; leopards and hyenas have spotted patterns; giraffes are very blotchy (as well as very tall), while butterflies and moths may possess them all: spots, blotches, bands, and stripes. Everyone notices the wave patterns that move across oceans, lakes, ponds, and puddles, but fewer perhaps realize that waves move slowly across deserts in the form of sand dunes. In the sky brightly colored circular arcs—rainbows—beautify the sky after rain showers; from an airplane or a high peak there may sometimes be seen "glories"—small colored concentric rings surrounding the shadow of the airplane or observer—a phenomenon often called "the Specter of the Brocken" because it is frequently observed by climbers on the Brocken peak in Germany. On occasion circular and noncircular halos around and about the sun can be produced by ice crystals. If still we look up, we may sometimes observe parallel bands of cloud spreading across the sky; these may be billows or lee waves depending on the mechanism producing them—the latter are sometimes called mountain waves for this very reason. We may also see hexagon-like clouds hanging suspended from a blue ceiling: this is a manifestation of cellular convection. Patterns of light scattering are exhibited in the pre-dawn and twilight skies, at sunrise and sunset; blue sky and iridescent clouds. . . .

If we look *down* and around in a well-tended garden, we may become aware upon further investigation that arrangements of leaves, petals, seeds, and florets are intimately associated with spiral patterns, the related sequence of numbers 1, 1, 2, 3, 5, 8, 13, 21, 34, 55, 89, 144, . . . , and also with an *angle* of about 137.5° (or its complement, 222.5°). Moving on to bigger plants, we may note that the heights of trees are closely related to their diameters, following sound engineering principles. Spirals in three dimensions (helices) with interesting geometric properties plus striped patterns combine to make exquisite sea shells. Branching patterns in trees, leaves, river networks, lungs, and blood vessels exhibit similar (fractal-like) features; there is an amazing unity (without uniformity) in nature. There are also to be found some fascinating geometric properties associated with mud cracks and patterns in the bark of trees.

Although it developed out of a university course, this is not a textbook per se. It will be very useful as such, I hope, but it can also be dipped into at leisure; simple examples are scattered liberally throughout the book, and especially in the early chapters. For whom, then, is the book written? The answer is that it is for a mixture of communities, academic and otherwise. Certainly I have in mind the college population of undergraduate students in mathematics, science, and engineering (and their professors), who may be able to use it as a supplement to their standard texts in various courses, particularly those in *mathematical modeling*. The material covered here is very broad in its scope, and I hope that it will be of considerable interest to professors and students engaged in both these and interdisciplinary courses. My further hope is that this material will appeal to high school teachers and their students who may have the opportunity for "mathematical enrichment" beyond the normal syllabus, if time permits.

Anyone interested in the beauty of nature, regardless of mathematical background will also (I trust) enjoy much in this book. Although the mathematical level ranges across a broad swath, from "applied arithmetic" to partial differential equations, there is a measure of nonmathematical discussion of the basic science behind the equations that I hope will also appeal to many others who might wish to ignore the equations (but not at their peril). Thus those who have no formal mathematical background will find much of value in the descriptive material contained here.

The level of mathematics used varies from basic arithmetic, geometry, and trigonometry through calculus of a single variable (and a smidgin of linear algebra) to the *occasional* senior or first-year graduate level topic in American universities and colleges. A background in geometry, trigonometry, and single-variable calculus will suffice for most purposes; familiarity with the theory of linear ordinary and partial differential equations is useful but definitely not necessary to appreciate the contents of this book. It should therefore be accessible almost in its entirety to students of mathematics, science, and engineering, the occasional advanced topic notwithstanding.

One of the other major reasons for writing this book is to bring together different strands from the many fascinating books and scientific articles, both technical and popular, that I have collected, read and used, or just dipped into over the last twenty-five to thirty years. Some of the books are out of print, though fortunately many are not. I have been richly blessed and stimulated by the writings of many scientists and mathematicians during this time, and in one sense, therefore, this book is the result of having carried out some "intellectual janitorial activity," a phrase that I encountered in Blair Kinsman's book *Wind Waves* many years ago and have adopted as my own. My hope is that this book will be a valuable resource for you, the reader; it may provide details of previously unknown sources that I encourage you to search out for yourself if time permits, but failing that, may it be a useful introduction to some of the fascinating and varied research that has been carried out by some very clever people!

Continuing this personal theme, I would like to share my philosophy for both writing about and teaching applied mathematics. It is a simple one: (i) try to understand the material to be presented at as many levels of description as is reasonable, and (ii) attempt to communicate that understanding with enthusiasm, gentleness, and humor. Like most others in my profession, I continue to be fascinated by the beauty, power, and applicability of mathematics, and I try to induce that fascination in others (often with mixed success in the classroom). Mathematics is a subject that is misperceived, sadly, by many both inside and outside the academic world. It is thought either to involve "doing long sums" or to be a cold, austere subject of little interest in its own right and no practical application whatsoever. These extremes could not be further from the truth, and one of my goals in teaching mathematics is to try to open students' minds to the above-mentioned triad of beauty, power, and applicability (even one out of three would be valuable!). My goals are the same in writing.

Nowadays a great deal of what is taught in universities and colleges by applied mathematicians comes under the general description *mathematical modeling*. Mathematical modeling is as much "art" as "science": it requires the practitioner to (i) identify a so-called "real world" problem (whatever the context may be); (ii) formulate it in mathematical terms (the "word problem" so beloved of undergraduates); (iii) solve the problem thus formulated (if possible; perhaps approximate solutions will suffice, especially if the complete problem is intractable); and (iv) interpret the solution in the context of the original problem. (What does this answer tell me? What does it really mean? Is it consistent with what I know already about the problem? What predictions can be made?) The formulation stage is often the most difficult: it involves making judicious simplifications to "get a handle" on the salient features of the problem. This in turn provides the basis in principle for a more sophisticated model, and so on. Whether the class is at an elementary, intermediate, or advanced level, it is important to convey aspects of the modeling process that are relevant to a particular mathematical result or technique that may be discussed in that class. Often this is most easily accomplished by illustrating the result in the context of a particular application. Consequently there are plenty of applications in this book.

Stephen Wolfram's recent book *A New Kind of Science* offers a fascinating and thought-provoking alternative to classical applied mathematics. He writes: "The traditional mathematical approach to science has historically had its great success in physics—and by now it has become almost universally assumed that any serious physical theory must be based on mathematical equations... we will see that some extremely simple programs seem able to capture the essential mechanisms for a great many physical phenomena that have previously seemed completely mysterious. .... Existing methods in theoretical physics tend to revolve around ideas of continuous numbers and calculus—or sometimes probability. Yet most of the systems in this book involve just simple definite discrete elements with definite rules."

The question arises: does this approach to modeling natural phenomena really explain the mechanisms underlying the observed patterns? Or do such cellular automata reproduce and describe them without explaining them? No doubt time will tell, as the implications of this challenging book are assimilated and debated both within and outside the scientific community.

Some will wonder why I have not included a chapter on "Fractals and Chaos in Nature," since fractals, chaos, and "complexity" are of such interest within the scientific community. There are two basic reasons: (i) many others have done an excellent job already on this topic, and (ii) it is a huge subject that could occupy several volumes if carried out properly. Fractal geometry has been characterized by some as the only realistic way to mimic nature and describe it in mathematical terms, and without wishing to question the foundations upon which this book is based (non-fractal mathematics), there is a measure of truth to that statement. But it has always been a strongly held opinion of mine that, in order to gain the most understanding of a physical phenomenon, it is necessary to view it with as many complementary levels of description and explanation as possible (but see p. 340 for a different perspective). The "classical applied mathematics" utilized in this book represents some of these levels, of course, and fractal geometry represents another, very profound approach. Furthermore, the latter is always "lurking in the background" in a book of this type, and to this end I have provided a very brief and cursory appendix on the topic of fractals and chaos, with some quotes from experts in the field and references for further reading. Indeed, the bibliography contains chapter numbers after each reference to indicate which chapter(s) draw on this reference, either as source material or as a valuable place to go deeper into a topic. There are many valuable sources in the literature, and I consider it a privilege if I can point readers to some of these in their quest for a better appreciation of "mathematics in nature." Yet another approach to the scientific and mathematical description of nature is via statistics and probability theory, and while those subjects do not constitute a major thread in this book, they are discussed in a little more detail in chapter 1.

A word about the epigraphs at the beginning of each chapter: they are from the New International Version (NIV) of the Old Testament, with the exceptions of those used in chapters 8, 10, and 11 which are from the King

James Version (KJV), and that in chapter 14, from the Revised Standard Version (RSV). As many theologians have said wisely, "A text without a context is a pretext." These verses lie in the intersection of two different contexts, and so I urge the interested reader to investigate both. While they may seem unusual in a book of this nature, I have chosen them because I think they have bearing on the subjects of each chapter, but sometimes you will have to think about that.

ACKNOWLEDGMENTS

I am grateful for the opportunity provided by Old Dominion University (and the Department of Mathematics and Statistics in particular) to develop the course, teach it, and write the book based on my classroom notes and literally hundreds of other sources (see the bibliography). Much of that task was accomplished during a one-semester study leave in the fall of 2001. All concerned have been very supportive of these my goals, and while it may have been merely aquiescence as a result of sheer exhaustion on their part ("Just go ahead and do it, John, and leave us in peace"), I rather think it was based on genuine and mutual collegial cooperation and respect. Mark Leslie has frequently dropped mathematical "snippets" in my mailbox or enthusiastically stopped by my office with a valuable reference, article, or "discovery." To him and all my colleagues I say thank you! Our hard-working secretaries Barbara Jeffrey and Gayle Tarkelson keep our department running smoothly, and my life would have been unnecessarily complicated without their constant and cheerful support. Bill Drewry in the Department of Civil and Environmental Engineering has imparted encouragement and wise advice to me over many a cup of coffee. I am especially grateful to Debbie Miller, who painstakingly drew the many figures in this book. There are many other friends and acquaintances who by words and emails have warmly wished me well in the writing of this book; I thank you all—you know who you are.

I am extremely appreciative of the sound advice and encouragement I have received from Vickie Kearn, senior mathematics editor at Princeton University Press. Right from the start of this project Vickie "caught the vision" I had for this book and kept me on track with her patience, enthusiasm, and humor. I have also greatly benefited from conversations (both electronic and acoustic) with her counterpart in the European PUP office, David Ireland. Thank you both. I would also like to thank Alison Anderson, who copyedited the typewritten manuscript, and Anne Reifsnyder, production editor at the Press, who gallantly dealt with my corrections at the proof stage. Frankly, it was a humbling experience to find I had made so many mathematical errors when typing the manuscript (with all of two fingers).

Perhaps it is too much to hope that I caught them all at the final proof stage. In the event that errors remain, I apologize for them and ask the reader's forbearance; they are my responsibility alone.

My wife Susan and children Rachel, Matthew, and Lindsay have been tremendously supportive and increasingly exicited as this project has evolved, and to each of them I say thank you for your continued love and encouragement.

Finally, a gentle plea of mine to the reader is as follows: *be observant*. There are many optical and fluid dynamical phenomena (particularly the latter) taking place in the everyday world around you—in the sky, clouds, rivers, lakes, oceans, puddles, faucets, sinks, coffee cups, and bathtubs. I hope that as a result of reading this book you will be better able to understand such phenomena, both mathematically and physically—it is a fascinating interdisciplinary area of applied mathematics, which richly rewards those who invest some time and effort to study it.

CREDITS

I am grateful to the following for granting me permission to reproduce or paraphrase material from their archives, books, or published articles.

To the National Science Teachers' Association for material in chapter 2, reprinted from Adam (1995) "Educated Guesses," *Quantum*. September/October: 20–24.

To Elsevier for material in chapter 5, reprinted from *Physics Reports*, vol. 356, John A. Adam, "The Mathematical Physics of Rainbows and Glories." pp. 229–365, © 2002, with permission from Elsevier Science.

To Wayne Armstrong for material in chapter 6, reproduced from http://waynesword.palomar.edu/ww0704.htm.

To Ned Mayo for material in chapter 6, paraphrased from Mayo, N. (1994), "A Hurricane for Physics Students," *The Physics Teacher*. 32: 148–154.

To Neil Shea for material in chapter 9, paraphrased from Shea, N. M. (1987), "Estimating the Power in the Tides," *The Physics Teacher*. October: 426.

PROLOGUE

Why I Might Never Have Written This Book

At about eleven years of age, I developed a passion for astronomy. I read everything about it I could get my hands on. My parents were very supportive, but my father, being a farmworker, had a very meager weekly income, about $30 per week at that time, so not a lot of money was available to support my astronomy habit. However, they did have a small insurance policy on my life, which they cashed in for about the equivalent of $50, as I recall. With this I bought a beautiful but somewhat dented old brass-tubed telescope: a 3-inch refractor (with an army-surplus tripod) that I have to this day. I spent many evening hours outside with it, observing the sky (most of which was cloudy—this was England, after all!). That set my career path, or so I thought. I had just started what in England we called secondary school, which spans the age range eleven to eighteen. At that time we were all streamed into different ability groups; there were three at Henley-on-Thames Grammar School, where I attended. Being something of a plodder, I was placed in the middle group and remained there for most of my years at the school. I remember very distinctly my algebra teacher asking us all what we wanted to be when we grew up. When Mr. Archibald Chanter (Arch) got to me, I said proudly, "An astronomer, Sir."

Immediately a rather worried look crossed his face, as he frowned and said, "But Adam, astronomers need to know an awful lot of maths, and you are very near the bottom of the class in this subject." Looking back, that was something of a turning point for me. I had two choices, though I didn't appreciate it at the time: to get discouraged and give up, or to use this as an opportunity to rise to the challenge and work my tail off to learn and understand the mathematics I needed to become an astronomer. Providentially, I chose the latter. I'll never know exactly what was in Arch's mind at the time, but I think he was genuinely concerned that I just didn't have the smarts to be an astronomer, even though I had the telescope!

Well, things didn't exactly change overnight. I buckled down and persevered throughout the next six or seven years, sometimes spending complete weekends working on my math(s) homework. I remember another mathematics teacher writing several years later in a report card that "by dint of sheer hard work, Adam is sometimes able to achieve far more than his brighter peers." And that was the secret: plod, plod, plod on my part and the blessing of having very supportive parents. While some of those brighter peers were groomed for Oxford and Cambridge Universities, the rest of us just applied to universities elsewhere, and, to cut a long story short, in the process of gaining my Ph.D. in theoretical astrophysics from the University of London, I became so enamored of applied mathematics that I resolved to turn my career, if possible, in the direction of becoming a lecturer (professor) in that field. I am also very grateful to a famous British amateur astronomer: Sir Patrick Moore, who is something of a cult figure in the UK, and who has hosted the television program *The Sky at Night* for over 45 years. It was he who answered my letters about becoming an astronomer, encouraged me to pursue my dream, and even invited my parents and me to visit him in Armagh, Northern Ireland, during his tenure as Director of the Armagh Planetarium. That was in 1966. Thank you, Sir Patrick! Although I am not now an astronomer as I had aspired to be all those years ago, on starry nights I still contemplate what are sometimes called the astronomer's psalms—Psalms 8 and 19. And were my father and Arch alive today, I'm sure they would appreciate the fact that in their own very different ways they helped me very, very much.

So, if I had chosen a nonscientific career, I would not have written this book. I'm thankful to have the privilege and joy of dedicating it to my mother and the memory of my father. My hope is that it will not receive the kind of review that the mathematician Mark Kac allegedly wrote—"This book fills a much-needed gap in the literature. The publishers should have opted for the gap."

Mathematics in Nature

Were the whole realm of nature mine
That were an offering far too small
Love so amazing, so divine
Demands my soul, my life, my all.

ISAAC WATTS (1674–1748)

CHAPTER ONE

The Confluence of Nature and Mathematical Modeling

> Great are the works of the Lord; they are pondered
> by all who delight in them.
>
> —Psalm 111:2

CONFLUENCE . . .

In recent years, as I have walked daily to and from work, I have started to train myself to observe the sky, the birds, butterflies, trees, and flowers, something I had not done previously in a conscious way (although I did watch out for fast-moving cars and unfriendly dogs). Despite living in suburbia, I find that there are many wonderful things to see: clouds exhibiting wave-like patterns, splotches of colored light some 22 degrees away from the sun (sundogs, or parhelia), wave after wave of Canada geese in "vee" formation, the way waves (and a following region of calm water) spread out on the surface of a puddle as a raindrop spoils its smooth surface, the occasional rainbow arc, even the iridescence on the neck of those rather annoying birds, pigeons, and many, many more nature-given delights. And so far I have not been late for my first class of the morning!

The idea for this book was driven by a fascination on my part for the way in which so many of the beautiful phenomena observable in the natural realm around us can be described in mathematical terms (at least in principle). What are some of these phenomena? Some have been already mentioned in the preface, but for a more complete list we might consider rainbows, "glories," halos (all atmospheric occurrences), waves in air, earth, oceans, rivers, lakes, and puddles (made by wind, ship, or duck), cloud formations (billows, lee waves), tree and leaf branching patterns (including phyllotaxis), the proportions of trees, the wind in the trees, mud-crack patterns, butterfly markings, leopard spots, and tiger stripes. In short, if you can see it outside, and a human didn't make it, it's probably described here! That, of course, is an exaggeration, but this book does attempt to answer on varied levels the fundamental question: what kind of scientific and mathematical principles undergird these patterns or regularities that I claim are so ubiquitous in nature?

Two of the most fundamental and widespread phenomena that occur in the realm of nature are the scattering of light and wave motion. Both may occur almost anywhere given the right circumstances, and both may be described in mathematical terms at varying levels of complexity. It is, for example, the scattering of light both by air molecules and by the much larger dust particles (or more generally, aerosols) that gives the amazing range of color, hues, and tints at sunrise or sunset that give us so much pleasure. The deep blue sky above and the red glow near the sun at the end of the day are due to molecular scattering of light, though dust or volcanic ash can render the latter quite spectacular at times.

The rainbow is formed by sunlight scattered in preferential directions by near-spherical raindrops: scattering in this context means refraction and reflection (although there many other fascinating features of light scattering that will not be discussed in great detail here). Using a simple mathematical description of this phenomenon, René Descartes in 1637 was able to "hang the rainbow in the sky" (i.e., deduce its location relative to the sun and observer), but to "paint" the rainbow required the genius of Isaac Newton some thirty years later. The bright primary and fainter secondary bows are well described by elementary mathematics, but the more subtle observable features require some of the most sophisticated techniques of mathematical physics to explain them. A related phenomenon is that of the "glory," the set of colored, concentric rainbow-like rings surrounding, for example, the shadow of an airplane on a cloud below. This, like the rainbow, is also a "backscatter" effect, and, intriguingly, both the rainbow and the glory have their counterparts in atomic and nuclear physics; mathematics is a unifying feature between these two widely differing contexts. The beautiful (and commonly circular) arcs known as halos, no doubt seen best in arctic climes, are formed by the refraction of sunlight through ice crystals of various shapes in the upper atmosphere. Sundogs, those colored splashes of light often seen on both sides of the sun when high cirrus clouds are present, are similarly formed.

Like the scattering of light, wave motion is ubiquitous, though we cannot always see it directly. It is manifested in the atmosphere, for example, by billow clouds and lee-wave clouds downwind from a hill or mountain. Waves on the surface of puddles, ponds, lakes, or oceans are governed by mathematical relationships between their speed, their wavelength, and the depth of the water. The wakes produced by ships or ducks generate strikingly similar patterns relative to their size; again, this correspondence is described by mathematical expressions of the physical laws that govern the motion. The situation is even more complex in the atmosphere: the "compressible" nature of a gas renders other types of wave motion possible. Sand dunes are another complex and beautiful example of waves. They can occur on a scale of centimeters to kilometers, and, like surface waves on bodies of water, it is only the waveform that actually moves; the body of sand is stationary (except at the surface).

In the plant world, the arrangement of leaves around a stem or seeds in a sunflower or daisy face shows, in the words of one mathematician (H.S.M. Coxeter), "a fascinatingly prevalent tendency" to form recurring numerical patterns, studied since medieval times. Indeed, these patterns are intimately linked with the "golden number" ($(1 + \sqrt{5})/2 \approx 1.618$) so beloved of Greek mathematicians long ago. The spiral arrangement of seeds in the daisy head is found to be present in the sweeping curve of the chambered nautilus shell and on its helical counterpart, the *Cerithium fasciatum* (a thin, pointy shell). The curl of a drying fern and the rolled-up tail of a chameleon all exhibit types of spiral arc.

In the animal and insect kingdoms, coat patterns (e.g., on leopards, cheetahs, tigers, and giraffes) and wing markings (e.g., on butterflies and moths) can be studied using mathematics, specifically by means of the properties and solutions of so-called reaction-diffusion equations (and other types of mathematical models). Reaction-diffusion equations describe the interactions between chemicals ("activators" and "inhibitors") that, depending on conditions, may produce spots, stripes, or more "splodgy" patterns. There are fascinating mathematical problems involved in this subject area, and also links with topics such as patterns on fish (e.g., angel fish) and seashells. In view of earlier comments, seashells combine both the effects of geometry and pattern formation mechanisms, and mathematical models can reproduce the essential features observed in many seashells.

Cracks also, whether formed in drying mud, tree bark, or rapidly cooling rock, have their own distinctive mathematical patterns; frequently they are hexagonal in nature. River meanders, far from being "accidents" of nature, define a form in which the river does the least work in turning (according to one class of models), which then defines the most probable form a river can take—no river, regardless of size, runs straight for more than ten times its average width.

Many other authors have written about these patterns in nature. Ian Stewart has noted in his popular book *Nature's Numbers* that "We live in a universe of patterns. . . . No two snowflakes appear to be the same, but all possess six-fold symmetry." Furthermore, he states that

> there is a formal system of thought for recognizing, classifying and exploiting patterns. . . . It is called mathematics. Mathematics helps us to organize and systemize our ideas about patterns; in so doing, not only can we admire and enjoy these patterns, but also we can use them to infer some of the underlying principles that govern the world of nature. . . . There is much beauty in nature's clues, and we can all recognize it without any mathematical training. There is beauty too in the

mathematical stories that ... deduce the underlying rules and regularities, but it is a different kind of beauty, applying to ideas rather than things. Mathematics is to nature as Sherlock Holmes is to evidence.

We may go further by asking questions like those posed by Peter S. Stevens in his lovely book *Patterns in Nature*. He asks,

> Why does nature appear to use only a few fundamental forms in so many different contexts? Why does the branching of trees resemble that of arteries and rivers? Why do crystal grains look like soap bubbles and the plates of a tortoise shell? Why do some fronds and fern tips look like spiral galaxies and hurricanes? Why do meandering rivers and meandering snakes look like the loop patterns in cables? Why do cracks in mud and markings on a giraffe arrange themselves like films in a froth of bubbles?

He concludes in part that "among nature's darlings are spirals, meanders, branchings, hexagons, and 137.5 degree angles.... Nature's productions are shoestring operations, encumbered by the constraints of three-dimensional space, the necessary relations among the size of things, and an eccentric sense of frugality.

In the book *By Nature's Design*, Pat Murphy expresses similar sentiments, writing,

> Nature, in its elegance and economy, often repeats certain forms and patterns ... like the similarity between the spiral pattern in the heart of a daisy and the spiral of a seashell, or the resemblance between the branching pattern of a river and the branching pattern of a tree ... ripples that flowing water leaves in the mud ... the tracings of veins in an autumn leaf ... the intricate cracking of tree bark ... the colorful splashings of lichen on a boulder.... The first step to understanding—and one of the most difficult—is to see clearly. Nature modifies and adapts these basic patterns as needed, shaping them to the demands of a dynamic environment. But underlying all the modifications and adaptations is a hidden unity. Nature invariably seeks to accomplish the most with the least—the tightest fit, the shortest path, the least energy expended. Once you begin to see these basic patterns, don't be surprised if your view of the natural world undergoes a subtle shift.

Another fundamental (and philosophical) question has been asked by many—How can it be that mathematics, a product of human thought independent of experience, is so admirably adapted to the objects of reality? This fascinating question I do not address here; let it suffice to note that, hundreds of years ago, Galileo Galilei stated that the Universe "cannot be read until we have learnt the language and become familiar with the characters in which it is written. It is written in mathematical language." Mathematics is certainly the language of science, but it is far, far more than a mere tool, however valuable, for it is of course both a subject and a language in its own right. But lest any of us should balk at the apparent need for speaking a modicum of that language in order more fully to appreciate this book, the following reassuring statement from Albert Einstein, when writing to a young admirer at junior high school, should be an encouragement. He wrote "Do not worry about your difficulties in mathematics. I can assure you that mine are still greater." Obviously anyone, even scientists of great genius, can have difficulties in mathematics (one might add that it's all a matter of relativity in this regard).

Obviously a significant component of this book is the application of elementary mathematics to the natural world around us. As I have tried to show already, there are many mathematical patterns in the natural world that are accessible to us if we keep our eyes and ears open; indeed, the act of "asking questions of nature" can lead to many fascinating "thought trails," even if we do not always come up with the correct answers. First, though, let me remind you (unnecessarily, I am sure) that no one has all the answers to such questions. This is true for me at all times, of course (not just as a parent and a professor), but especially so in a subject as all-encompassing as "mathematics in nature." There will always be "displays" or phenomena in nature that any given individual will be unable to explain to the satisfaction of everyone, for the simple reason that none of us is ever in possession of all the relevant facts, physical intuition, mathematical techniques, or other requirements to do justice to the observed event. However, this does not mean that we cannot appreciate the broad principles that are exemplified in a rainbow, lenticular cloud, river meander, mud crack, or animal pattern. Most certainly we can.

It is these broad principles—undergirded by mathematics, much of it quite elementary—that I want us to perceive in a book of this admittedly rather free-ranging nature. My desire is that by asking mathematical questions of the phenomena we will gain both some understanding of the symbiosis that exists between the basic scientific principles involved and their mathematical description, and a deeper appreciation for the phenomenon itself, its beauty (obviously rather subjective), and its relationship to other events in the natural world around us. I have always found, for example, that my appreciation for a rainbow is greatly enhanced by my understanding of the mathematics and physics that undergird it (some of the mathematics can be extremely advanced; some references to this literature are provided in the bibliography). It is important to remember that this is a book on aspects of *applied mathematics*, and there will be at times some more advanced and even occasionally rigorous mathematics (in the form of theorems and sometimes proofs); for the most part, however, the writing style is intended to be informal. And so now, on to

... MODELING

An important question to be asked at the outset is *What is a mathematical model?* One basic answer is that it is the formulation in mathematical terms of the assumptions and their consequences believed to underlie a particular "real world" problem. The aim of mathematical modeling is the practical application of mathematics to help unravel the underlying mechanisms involved in, for example, economic, physical, biological, or other systems and processes. Common pitfalls include the indiscriminate, naïve, or uninformed use of models, but, when developed and interpreted thoughtfully, mathematical models can provide insight into the nature of the problem, be useful in interpreting data, and stimulate experiments. There is not necessarily a "right" model, and obtaining results that are consistent with observations is only a first step; it does not imply that the model is the only one that applies, or even that it is "correct." Furthermore, mathematical descriptions are not explanations, and never on their own can they provide a complete solution to the biological (or other) problem—often there may be complementary levels of description possible within the particular scientific domain. Collaboration with scientists or engineers is needed for realism and help in modifying the model mechanisms to reflect the science more accurately. On the other hand, workers in nonmathematical subjects need to appreciate what mathematics (and its practitioners) can and cannot do. Inevitably, as always, good communication between the interested parties is a necessary (but not sufficient) recipe for success.

In the preface mention was made of fundamental steps necessary in developing a mathematical model (see figure 1.1): formulating a "real world" problem in mathematical terms using whatever appropriate simplifying assumptions may be necessary; solving the problem thus posed, or at least extracting sufficient information from it; and finally interpreting the solution in the context of the original problem (which as noted above may include validation of the model by testing both its consistency with known data and its predictive capability). Thus the art of good modeling relies on (i) a sound understanding and appreciation of the scientific or other problem; (ii) a realistic mathematical representation of the important phenomena; (iii) finding useful solutions, preferably quantitative ones; and (iv) interpretation of the mathematical results—insights, predictions, and so on. Sometimes the mathematics used can be very simple. The usefulness of a mathematical model should not be judged by the sophistication of the mathematics, but by different (and no less demanding) criteria.

Although techniques of statistical analysis may frequently be used in portraying and interpreting data, it is important to note two fundamentally distinct approaches to mathematical modeling, which differ somewhat in

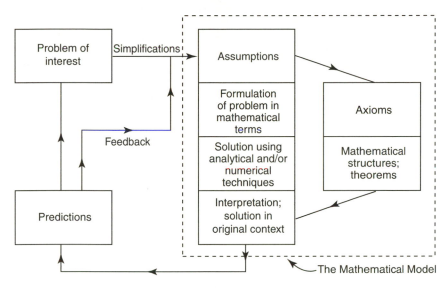

Figure 1.1. A generic flow chart illustrating stages and levels of mathematical modeling. Reprinted with permission of Birkhäuser, copyright 1997, from the book "A Survey of Models for Tumor-Immune System Dynamics," edited by J. A. Adam and N. Bellomo.

both mathematical and philosophical characteristics. These are deterministic models and probabilistic or statistical (sometimes referred to as stochastic) models. One's preference often depends upon the way one has been trained, which usually determines the way one looks at the world mathematically. Most of this book is written from the deterministic perspective. A general summary of the philosophy and methodology of this approach (with many references to applications in cancer biology) may be found in chapter 2 of the book edited by Adam and Bellomo (1997).

Deterministic models frequently possess the property that a system of interest (e.g., a population of cells, a mixture of chemicals or enzymes, a density or pressure imbalance in a gas) can be "observed" mathematically to evolve away from some initial configuration.

Frequently in deterministic models the assumption is made that the dependent variables are at least differentiable functions of their argument(s), and hence continuous. This assumption is quite reasonable when the magnitude of (say) the population of cells is very much larger than a typical change (increase or decrease) in the population. For a small tumor composed of about a billion cells, the individual cell cycles have become asynchronous, and so on the basis of a simple model the evolution of the tumor can be reasonably described using the techniques of differential and integral calculus. The situation is inevitably much more complex than this, and many other factors need to be included to derive a realistic model of tumor growth. Regardless of the complexity, deterministic models may be used to predict the size (or cell number) of the tumor as a function of time, in particular, predicting that the tumor will have $N(t)$ cells at time t. In practice, however, there is an element of uncertainty in every event, from the growth of a tumor to catching a bus. External factors, usually beyond our control, play a role in determining the outcome of the event. Thus diet, fitness, treatment regimen, and mental attitude may contribute to the eventual outcome for the cancer patient, while the faulty alarm clock used by the bus driver may influence whether the bus arrives on time. Similarly, in studying the relation between stature and heredity, differences in environment and nutrition may be sources of uncertainty in the results. Stochastic models enable researchers to identify and study many of these uncertainties.

Note also that in an experiment errors that arise may be classified as random or systematic. The word *error* as used here does not mean *mistake*; it is introduced into the results by external influences beyond the control of the experimenter. The static heard when listening to a radio with poor reception is generally random; the discovery of the 4°K microwave background radiation by Arno Penzias and Robert Wilson in 1964 resulted from *systematic* error (i.e., no matter what they tried, they could not eradicate it); and this in turn resulted in confirmation of a prediction by the scientist George Gamow (and others) in 1948. Systematic errors can lead to new discoveries! Probabilistic or stochastic models incorporate a measure of uncertainty; for example, they predict the *probability* that the tumor will have $N(t)$ cells at time t. Furthermore, if the cell population of interest is rather small, a typical change in cell number may well be a significant fraction of the total population, and so the "state" of the population must be represented, say, by an integer-valued random variable.

It has already been noted that mathematical models are not necessarily "right" (though they may be wrong as a result of ignoring fundamental processes). One model may be better than another in that it has better explanatory features: more specific predictions can be made that are subsequently confirmed, at least to some degree. Some of the models presented in this book are still controversial; in particular, the reaction-diffusion models of pattern formation presented in chapter 14 are not universally accepted by biomathematicians. There are other models, for example in the study of wound healing, that utilize more of the mechanical properties of the medium (skin in this case) and are therefore designated *mechanochemical* models (they are not discussed in this book). Some of the models cited are somewhat elderly or incomplete (or both), examples being those of sand dune formation and river meanders presented in chapters 6 and 12, respectively. Indeed, I venture to suggest that all mathematical models are flawed to some extent: many by virtue of inappropriate assumptions made in formulating the model, or (which may amount to the same thing) by the omission of certain terms in the governing equations, or even by mis-interpretation of the mathematical conclusions in the original context of the problem. Occasionally models may be incorrect because of errors in the mathematical analysis, even if the underlying assumptions are valid. And, paradoxically, it can happen that even a less accurate model is prefer-able to a more mathematically sophisticated one; it was the mathematical statistician John Tukey who stated that "it is better to have an approximate answer to the right question than an exact answer to the wrong one."

This is well illustrated by Lee and Fraser in their comparison of the less accurate Airy theory of the rainbow with the more general and powerful Mie theory. They write, "Our point here is not that the exact Mie theory describes the natural rainbow inadequately, but rather that the approxi-mate Airy theory can describe it quite well. Thus the supposedly outmoded Airy theory generates a more natural-looking map of real rainbow colors than Mie theory does, even though Airy theory makes substantial errors in describing the scattering of monochromatic light by isolated small drops. *As in many hierarchies of scientific models, the virtues of a simpler theory can, under the right circumstances, outweigh its vices*" (italics added).

With such provisos in mind, the aim of this book is *not* to present thor-ough mathematical descriptions of many naturally occurring phenomena—an impossible task—but instead to try to present a compilation and syn-thesis of several mathematical models that have been developed within these contexts. The extensive set of references in the bibliography is pro-vided to encourage the inquisitive reader to pursue the original articles and books from which these models were first presented. There are undoubtedly many published papers on these topics of which I am unaware, and which certainly would have enhanced this book, and for that I must point out, regrettably, that one has to stop somewhere (the publisher requires it).

The organization of the book is as follows. Chapters 2 and 3 constitute a rather gentle introduction to the importance and usefulness of estimation (chapter 2) and the problem of shape, size, and scale, plus an introduction to the methods of dimensional analysis (chapter 3). It is important to note that most of the material in these two chapters is mathematically "fuzzy"; the conclusions drawn are not exact, and cannot be. They are intended to provide the reader with some guidelines, domains of validity, and basic principles to be borne in mind when constructing the next level of model, by which is meant a rather more sophisticated mathematical approach to the problem of interest, assuming this is appropriate. However, there may on occasion be specific counterexamples to the conclusions drawn, of the genre "My uncle smoked two packs of cigarettes every day for seventy years, and he never developed cancer"; "That tree must have had at least 10^5 leaves:

I just raked them all up"; or, perhaps more relevant to chapter 3, "We owned a horse that could beat any dog running uphill." Obviously such anecdotal comments, while perhaps true, do not vitiate broad conclusions based on solid statistical evidence (in the case of cigarette smoking) or general principles of bioengineering, provided that in each case we operate within the domains of validity of the underlying assumptions and procedures. The reader is reminded of the Tukey quote earlier in this section.

The next three chapters (4–6) are of a meteorological character and include straightforward mathematical (and physical) descriptions of shadow-related phenomena, rainbows, halos, mirages, and some aspects of clouds, hurricanes, and sand dunes. Chapters 7–9 are fluid dynamical in nature, introducing aspects of linear and nonlinear wave motion, respectively (chapters 7 and 9), separated by a chapter dealing with the "other side of the linear coin," so to speak, namely instability in some of its various forms. Chapter 10 examines some of the many properties of the golden ratio—an irrational number with fascinating connections to the plant world in particular. The topics of chapter 11 include honeycombs, soap bubbles (and foam), and mud cracks; they are loosely connected by the continual search for optimal solutions and principles of minimization that has been a major theme in mathematical research for millennia.

A not-unrelated theme meanders through part of chapter 12: river meanders and branching patterns, followed by some arboreal mathematical models—the application of engineering principles to establish the height/width relationship for a generic tree, and the "murmur" of the forest. Basic principles of bird flight are discussed in chapter 13, as is the underlying fluid dynamical theorem of Bernoulli. The final chapter contains selected models of pattern formation based on an examination of the diffusion equation (and related equations). Applications of these models to animal, insect, and seashell patterns and plankton blooms are briefly considered, followed by some applications of the diffusion equation that are primarily of historical interest. A short appendix on fractals is designed to whet the reader's appetite for more.

APPENDIX: A MATHEMATICAL MODEL OF SNOWBALL MELTING

The purpose of this appendix is to illustrate some features of mathematical modeling by means of a simple (some would say silly) example. Before doing so, however, readers may find helpful the following quotation from the above-mentioned book, *Wind Waves* by Blair Kinsman. It identifies both the importance and the ubiquity of assumptions made in the process of mathematical modeling.

In the derivation to follow, it will be *assumed* that the Earth is flat, the water is of constant density, that the Coriolis force is negligible, that the density of the air can be neglected compared with the density of water, that the body of water is of infinite extent and completely covered by waves, and that viscosity and surface tension can be neglected. Moreover, *to simplify the problem*, it will be *assumed* that there is no variation in the wave properties in the y-direction. A few more *assumptions* will turn up along the way in deriving the equations to be solved, but the equations will still be unsolvable in closed form as they will be nonlinear. (italics added)

The following simplistic problem is expressed in the form of a question and an unhelpful answer to illustrate the fact that many problems we might wish to model may be, at best, ill-defined. It is a case study that may be helpful to the reader in revealing some of the more intuitive aspects of mathematical modeling.

Q: Half of a snowball melts in an hour. How long will it take for the remainder to melt?

A: I don't know.

Why may such a response be justified, at least in part? Because the question stated is not a precise one; it is ambiguous. Half of what? The mass of the snowball or its volume? Under what kinds of assumptions can we formulate a mathematical model and will it be realistic? This type of problem is often posed in "Calculus I" textbooks, and as such requires only a little basic mathematical material, for example, the chain rule and elementary integration. However, it is what we do with all this that makes it an interesting and informative exercise in mathematical modeling. There are several reasonable assumptions that can be made in order to formulate a model of snowball melting; however, unjustifiable assumptions are also a possibility! The reader may consider some or all of these to be in the latter category, but ultimately the test of a model is how well it fits known data and predicts new phenomena. The model here is less ambitious (and not a particularly good one either), for we merely wish to illustrate how one might approach the problem. It can lead to a good discussion in the classroom setting, especially during the winter. Some plausible assumptions (and the questions they generate) might be as follows:

i. Assume that the snowball is a sphere of radius $r(t)$ at all times. This is almost certainly never the case, but the question becomes one of simplicity. Is the snowball roughly spherical initially? Subsequently? Is there likely to be preferential warming and melting on one side even if it starts life as a sphere? The answer to this last question is yes: preferential melting will probably occur in the direction of direct sunlight unless the snowball is in the shade or the sky is uniformly overcast. If we can make this assumption, then the resulting surface area and volume considerations involve only the one spatial variable r.

ii. Assume that the density of the snow/ice mixture is constant throughout the snowball, so there are no differences in "snow-packing." This may be reasonable for small snowballs (hand-sized ones), but large ones formed by rolling will probably become more densely packed as their weight increases. A major advantage of the constant density assumption is that the mass (and weight) of the snowball is then directly proportional to its volume.

iii. Assume that the mass of the snowball decreases at a rate proportional to its surface area, and only this. This appears to make sense since it is the outside surface of the snowball that is in contact with the warmer air that induces melting. In other words, the transfer of heat occurs at the surface. This assumption in particular will be examined in the light of the model's prediction. But even if it is a good assumption to make, is the "constant" of proportionality really constant? Might it not depend on the humidity of the air, the angle of incidence and intensity of sunlight, the external temperature, and so on?

iv. Assume that no external factors change during the "lifetime" of the snowball. This is related to assumption (iii), and is probably the weakest of them all; unless the melting time is very much less than a day it is safe to say that external factors will vary! Obviously the angle and intensity of sunlight will change over time, and possibly other factors as noted above. Let us proceed, nevertheless, on the basis of these four assumptions, and formulate a model by examining some of the mathematical consequences of these assumptions. We may do so by asking further questions, for example:

i. What are expressions for the mass, volume, and surface area of the snowball?

ii. How do we formulate the governing equations? What are the appropriate initial and/or boundary conditions? How do we incorporate the information provided?

iii. Can we obtain a solution (analytic, approximate, or numerical) of the equations?

iv. What is the physical interpretation of the solution and does it make sense? That is, is it consistent with the information provided and are the predictions from the model reasonable?

v. Does a unique solution exist?

We will answer questions (i)–(iv) first, and briefly comment on (v) at the end. It is an important question that is of a more theoretical nature than the rest, and its consequences are far reaching for models in general. Let the initial radius of the snowball be $r(0) = R$. If we denote the uniform density of the snowball by ρ, its mass by $M(t)$, and its volume by $V(t)$, and measure time t in hours, then the mass of the snowball at any time t is

$$M(t) = \frac{4}{3}\pi\rho r^3(t). \tag{1}$$

It follows that the instantaneous rate of change of mass or time derivative is

$$\frac{dM}{dt} = 4\pi\rho r^2 \frac{dr}{dt}. \tag{2}$$

By assumption (iii)

$$\frac{dM}{dt} = -4\pi r^2 k, \tag{3}$$

where k is a positive constant of proportionality, the negative sign implying that the mass is *decreasing* with time! By equating the last two expressions it follows that

$$\frac{dr}{dt} = -\frac{k}{\rho} = -\alpha, \text{ say.} \tag{4}$$

Thus, according to this model, the radius of the snowball decreases uniformly with time. Upon integrating this differential equation and invoking the initial condition we obtain

$$r(t) = R - \alpha t = R\left(1 - \frac{t}{t_m}\right) = 0 \quad \text{when } t = \frac{R}{\alpha} = t_m, \tag{5}$$

where t_m is the time for the original snowball to melt, which occurs when its radius is zero! We do not know the value of α, since that information was not provided, but we *are* informed that after one hour half the snowball has melted, so we have from equation (5) that $r(1) = R - \alpha$. A sketch of the linear equation in (5) and use of similar triangles in figure 1.2a shows that

$$t_m = \frac{R}{R - r(1)}$$

and furthermore

$$\frac{V(1)}{V(0)} = \frac{1}{2} = \frac{r^3(1)}{R^3},$$

so that

$$r(1) = 2^{-1/3}R \approx 0.79R.$$

Hence $t_m \approx 4.8$ hours, so that according to this model the snowball will take a little less than 4 more hours to melt away completely. This is a rather long time, and certainly the sun's position will have changed during that

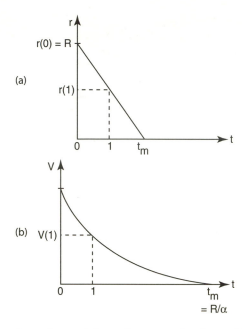

Figure 1.2. The time dependence of the radius $r(t)$ and volume $V(t)$ for the model of snowball melting.

time (through an arc of roughly 60°), so in retrospect assumption (iv) is not really justified. A further implication of equation (5) is that both the volume and mass of the snowball (by assumption (ii)) decrease like a cubic polynomial in t, i.e.,

$$V(t) = V(0)\left(1 - \frac{t}{t_m}\right)^3.$$

(See figure 1.2b). Note that $V'(t) < 0$ as required, and $V'(t_m) = 0$. Since $V''(t) > 0$ it is clear that the snowball melts more quickly at first, when $|V'|$ is larger, than at later times, as the value for t_m attests. I recall being told as a child by my mother that "snow waits around for more," but this model is hardly a "proof" of that, despite further revelations below! It may be adequate under some circumstances, but there are obvious deficiencies given the initial "data" (which to be honest, I invented). What other factors have been ignored here? Here are some:

We are all familiar with the fact that the consistency of snow varies depending on whether it is "wet" or "dry"; snowballs are more easily made with the former (at least, I have found it to be so). Wet snow can be packed more easily and a layer of ice may be formed on the outside. This can in turn cool a thin layer of air around the surface, which will insulate (somewhat) the snowball from the warmer air beyond that. A nice clean snowball, as opposed to one made with dirty snow, may be highy reflective of sunlight (it has a high *albedo*), and this will reduce the rate of melting further. There are no doubt several other factors missing.

Some other aspects of the model are more readily appreciated if we generalize the original problem by suggesting instead that "a fraction β of a snowball melts in h hours . . . ". The melting time is then found to be

$$t_m = \frac{h}{1 - \sqrt[3]{1 - \beta}}, \tag{6}$$

which depends linearly on h and in a monotonically decreasing manner on β. The dependence on h is not surprising; if a given fraction β melts in half the time, the total melting time is also halved. For a given value of h, the dependence on β is also plausible: the larger the fraction that melts in time h, the shorter the melting time.

A final point concerns question (v) on the existence and uniqueness of the solution to this mathematical model. While the existence of a solution is clear in this rather trivial example, it is certainly of fundamental importance in general terms, as is the uniqueness (or not) of a solution and the *stability* of such solution(s) to small variations in the initial and/or boundary conditions. Such considerations are outside the scope and theme of this book, but the interested reader can find information on these topics by consulting many undergraduate and most graduate texts on ordinary and partial differential equations. For completeness, we state and apply the relevant theorem for this particular example, namely:

Theorem: there exists a unique solution to the ordinary differential equation

$$\frac{dy}{dx} = f(x, y)$$

satisfying the initial condition

$$y(x_0) = y_0,$$

provided that both the function $f(x, y)$ and its partial derivative $\partial f / \partial y$ are continuous functions of x and y in a neighborhood of the initial point (x_0, y_0).

In many cases of practical interest, these continuity conditions are satisfied for all values of (x_0, y_0).

So what has this to do with our snowball? Everything! Our problem has reduced to

$$\frac{dr}{dt} = f(r, t) \equiv -\alpha, \quad r(0) = r_0 \equiv R,$$

where α is a positive constant. Since constants and their derivatives are continuous everywhere (!), the theorem applies, and hence a unique solution exists for the problem as posed. Of course, like the familiar saying about snowflakes, the snowball is probably unique as well.

As a final comment, it should be pointed out that in a majority of the mathematical models that follow in this book, the assumptions and their consequences will not be formally laid out as they have been in the above pedagogic example. Indeed, that organizational style is not generally used (in the author's experience) even in the modeling literature.

CHAPTER TWO

Estimation: The Power of Arithmetic in Solving Fermi Problems

> Who has measured the waters in the hollow of his
> hand, or with the breadth of his hand marked off the
> heavens? Who has held the dust of the earth in a
> basket, or weighed the mountains on the scales and
> the hills in a balance?
>
> —Isaiah 40:12

> It is the mark of an instructed mind to rest satisfied
> with the degree of precision which the nature of the
> subject permits and not to seek an exactness where
> only an approximation of the truth is possible.
>
> —Aristotle

The second quotation makes much the same point as that by John Tukey over two millennia later (see chapter 1). We may not be as erudite as Aristotle or as brilliant as Enrico Fermi, but we can learn to apply elementary reasoning to obtain "ballpark estimates" to problems (subsequently named *Fermi problems*) in the manner attributed to that great physicist. The rationale behind such estimates has been well described by Hans Christian von Baeyer, who wrote in *The Fermi Solution*,

> Fermi's intent was to show that although, at the outset, even the answer's order of magnitude is unknown, one can proceed on the basis of different assumptions and still arrive at estimates that fall within range of the answer. The reason is that, in any string of calculations, errors tend to cancel one another out. . . . It is as improbable that all of one's errors will be underestimates (or overestimates) as it is that all the throws in a series of coin tosses will be heads (or tails). The law of probabilities dictates that deviations from the correct assumptions will tend to compensate for one another, so the final results will converge toward the right number.

Several years ago a short article, written in the same spirit by David Halliday, appeared in *Quantum* (May 1990). In the context of a specific problem, Halliday showed how to obtain "order of magnitude" answers to problems by breaking them down into their components and making appropriate commonsense estimates. Of course, the ideas expressed and methods used in such Fermi problems go far beyond specific scientific applications into the realm of everyday activities. Two excellent resources that I have enjoyed reading and using are *Innumeracy* by John Allen Paulos and *Consider a Spherical Cow* by John Harte. You will recognize some of the problems cited here if you have encountered these books before. After a while you will get comfortable with posing and estimating answers to your own Fermi problems. The book by Paulos has been an eye-opener to many: in particular, he provides many examples of the power of plausible assumptions coupled with simple calculations. The book by Harte is a good introduction to mathematical modeling (particularly environmental problem solving) with little or no use of calculus. While on the subject of relevant books, *The Universe Down to Earth* by Neil de Grasse Tyson has some chapters (1 and 3) relevant to the present chapter, and von Baeyer's *The Fermi Solution* (containing the above quotation) has its opening chapter devoted to a clear exposition of Fermi problems and is well worth reading.

In much of what follows, letters are used to represent typical dimensions or other quantities. This will enable readers to obtain their own estimates, though they should resist the temptation just to "plug in" their numbers in the formula without following the prior reasoning. Almost certainly we will differ on typical sizes of objects (e.g., grains of sand). But almost as certainly we will choose typical dimensions in the range (for this example) of 10^{-1} mm $\leq d \leq 2$ mm, and will probably not, therefore, differ significantly in our subsequent order-of-magnitude answers. Remember that it is to be understood that, whenever ratios of dimensional quantities are to be sought, a conversion of units may be necessary in order to compare like quantities. For completeness, actual numerical estimates are given; some of their values may surprise you.

Needless to say, the question will be asked—What is the point of knowing how to estimate the number of grains of sand that would fill Buckingham Palace? Apart from a spell in prison for attempting to verify such an estimate, it is a great encouragement to realize that this type of "back of the envelope" calculation can be carried out with a modicum of salient information for a "real-world problem." Not only might this save a considerable amount of money and computer time on occasion, but also it can greatly enhance the scope of the teaching resources available to teachers of mathematics at the middle school/high school level. There is an excitement that comes from realizing how powerful simple arithmetic can be! And this can be passed on to receptive students. I have observed the "lights go on" when intelligent, educated people realize the real distinction between 10^6 seconds ($\approx 11\frac{1}{2}$ days) and 10^9 seconds (≈ 32 years). Sometimes all we need are the right "pegs" to hang numbers (and concepts) on.

Let us begin by considering some rather straightforward but eclectic examples. Among the simplest estimation problems are those arising from ratios of lengths, areas, and volumes. Thus, if D is a typical linear dimension of a given object (e.g., a classroom), and $d < D$ is a typical linear dimension of a smaller object (e.g., a piece of popcorn, popped), then $N = D^3/d^3$ is the approximate number of smaller objects that would fill the latter. Thus, by using appropriate choices of D and d we may estimate, for example, the number of (1) *golf balls required to fill a suitcase*; (2) *pieces of popcorn to fill a room*; (3) *soccer balls to fill an average-sized house*; (4) *cells in a human body*; (5) *grains of sand to fill the earth*. Related problems involve volumetric measures of fluids, e.g., (6) *the volume of human blood in the world*; (7) *the number of one-gallon buckets required to empty Loch Ness* (and hence expose the monster). Others will be found below, and still more in the article by the author (1995).

Sometimes everyday objects are obviously *misrepresented* by cubes. Thus for an object with linear dimensions a, b, c, $N = abc/d^3$ is more appropriate. For problem (1) above, we might suggest that $a = 20$, $b = 24$, $c = 8$, and $d = 1.5$ inches, respectively, so $N \approx 10^3$. For problem (2), suppose $a = 10$, $b = 20$, $c = 15$ (all in feet: a typical classroom size), and $d = 1$ cm; then, on conversion to cubic cm, $N \approx 3000 \times 30^3 \approx 10^8$. For (3), consider $D = 30$ ft and $d = 1$ ft, respectively, yielding $N \approx 10^4$. Problem (4) yields 10^{14}, and the answer to problem (6) is less than $1/200$ of a cubic mile; both of these are discussed below. For problem (5), $D \approx 10^4$ km and $d = 1$ mm yields $N \approx (10^4 \times 10^3 \times 10^2 \times 10)^3 = 10^{30}$. A cubic earth, you ask? To this degree of approximation, that is not a problem (see the related comment on problem 12). Using the fact that 1 ft^3 of liquid (water, soup, blood, etc.) is about 7.5 gallons, we arrive at $N \approx 10^{12}$ buckets to empty Loch Ness (problem 7: the Loch has a volume of approximately 2 cubic miles, so $2 \times 5280^3 \times 7.5 \approx 10^{12}$). And while we are talking about gallons, here is problem 8: estimate the thickness of paint if only one gallon *could* be used to paint a building of surface area A. Clearly, if A is in ft^2, then the thickness $d = (7.5A)^{-1}$ ft. For the "cubical house" of problem 3 (full of soccer balls by now, you will recall), $A \approx 6 \times 30^2 \approx 5 \times 10^3$ ft^2, so $d \approx 10^{-5}$ ft $\approx 10^{-4}$ in. In practice we would need much more paint to do the job, even if we ignored the windows (!). And it would no doubt be impossible to spread paint this thinly (for me, anyway). The point here is to *play* with estimates.

From paint we now go back to the blood problem (6). For a population of 6×10^9 with an *average* of 1 gallon of blood per person, then $V \approx 6 \times 10^9/7.5 = 8 \times 10^8$ ft^3. This could be contained in a cube of side length $(8 \times 10^8)^{1/3} \approx 930$ feet. Putting things a little more picturesquely, Central Park has an area of 1.3 mi^2 so all this blood would cover the park to a depth of approximately $(8 \times 10^8)/[1.3 \times (5280)^2] \approx 22$ ft. Interesting.

Questions of a more sophisticated nature require, not surprisingly, more terms in the estimation formulae. Thus we have the following problems:

9. *How much dental floss does a convict really need?* A recent newspaper article featured the story of an inmate at a correctional center in West Virginia who escaped from the prison grounds by using a rope made from dental floss to pull himself over the courtyard wall. The rope was estimated to be the thickness of a telephone cord, and the wall was 18 ft high. Taking 4 mm for the diameter of the telephone cord, and $\frac{1}{2}$ mm for the floss diameter, then the number of floss fibers in a cord cross section is $(4/\frac{1}{2})^2 \approx 60$, and if each packet contains the standard length of 55 yds, the number of packets required is $N \approx (20 \times 60)/(55 \times 3) \approx 7$. I don't know how many he actually used, nor if his teeth were subsequently in good condition.

10. *Estimate the number P of piano tuners in a certain city or region.* Consider a population in the region totaling N, with an average of p pianos per family (generally $p < 1$). Suppose that pianos are tuned on average b times per year (generally we expect $0 < b < 2$, so the number tuned per year is Npb/n_1, where n_1 is the average number of individuals in a household. If each tuner tunes n_2 pianos per day (where $0 < n_2 < 4$ in general), this corresponds to $250n_2$ per year (for a reasonable working year of 50×5 days). So the number of tuners in the region (city, town, or county) is approximately $Npb/250n_1n_2$. Let's put some numbers in here. If for New York City, say, $N \approx 10^7$, $n_1 = 5$, $b = 0.5$, $p = 0.2$, $n_2 = 2$, then $P \approx (10^7 \times 10^{-1})/(250 \times 10) = 4 \times 10^2$, that is, within the order of magnitude range 10^2 to 10^3. Please note that, as in all these examples, the real object of the exercise is to be able to estimate *any* particular quantity of interest; it is unlikely that the number of piano tuners in a given city is of prime importance to people other than the tuners themselves!

11. *Estimate how fast human hair grows (on average) in mph.* If the hair is cut every n months (usually $n < 2$) and the average amount cut off is x inches, then x/n inches per month $\approx x/n \times 1/(5280 \times 12) \times 1/(30 \times 24)$ mph $\approx 10^{-8}(x/n)$ mph. If $n = 2$ and $x = 1$, then the rate of hair growth is approximately 10^{-8} mph.

12. *The asteroid problem.* In light of the impact(s) of ex-comet Shoemaker-Levy on Jupiter's outer atmosphere, the question has been raised: could it happen here on earth? It may have happened before—one theory for dinosaur extinction (not Gary Larson's; see below) is that about 65 million years ago such an encounter

occurred, this time with an asteroid. Eventually dust from the impact settled back to the surface of the earth, having done a superb job of blocking sunlight and thus devastating plant and animal life. According to one hypothesis, about 20 percent of the asteroid's mass was uniformly deposited over the (now rather inhospitable) surface of the earth—about 0.02 gm/cm^2. *Question:* how large was the asteroid? The mass is clearly about $4\pi r^2 \times 0.02 \times 5$ if r is the radius of the earth in cm. This must be equated to density $\times$ volume for a cube of side L (this is obviously the simplest geometry to consider: the largest sphere that can be inscribed in a cube of side L differs in volume from that cube by a factor $\pi/6 \approx \frac{1}{2}$, so this will not affect our order-of-magnitude estimate). Suppose that we take a typical rock-type density of 2 gm/cm^3, so that $2L^3 \approx 0.4\pi r^2$, whence $L \approx (0.2\pi r^2)^{1/3}$. Since $r \approx 4000 \times 1.6 \times 10^5$ cm (converting miles to centimeters) $= 6.4 \times 10^8$ cm, then $L \approx 6 \times 10^5$ cm, or 6 km (10 km by order of magnitude). This is not unreasonable for an asteroid (even though the dinosaurs may disagree). By the way, in a Far Side cartoon, Gary Larson shows some dinosaurs standing around smoking cigarettes; the caption explains that this is the real reason they became extinct.

13. *Thickness of an oil layer.* As rumor has it, Benjamin Franklin noted that 0.1 cm^3 of oil dropped on a lake spreads to a maximum area of 40 m^2. If d is the thickness of the layer in meters, then $40d = 10^{-7}$, so $d = 25 \times 10^{-10}$ m, or 25 angstroms. Interestingly this corresponds to a "monomolecular layer" of 10–12 atoms (with atom-space-atom-... for a molecule), which is about right for a molecule of "light" oil.

14. *The number of leaves (N) on a tree.* A *very* simplistic argument follows. If r is the typical radius for a tree's leaf "canopy," the "surface area" of the canopy is $4\pi r^2$, and d is (in the same units as r) a typical leaf size, an estimate for the number of leaves is $4\pi r^2/d^2$. Clearly, leaves do not continuously cover the "surface" of the canopy; this does compensate, however, for the fact that there are generally many leaves on branches interior to the canopy (although this will depend on the type of tree). For a small tree (for example a 15- to 20-year-old yew), the leaf canopy has a radius $r \approx 4$ ft, and $d \approx 1$ in, so $N \approx 4 \times 3 \times (50)^2 = 30,000$—that is, an order of magnitude of 10^4–10^5 in general, if we include larger trees as well. Alternatively we could consider a cubical tree canopy of side r and perform the estimate for "surface area" $6r^2$; the result will differ from the former by about a factor of two.

15. *The number of blades of grass on the earth (N).* This, I must admit, is stretching things somewhat, but it's certainly worth a try. Anyway, if 40 percent of the surface of the earth is covered by land, a fraction f of this land will be covered by grass. If the average number of blades of grass per square inch is n, then $N \approx 0.4 \times 4\pi R^2 \times f \times n$ for R measured in inches. Thus for $R \approx 4000 \times 5280 \times 12$, $f \approx 10^{-1}$ or 10^{-2} (this is rather difficult to estimate without a little research), and $n \approx 20$, $N \approx 10^{16}$ or 10^{17}. But I have enough difficulty trying to keep the $\approx 10^6$–10^7 blades in my yard under control.

16. *Population square.* If each person on earth were given enough space to stand comfortably on the ground without touching anyone else, estimate the length of the side of a square that would contain everybody in this fashion. If we give everyone a square $\frac{1}{2}$ meter on a side, then the side of the large square is $L \approx (6 \times 10^9)^{1/2} \times \frac{1}{2} \times 10^{-3}$ km ≈ 40 km.

17. *Human surface area and volume.* To estimate these quantities crudely but quickly, consider a cylinder of radius r and height h: if $r \approx \frac{1}{2}$ ft and $h \approx 6$ ft, then the volume $V = \pi r^2 h \approx 4$–5 ft^3, and the surface area $S = 2\pi r h \approx 20$ ft^2. Since 1 ft ≈ 0.3 m, $V \approx 0.1$ m^3, and $S \approx 2$ m^2. This may be something of an overestimate, and one could use a rectangular box (or parallelepiped!) rather than a cylinder as an alternative. Robert Ehrlich notes that, since most individuals float in water, the average density of a human is about the same as that of water, or 1 gm/cm^3. So a kilogram of you or me occupies about 1000 cm^3 or 1 liter. A person weighing 170 pounds (77 kg) thus has a volume of about 77 liters or roughly 0.08 m^3. For simplicity we will take the former estimates based on the cylinder, but note that whichever value we choose, we can estimate the answer to Ehrlich's question—*How long a hot dog would you make?* The answer may surprise you. He further points out, aided and abetted by a cartoon from Gary Ehrlich, that there *is* one profession that regularly approximates people by rectangular boxes.

Now we are in a position to return to problem 4. If we assume an average cell diameter of 10 microns or 10^{-5} m, then, since, as noted above, 1 ft ≈ 0.3 m, using the estimate of V from problem (17) above, we find that $N \approx 10^{-1}/(10^{-5})^3 = 10^{14}$ cells in the human body. Remember that, even though we know cells are not tiny cubes, we are probably not *too* far off the correct value by assuming this.

18. *The average rate of growth of a child from birth to 18 years.* Over this time span (if h_n denotes height at age n), the average "speed"

is approximately equal to

$$\frac{\Delta h}{\Delta t} = \frac{h_{18} - h_0}{18},$$

that is, $v \approx 1/18$ m/yr $\approx 10^{-3}/(20 \times 400 \times 20)$ km/hr $\approx 10^{-8}$ km/hr—about the same order of magnitude as the speed of hair growth! Perhaps we could label children as super- or sub-follicular depending on whether or not they grow faster than their hair!

19. *Saturday morning reading.* The value of π was calculated to 2 billion decimal places several years ago (probably more by now). Estimate the thickness in inches of your local newspaper if it were solely devoted to printing π (without commas) to this number of places. No doubt the future of the paper's editor would be uncertain for so doing, and subsequent "letters to the editor" would make interesting reading. We shall assume there are no advertisements or photographs or any other extraneous material. According to my estimate, there would be about 3×10^4 digits on each side of a typical page (double that per sheet; your estimate may differ from mine depending on the font size), so that the number of pages needed is $\approx 2 \times 10^9/6 \times 10^4 \approx 3 \times 10^4$. A sheet of newspaper is thin, say about $1/300$ inch thick, so the newspaper would be about 100 inches thick! Imagine being woken up (or perhaps injured) by an 8-foot-high newspaper being thrown onto your porch.

The next set of problems deal more specifically with applications to the Earth as our abode.

20. *The weight of our atmosphere.* At first sight this might seem to be a very difficult problem, but noting that atmospheric pressure at sea level is about 14.8 lb/in^2, all we need to do is calculate the surface area of the Earth in square inches, because each square inch supports a column of air weighing 14.8 pounds. We shall of course conveniently assume that we live on a perfectly smooth sphere of radius 3960 miles! Therefore, if W denotes the weight of the atmosphere in pounds, then

$$W = 4\pi(3960 \times 5280 \times 12)^2 \times 14.8 \approx 1.19 \times 10^{19} \text{ lb.}$$

A similar type of calculation, though more of an estimate, for the weight of a cloud, can be found in chapter 6.

21. *A tunnel through the earth.* Consider the earth to be a perfect sphere (yet again) of radius r. Imagine a straight tunnel $2a$ miles long bored through it perpendicular to the north-south diameter, with midpoint

a distance b miles below the north pole, and let $c = 2r - b$ (see figure 2.1). By the *sagittal theorem*, which is easily derived using similar triangles, $a = \sqrt{bc}$, and since $2a \leq b + c$, we have a geometrical interpretation of the *arithmetic mean/geometric mean inequality* for two non-negative quantities b and c, that is,

$$\frac{b+c}{2} \geq \sqrt{bc}$$

(see chapter 7 for a simple algebraic proof of this result). Let us now apply this to the tunnel, with endpoints at A and B. Suppose they are one mile apart; how far below the north pole is the middle of the tunnel? Equivalently, how big is the "bulge" b? Since $b = a^2/c$, the diameter of the earth is approximately 7920 miles, and $c \gg b$, it follows that

$$b \approx \frac{(1/2)^2}{7920} \text{ miles or } \approx 2 \text{ inches.}$$

Given any tunnel for which the length is small relative to r, c is approximately constant and so b is proportional to a^2. If the tunnel is 10 miles long ($a = 5$ mi) then b is 100 times larger, or about 17 feet. In other words, the center of the tunnel is 17 feet closer to

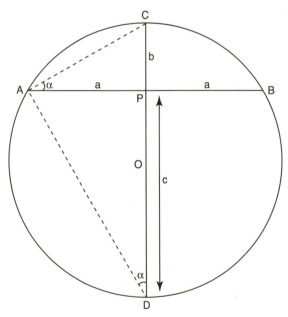

Figure 2.1.   The geometry for the application of the sagittal theorem.

the center of our idealized earth than are the ends. It is not surprising, therefore, that water has a tendency to collect in the middle of such feats of engineering! As an afterthought, perhaps the sagittal theorem is so called because the diagram for this problem looks rather like a crossbow ready to be used.

Let us now examine some problems that I found rather surprising and counterintuitive when I first encountered them. We will go a little beyond the arithmetic of estimation, but not by much. The first example below (unnumbered) will set the scene for the second and more important one. They are closely related in that they both involve expansion due to a small increase in temperature of a solid horizontal strip, but for the first problem the strip is rigidly fixed at each end. Suppose that the strip is initially $2D$ miles in length, and expands a total of $2\Delta D$ inches while rigidly attached at each end (the factors of 2 are introduced for computational simplicity). Since it has nowhere else to go, the strip will "bow" or buckle upward, *but by how much?* The question requires us to find the maximum height d of the strip above ground level, and since by assumption the expansion due to heating will be taken as uniform along the strip, all we need to do is calculate the height of the arch midway between the fixed ends. The shape of the arch itself is not relevant here, so we represent it in our "model" by an isosceles triangle with vertices at the fixed ends and the top of the arch (sketch the picture for yourself). Obviously, this will induce some error, in that we are now replacing each half of the arch by the hypotenuse of a right triangle, but it can be shown that the curvature of the arch is so small that the latter is well approximated by two such line segments. This will be the case provided the relative expansion $\delta D/D \ll 1$, where δD is defined as $\Delta D/(12 \times 5280)$ by converting ΔD to units of miles. It follows from the Pythagorean theorem that

$$d = \sqrt{(D + \delta D)^2 - D^2} \approx \sqrt{2D \times \delta D} \text{ mi}$$

if $\delta D/D \ll 1$. Suppose now that the strip is one mile in length and it expands by 2 inches, that is, $2D = 1$ and $\Delta D = 1$, or $\delta D \approx 1.58 \times 10^{-5}$, so that the relative expansion is tiny; then $d \approx 3.97 \times 10^{-3}$ mi or about 21 feet! This is quite a surprising result, at least to me. Note that for a given strip the length d is proportional to the square root of the extension, so quadrupling the extension results in a twofold increase in the buckling height. If the extension is only one inch, the buckling height is nearly 15 feet, which is still quite impressive. Obviously this is one reason why rails and other constructs have expansion gaps built into them. Next we proceed with what at first sight appears to be another "impractical" problem!

22. *Band tectonics.* The polar and equatorial circumferences of the earth differ by little more than one hundred miles, with each circumference being slightly less than 25,000 miles. We shall adopt this figure

for simplicity, although it will not make any difference to the conclusions drawn. Consider then the earth to be a perfectly smooth sphere (no mountains or valleys!) of radius r such that the circumference $C = 2\pi r = 25{,}000$ mi. Now imagine a metal band to be wrapped tightly around the globe, so its length is also equal to C. Then cut the band, insert an extra strip ΔC feet in length into it, and arrange the now rather loose band into a circle concentric with that of the earth, that is, with the same "gap" everywhere. If this gap is d feet high, how large is d? Converting ΔC and d to miles, it follows that

$$\left(C + \frac{\Delta C}{5280}\right) = 2\pi\left(r + \frac{d}{5280}\right).$$

Subtracting the equation $C = 2\pi r$ from this yields the result for the width of the gap as

$$d = \frac{\Delta C}{2\pi}.$$

Putting in ΔC as 1, 10, 20, and 100 feet, respectively, gives the corresponding set of gaps in feet as approximately 0.16 (about 1.9 inches), 1.6, 3.2, and 16, really very surprising! To put it another way, adding an extra foot to a band 25,000 miles in length results in a gap that a rat could squeeze under!

Notice that the final result is *independent* of the radius of the earth, or any other spherical object we choose to circumscribe, so these values hold for *any* such object—Jupiter (also not a sphere, but for our purposes $r = 88{,}000$ mi), or the sun ($r = 864{,}000$ mi)! Of course, for decreasingly small objects (such as soccer balls, billiard balls, or even a mathematical point!) the gap is the same, but the result does not seem so impressive.

Another point deserves consideration: really just how "smooth" is the earth, in comparison with a billiard ball, for example? A crude estimate can be made, still assuming that the earth is a sphere of radius r (≈ 3960 mi), by computing the "smoothness factor"

$$s = \frac{h_m - d_t}{r} \equiv \frac{\Delta r}{r},$$

where h_m and d_t are the height of the highest mountain and depth of the deepest trench, respectively, both relative to sea level. The height of Mt. Everest is about 29,000 feet, or 5.5 miles, while the Mariana trench is about 6 miles deep (or -6 miles high). This means that

$$s = \frac{5.5 - (-6)}{3960} \approx 3 \times 10^{-3},$$

or three parts in one thousand, probably at least as smooth as a billiard ball!

At this juncture we may well ask: so what? This is a well-known example of a simple counterintuitive result that is easily derived, but does it have any realistic implications? Ernest Zebrowski, Jr., in his book on the history of the circle, certainly thinks so. He examines what this example implies about the "real tectonic processes that have wrinkled the crust of our planet." He considers the now-solidified crust of the earth as being composed of many such tight circumferential bands. In the present era this crust is a very thin layer between 4 and 40 miles thick overlying an interior of molten rock. If one of these "bands" expands by 20 feet as a result of even a tiny increase in temperature (Zebrowski suggests 0.015°F), then, as we have seen above, the result could be a "wrinkle" in the crust over 3 feet high! In general the situation would be one of uneven heating and cooling of the crust resulting in irregular regions where expansion and contraction occur.

We end this particular example by relating Zebrowski's conclusions in his own words:

> The crust is thrust upward in some places, while in other places it is drawn apart. The result is a spectrum of geophysical phenomena: earthquakes, volcanoes, and the growth of mountain ranges (Mt. Everest, for instance, grew about 26 feet between 1850 and 1950, and in the Alaskan earthquake of 1964, the surface in some locations was thrust upward by as much as 30 feet). The temperature fluctuations that drive this mechanism are very small and can't easily be measured, for we are talking about temperature changes *within* the crustal bands, at depths of several miles. The conclusion is nevertheless inescapable that significant geophysical surface phenomena result from tiny irregular changes in the planet's circumference. The algebraic calculation does not predict where such events will occur, but it does give us a mathematical analogy to explain them.

23. *The weight of a hypothetical mountain.* No real mountain is a perfect cone, but ours will be, and, following the choice made by COMAP, it will be made of granite! One cubic foot of granite weighs about 165 lb, and we shall take our Mt. Neverest to be 5.5 miles high with a base diameter of 5 miles. Using the formula for the volume of a right circular cone, the volume of the mountain is about 36 cubic miles and the weight is found to be

$$W = \frac{1}{3}\pi(2.5 \times 5280)^2 \times (5.5 \times 5280) \times 165 \text{ lb} \approx 8.7 \times 10^{14} \text{ lb},$$

or approximately one ten-thousandth the weight of the atmosphere!

A related topic is the question: how high can a mountain be? This is no doubt determined by a combination of many geophysical factors, but the most significant is probably the pressure at the base of the mountain. If this pressure exceeds what is called the yield strength, then the base rock will undergo a phase change and become plastic, and flow out under the weight of the rock above it. This will reduce the height and hence the weight of the mountain, so if the base area remains about the same, the base pressure is correspondingly reduced, and the base of the mountain will revert to the solid phase again. This places a natural restriction on the height of a mountain. For Mt. Neverest the pressure is

$$\frac{\text{weight}}{\text{base area}} = \frac{8.7 \times 10^{14}}{\pi(2.5 \times 5280)^2} \text{ lb/sq. ft} \approx 1.6 \times 10^6 \text{ lb/sq. ft.}$$

The yield strength of granite is much higher than this, being close to 30 million pounds per square foot, so our mountain is nowhere near the limiting size. What might this size be? On the basis of dimensional arguments alone (see chapter 3), we can deduce an upper bound that is probably excessively large, for no account has been taken of structural features of the earth's crust, volcanic activity, and other mechanisms of mountain formation, and so forth.

Proceeding, then, with a large pinch of sodium chloride, let us generalize the above ideas to include other worlds, for example, the moon and the planet Mars. At this point the reader's forbearance is requested. In the following chapter a discussion will be found as to how various physical attributes such as area, volume, weight, and strength, vary with size for geometrically similar objects. Since the present example uses some of those ideas while being more appropriately placed in *this* chapter, we proceed by supposing for simplicity that the mountains in each location are geometrically similar and are made of the same material, in which case we can use the scalings that weight $\sim h^3 g$ and base area $\sim h^2$, where h is the maximum permitted height of the mountain and g is the local gravitational acceleration. Then, since the yield pressure Y must be the same on each "planet," we may write

$$Y = \frac{\text{weight}}{\text{base area}} \sim \frac{h^3 g}{h^2} \sim hg \sim \text{constant},$$

from which we conclude that maximum mountain height is inversely proportional to local gravitational acceleration for a given yield strength. This means that the height limit of mountains on both the moon and Mars should be higher than on earth (if the assumptions made above are not seriously violated). The fact that the Leibnitz range on the moon has mountains comparable in height to the Himalayas and Olympus Mons on Mars is about three times as high as Mt. Everest is consistent with this claim but certainly not proof for it. Returning to our planet (for which g may be taken as constant), the above result also implies that $Y \sim h$. Thus by simple proportion the limiting height for our Mt. Neverest is determined by the ratio of yield strength to base pressure, times the height—

$$h_{max} \approx \frac{30 \times 10^6}{1.6 \times 10^6} \times 5.5 \approx 100 \text{ miles!}$$

As indicated above, due to the crudeness of the argument, this is excessively high. But an extreme optimist might argue that, since there are no mountains that high, the theory must be correct! Note that, if our mountain were made of something other than granite, then yield strength would also be different, and the maximum height correspondingly so.

24. *The speed of descent of a cloud droplet in still air.* The behavior of small particles like aerosols, or tiny cloud droplets falling slowly through air, or sediment settling down to the bottom of a lake is well described by *Stokes's law*, provided their speed of descent is small enough that no turbulence is generated in their wake. This will occur if the Reynolds number $\text{Re} = v r \rho_a / \mu < 1$, where under steady conditions v is the (constant) speed of descent, ρ_a is the density of the air, and μ is the dynamic viscosity of the air. For a spherical object of radius r the law states that the drag force F encountered by the object is

$$F = 6\pi r v \mu.$$

The constant speed, if it occurs, is called the terminal speed and is a consequence of the balance between the downward direction of the weight of the particle and the upward forces of buoyancy and drag (the former, being tiny, may usually be ignored in air for small particles). Therefore, if m is the mass of the particle,

$$6\pi r v \mu = mg,$$

and so

$$v = \frac{2\rho_p g r^2}{9\mu},$$

where ρ_p is the density of the particle. Let us apply this result to calculate the terminal speed of a typical cloud droplet of diameter 20 microns (2×10^{-5} m) in still air at 5°C at an altitude of 1000 m, given that $g \approx 9.804$ m/s^2, $\rho_p \approx 1000$ kg/m^3, and $\mu \approx 1.758 \times 10^{-5}$ Ns/m^2 (see table C2 in Munson et al.). Therefore

$$v = \frac{2 \times 10^3 \times 9.804 \times (10^{-5})^2}{9 \times 1.758 \times 10^{-5}} \approx 1.2 \times 10^{-2} \text{ m/s},$$

or about one cm/s. The Reynolds number for this droplet is

$$\text{Re} \approx \frac{1.2 \times 10^{-2} \times 10^{-5} \times 1.1}{1.8 \times 10^{-5}} \approx 7 \times 10^{-3} < 1$$

(where the density of the air, $\rho_a \approx 1.1$ kg/m^3), so the use of Stokes' law is validated. Note that since $v \propto r^2$ a droplet one hundred times smaller (of radius 0.1 micron) falls at about 10^{-6} m/s. Elsewhere in this book a more generic form of the Reynolds number is used, namely $\text{Re} = ul/v$, where $u, l,$ and v represent a typical speed, length scale, and kinematic viscosity respectively.

Our final "problem" in this chapter is just for fun and uses a little differential calculus, but it still qualifies as an estimation problem. In the science fiction book *The Black Cloud* by the late Sir Fred Hoyle (a famous astrophysicist at the University of Cambridge), one of the characters in the story makes a neat little "back of the envelope" calculation, using observational data, to estimate the time of arrival T of a mysterious and seemingly intelligent cloud of dust and gas that is directly approaching the earth. At that point in time, at a distance D from the earth, the cloud subtends a small angle (a radians) at the earth, and its speed of approach is denoted by V. Clearly a and D are both functions of time (assumed to be differentiable!). In terms of the linear width d of the cloud and its distance D from the earth, $d \approx Da$, since a is small, from which it follows that

$$\frac{da}{dt} = -\frac{d}{D^2}\frac{dD}{dt} = \frac{dV}{D^2},$$

since $V = -dD/dt$. But if V is constant, then $V = D/T$, so that

$$T = \frac{D}{V} = \frac{d}{D}\left(\frac{da}{dt}\right)^{-1} = a\left(\frac{da}{dt}\right)^{-1}.$$

Therefore for small enough intervals of time (ΔT) and the corresponding increment in a (Δa) we may write

$$T \approx a\frac{\Delta t}{\Delta a} = \left(\frac{\Delta a}{a}\right)^{-1}\Delta t.$$

Over an interval of one month the concerned astronomers in the story noticed that the relative increase in the cloud's angular diameter ($\Delta a/a$) was about 5 percent, or 0.05, so they were able to estimate that the time until "contact" was about 20 months. To see what happened next you will have to read the book, but do try to finish this one first.

CHAPTER THREE

Shape, Size, and Similarity: The Problem of Scale

> What strength he has in his loins, what power in the muscles of his belly! His tail sways like a cedar; the sinews of his thighs are close-knit. His bones are tubes of bronze, his limbs like rods of iron.
>
> —Job 40:16-18

DIMENSIONAL ANALYSIS I

The phrase *dimensional analysis* will be used in two senses in this chapter. The first one is somewhat loosely defined, but essentially it concerns the way that physical characteristics of an object (such as surface area, volume, strength, power) vary with its size L. "Size" here means any representative linear dimension of the object—height, width, leg length, and so on—provided that dimension is used consistently in all comparisons made for a particular property. This is particularly useful (and valid) when the objects are geometrically similar, and that assumption will be made in much of this section. It is not particularly accurate for fauna in general, but as a first (or even zeroth) approximation to reality it can provide some useful insights into the way things "work."

Simply put, the problem of scale means—*What happens as things get bigger?* Consider a simple three-dimensional object: a cube of side L (units of length used—feet, centimeters, etc.—are of no significance here provided we use them consistently). The surface area of the cube is $6L^2$ square units and the volume is L^3 cubic units (from now on we will omit the cumbersome expressions "square units" and "cubic units"). Doubling the linear dimensions of the cube will result in a new area of $6(2L)^2$ or $24L^2$ and a new volume of $(2L)^3$ or $8L^3$. Thus the area has increased by a factor of 4 and the volume by a factor of 8. Furthermore, the *ratio* of the area to the volume has *decreased* from a numerical value of $6L^2/L^3 = 6L^{-1}$ to $24L^2/8L^3 = 3L^{-1}$ (the units being inverse feet, cm, etc. depending on the original units chosen). Indeed, if the cube dimension is changed by a factor α, the area changes by a factor α^2 and the volume by a factor α^3 (if the side of the cube is doubled, its area quadruples and its volume increases eightfold; if the side is halved, the area and volume are 1/4 and 1/8 of the respective original values). The corresponding area-to-volume ratio changes by a factor α^{-1} for a cube. These properties associated with such a change of scale are illustrated in table 1.

Dimension L	Area $6L^2$	Volume L^3	$\dfrac{Area}{Volume} = 6L^{-1}$
0.01	6×10^{-4}	10^{-6}	600
0.1	0.06	10^{-3}	60
0.5	1.5	0.125	12
1	6	1	6
2	24	8	3
3	54	27	2
4	96	64	1.50
5	150	125	1.20
10	600	1000	0.60
100	6×10^4	10^6	0.06

What if we were to change the object from a cube to a tetrahedron, say, or any of the other platonic solids, or even a sphere? Apart from a different numerical factor in place of the 6 for a cube, nothing new would occur. The surface area would "scale" in proportion to L^2 and the volume in proportion to L^3 as before, where L is the length of an edge (or the radius of the sphere). The surface area to volume ratio is therefore still proportional to L^{-1}, or if you prefer, inversely proportional to L. This is illustrated in table 2. The five platonic or regular solids, by the way, are

  (i) the tetrahedron, with four equilateral triangular sides,
 (ii) the cube, with six square sides,
(iii) the octahedron, with eight equilateral triangular sides,
 (iv) the dodecahedron, with twelve regular pentagonal sides, and
  (v) the icosahedron, with twenty equilateral triangular sides.

There can be no others in three spatial dimensions unless we are permitted to join regular plane figures of more than one type (see Stevens's or Huntley's books for a brief discussion of why this is so). In passing, let us note a very important result that is true for all polyhedra, not just regular ones. If v, e, and f respectively are the numbers of vertices, edges, and faces for a given polyhedron, then

$$v - e + f = 2.$$

Always. Check it out for yourself. This is known as Euler's formula (although there are many formulae thus designated: Euler was a very prolific mathematician!).

Solid	S. Area	Volume	S.A./Vol.	or $\approx$
Tetrahedron	$\sqrt{3}L^2$	$\sqrt{2}L^3/12$	$6\sqrt{6}/L$	$14.7/L$
Cube	$6L^2$	L^3	$6/L$	$6/L$
Octahedron	$2\sqrt{3}L^2$	$\sqrt{2}L^3/3$	$3\sqrt{6}/L$	$7.3/L$
Dodecahedron	$\frac{15L^2}{\tan(\pi/5)}$	$\frac{5L^3\cos^3(\pi/5)}{\sin^2(\pi/5)}$	$\frac{3\sin(\pi/5)}{L\cos^2(\pi/5)}$	$2.7/L$
Icosahedron	$5\sqrt{3}L^2$	$\frac{10L^3\cos^2(\pi/5)}{3}$	$\frac{3\sqrt{3}}{2L\cos^2(\pi/5)}$	$4.0/L$
Sphere	$4\pi L^2$	$4\pi L^3/3$	$3/L$	$3/L$

Straightforward considerations like this are important in the problem of scale, because differences in size affect structure (e.g., a large cube is relatively weaker than a small cube, and King Kong, as a scaled-up gorilla, simply could not exist; see later in this chapter) so to maintain the same structural characteristics a change in size is usually accomplished by a change in shape. Two examples illustrating this last point are worth considering: (i) very large animals have bones that are larger in proportion to their size than do small ones, as any visit to a museum of natural history will show; (ii) if we neglect the fact that the materials are very different, and assume for simplicity a square base for the Empire State Building, then with the same height/width ratio as a stalk of wheat it would be five feet wide! Clearly, and for good engineering reasons, the Empire State Building is somewhat wider! This little calculation is based on simple proportion: neglecting the television tower, the building is about 1,250 feet high; wheat can grow to about five feet in height, and the stalk is approximately 1/4 inch thick, so

$$\frac{w}{1250} = \frac{1}{4 \times 12 \times 5}$$

or $w \approx 5$ feet.

Returning to the first example: if you have not visited a natural history museum, ask yourself the following question—are elephants shaped like gigantic mice? Of course not, but why not? The surface area of an object— you, me, or a cube—is the fundamental means by which the object interacts with the environment. There is no other easy way for heat to get in or out, even for warm-blooded creatures that generate their own heat from within (from food). If the surface area to volume ratio is too small (L is too large), the animal must have a low enough metabolic rate in warm climates to allow the heat generated to escape adequately through the available surface area, or use some other means of cooling. That is one reason why elephants (for whom L is large) have large ears—to act as radiators! If the animal is small, the surface area to volume ratio is large, and in cold climates such an animal may lose heat too rapidly to survive without compensating insulation. Other things being equal, smaller creatures will avoid higher latitudes where larger ones fare better.

If objects are geometrically similar, then the ratio of their volumes is equal to their (ratio of heights)^3; if we assume that the mean densities of the objects are the same, then this ratio is also equal to the ratio of their weights. Similarly, the ratio of their surface areas is equal to their (ratio of heights)^2. While humans and most other animals do not grow in a geometrically similar fashion, it is nonetheless easy to see from these statements that one such object must have height $2^{1/3} \approx 1.26$ times that of the second in order to possess twice the volume, and a height $\sqrt{2} \approx 1.41$ times that of the second in order to have twice the surface area. Such arguments provide estimates for more realistic models of animal structure and development.

Many things vary with L^2, the square of their length (or other typical dimension such as width or height). Exchange of gas takes place through the surfaces of lungs, gills, or leaves. Food is assimilated in animals via the surface of the gut. Heat is primarily gained or lost (usually the latter) through the skin. The capacity to withstand compressive or bending forces—strength—is dependent on the cross-sectional area of the bone, beam, tree, or whatever is required to resist such forces. Drag, whether in air or water, for large Reynolds numbers is generally proportional to the area of the surface in contact with the fluid. (For small Reynolds numbers drag is proportional to the size or linear dimension of the object, as noted in problem 24 in the previous chapter. In intermediate regimes it is proportional to some combination of size and area.) A bird's ability to maintain lift is proportional to the area of its wings. Metabolic power consumed (and generated) by a muscle is related, for various reasons discussed below, to its cross-sectional area. Mass, the quantity of matter in an object, is proportional to its volume, which in turn means that weight (in a constant gravitational field) and mass are both proportional to L^3 (but remember that weight is a force). Buoyancy, by Archimedes' principle, also scales as L^3.

We have noted that the strength of a solid columnar beam (made of bone, steel, or any other material designed to support a load) in compression is proportional to its cross-sectional area, and so for geometrically similar beams of the same material their relative strengths are equal to the ratio of their cross-sectional areas. What about the relative strength of beams under a load acting perpendicular to them, causing them to sag? My office bookshelves are excellent examples of this phenomenon. We consider a slightly different problem, however: that of a uniform horizontal beam of length L, and weight W, supported at each end. What follows is a crude oversimplification of the problem, but consistent with the philosophy of this chapter, we are merely concerned with basic dimensional considerations. The weight of the beam is assumed to act effectively at the center of the beam. The *moment* of W about each end is $WL/2$, which is proportional to L^4, but the resistance of the beam to bending, it is reasonable to assume, is again proportional to the cross-sectional area of the beam, that is, to L^2. Thus the ability of the beam to resist a downward "sag" at its center is exceeded by the torque producing the sag in the ratio $L^4 : L^2$ and is therefore proportional to L^2 according to this admittedly crude argument. This implies that, given a two-inch-long matchstick and a six-foot-long geometrically similar beam made of the same material, the latter will sag by an amount $(36)^2 \approx 1300$ times as much. Thus a beam will eventually fracture and then fail under its own weight if its size is indefinitely increased in a geometrically similar manner.

Principles like those discussed here have implications for daily life for both the human race and the rest of the animal kingdom. Those of us who need sweetening stir our sugar into our coffee to increase the surface area by which the sugar comes in contact with the coffee and facilitate dissolving. For similar reasons we chew our food to increase its surface area so that ultimately the food energy associated with it may be absorbed easily. A log is difficult to ignite in the absence of a roaring fire; by chopping it into kindling wood we are increasing its surface area to improve our chances of starting a fire. Indeed, it has been pointed out that if we go one stage further and grind the wood into tiny particles of dust, these may, when mixed with air, ignite so rapidly that an explosion occurs.

In what follows, the symbol $O(\dots)$ is used to denote "order of magnitude." Thus, in connection with the size of cylindrical logs (as in the next paragraph), if the radius is $O(10 \text{ cm})$, it means the radius will be closer to 10 cm than either 1 m or 1 mm. Although the O symbol does have a precise mathematical meaning, it is used here as a rough measure of size that does not distinguish between, say, 7 cm and 20 cm because they are both $O(10 \text{ cm})$. On the other hand, 88 cm is $O(1 \text{ m})$. Essentially, the number of interest is rounded to the nearest power of ten. Additionally, the symbol $\sim$ used in this chapter means "varies as" or "scales as" the quantity it precedes. Thus surface area $\sim L^2$ and volume $\sim L^3$, where L is an appropriate linear dimension. It should not be confused with the commonly used symbol $\approx$ meaning "is approximately equal to."

We can go even further. Because of their small surface area to volume ratios, logs burn more slowly than kindling, which in turn burns more slowly than a gunpowder-impregnated fuse, as Herbert Lin remarks. Cylindrical logs (radius (10 cm)) should burn about 10 times more slowly than kindling (radius $O(1 \text{ cm})$) and about 100 times more slowly than a fuse (radius $O(1 \text{ mm})$). Experience with the first two at least (!) indicates that logs typically burn for an hour and kindling for minutes. A long fuse may burn for seconds (at least it always seems to in action movies).

Small animals huddle together to stay warm by reducing their combined surface area exposed to the air, hence losing less heat to their surroundings. To generate enough heat to stay warm, mice eat 25 to 50 percent of their body weight each day; hummingbirds take in more than their body weight each day; pigmy elephant shrews (who weigh about 1/5 ounce) eat almost constantly and if necessary resort to cannibilism. In many mammals, the length of the vocal chords and the size of the sound-receiving apparatus (ears and inner ears) are proportional to the size L of the animal, so the larger the animal, the deeper the voice and the lower the range of frequencies heard. Whale songs travel enormous distances in the ocean, partly because the sounds spread out cylindrically (rather than spherically) with consequently less geometric attenuation, but also because of their low pitch, which renders the sounds less susceptible to absorption mechanisms.

In a fascinating essay entitled "On Being the Right Size," J.B.S. Haldane has pointed out the advantages of being small (specifically, of weighing very little) as far as falls are concerned: "You can drop a mouse down a thousand-yard mineshaft; and, on arriving at the bottom, it gets a slight shock and walks away. A rat would probably be killed, though it can fall safely from the eleventh story of a building; a man is killed, a horse splashes." The consequences of being even larger are seen occasionally: a beached whale stranded in shallow water or on a sand bar can suffocate under its own weight. In subsequent sections we will be studying these and other dimensional considerations in a more quantitative manner, but already we may draw some qualitative inferences from these observations.

First, let us examine the Haldane example: *falling*. When an object falls, its energy of motion (kinetic energy) is proportional to its weight W (if the height fallen is h, then in appropriate units its kinetic energy is Wh). Upon impact this energy is mostly absorbed by the object, the small remainder being dissipated as sound. The kinetic energy is absorbed over part of the surface area of the body, so since the weight (and hence volume) of the object increases as L^3, the area of absorption increases also, but only as L^2. Thus, as size increases, so do the hazards of falling from the same height. Similar arguments apply for geometrically similar objects resting on a flat surface: pressure, or weight per unit area on the surface, increases linearly with L (L^3/L^2 for a cube). Obviously, if it is possible to increase the area in contact with the surface by means of a change of shape, the pressure can be reduced for a given weight.

What about *diving*? The diving ability of a mammal (whether human, whale, or hippo) depends on the volume of air (and hence oxygen) carried in the lungs, which in turn scales as the cube of the animal's size, L^3. A second factor is the rate at which oxygen is absorbed by the lungs, which at first sight is proportional to the lung area, scaling as L^2. Thus, crudely, we may estimate the duration of a dive, T, to be given by the following ratios:

$$T \sim \frac{\text{lung volume}}{\text{lung area}} \sim \frac{L^3}{L^2} = L.$$

As in all these arguments, there will be exceptions, even in more realistic models, but this does indicate a fundamental relationship between duration of dive and size. It is gratifying to note that this appears to be the case when comparing blue whales and humans. A blue whale (the largest living mammal) is typically 30 m in length; such whales can hold their breath for times on the order of half an hour, sometimes longer. Even the most skilled breath diver cannot sustain a dive longer than about two minutes. The ratio of durations (whale : human) is about fifteen, the ratio of the corresponding sizes.

Of course, there are other factors that we have neglected here. The oxygen-absorbing surface of a human lung (and presumably Moby Dick's also) is not flat. The bronchial tubes branch into smaller and smaller tubes (as does a tree limb by dividing into branches and twigs), ultimately opening into tiny air sacs (alveoli), perhaps more than 6×10^8 in number. If the walls of these air sacs alone could be spread out flat, they would cover an area typically of between 600 and 1000 ft^2 (approximately 50–100 m^2). This means that the rate of oxygen absorption scales more rapidly with size than L^2, possibly $L^{2.5}$, for example. This would render $T \sim \sqrt{L}$ instead of $T \sim L$, other things being equal. But they are not! The muscles of whales store much more oxygen than do those of other mammals (compare 40 percent for whales to about 13 percent of their supply for humans). During a whale's dive, the blood flow to the muscles is greatly reduced, and the heartbeat slows, both of which reduce oxygen consumption. These factors will increase T, which may well compensate for the surface area effect mentioned above. Needless to say, our simple scaling models cannot incorporate all the effects of this kind of "fine structure."

Before proceeding to other forms of mammal activity, note that the original argument about dive duration also applies to pulse rate. The pulse rate of a mammal is proportional to the rate at which the heart pumps blood and inversely proportional to the volume of the heart. For reasons identified below, the former scales as L^2, while the latter obviously scales as L^3. Hence, to a crude approximation, the pulse rate r

$$r \approx \frac{L^2}{L^3} = L^{-1} \propto M^{-1/3}.$$

We will improve on this later (in the appendix to this chapter), and show that $r \approx M^{-1/4}$ is a rather better description.

What about *jumping* (to conclusions, or otherwise)? A typical dog or human flea can jump about two feet into the air. Since for the flea this is many times its size ($L \approx 3$ mm, or about 1/8 inch), the thought may arise—just think how high a flea the size of a human could jump! Well, we will do just that (think, not jump). It turns out to be about the same height as the real flea can jump—nothing much changes. The muscle strength of the flea, human, kangaroo or other animal, is proportional to the cross-sectional area of the muscle. A jump is caused by the muscle contracting in a manner proportional to its length, so the ability to jump is proportional to the product of these two quantities, namely, to the volume of the muscle. This in turn is proportional to L^3, the volume of the "animal," as is the weight of the animal, so the weight and the energy available to jump go up in direct proportion. Thus the height of jumping is approximately independent of size, at least for geometrically similar objects. Somewhat more quantitatively, note that to jump a height h it must be that the leg muscles contract through a distance d, so that the potential energy gained is the work done—$mgh \sim Td$; since $T \sim L^2$ and $d \sim L$, we obtain the interesting result that

$$h \sim \frac{Td}{mg} \sim L^0 \sim \text{constant}, \qquad (T \text{ being tensile stress})$$

so that adult jerboas, kangaroos, and humans (and perhaps also Tigger) may be able to jump to about the same height. Similar details of the dimensional analysis for other examples follow below, but first we must examine the concept of *power*.

Power is defined as the rate of doing *work*, or for the dimensional purposes of this chapter, the work done by a force F in unit time. Since work (or energy) has dimensions of force times distance, it follows that power has dimensions of force times speed. This will be utilized below in a simple account of running on the level. In most animals, if L is a typical length scale, then their power output (or metabolic rate) is approximately proportional to their surface area, L^2. This is good news, because animal muscles are typically about 25 percent efficient, which means that the remainder is lost to the outside world as heat; if the power generated by muscles varied with an exponent greater than two, animals (including ourselves) would overheat on a regular basis! (Remember that all these arguments are imprecise at best; the intention is merely to establish some general size-dependent principles from which rough conclusions may be drawn. Such a process can be a very important prelude to formulating a mathematical model more precisely.)

The preceding comments notwithstanding, there are some good biological and physical reasons for this L^2-dependence: the rate of oxygen supply to muscles depends on the volume of blood reaching the muscle cells per unit time, which for a cylindrical vessel of radius r is (blood speed v) $\times \pi r^2$. For most animals it is known that the speed v is approximately independent of size (meaning here that it probably varies by no more than a factor of two or three at most in the spectrum of animal size from mouse to elephant). Since the radius of the vessel will be proportional to L, the rate of supply of life-giving oxygen will vary roughly as L^2. Another determinant of this power of L is the fact that muscular force is limited by the tensile stress (T) of muscles and tendons, which by our earlier arguments for column strength varies as L^2. When a muscle moves a limb of mass m, kinetic energy is supplied to the limb to move an average distance d with an average speed v, so, equating the kinetic energy to the work done, we have the approximate scaling:

$$\frac{1}{2}mv^2 \sim Td.$$

Since $m \sim L^3$ and $d \sim L$, it follows that

$$L^3 v^2 \sim L^3 \quad \text{or} \quad v \sim L^0 \sim \text{constant},$$

that is, the speed v is approximately size-independent. Now if the muscle contracts in a time t, where

$$t \sim \frac{d}{v} \sim \frac{L}{v} \sim L,$$

the power output of the muscle $\sim Td/t \sim L^2$, which also helps to justify the earlier assertion.

Now we examine the size-dependence of the maximum speed V for an animal running on level ground, and also uphill. When any object moves through a real fluid it experiences a drag force D. For the running (or indeed flying) animal this is air resistance, and the dependence of this retarding force on speed is rather complicated in practice. It is sufficient for our purposes to note that for "small" speeds (like that of a tiny cloud droplet falling through still air, as in chapter 2) the drag is proportional to speed, but for higher speeds (as considered here) it is more reasonable to take it as increasing like V^2. The drag experienced is also proportional to the surface area directed perpendicularly into the air, so in dimensional terms, $D \sim L^2 V^2$. We have noted above that the power (P) needed to "do the job" can be expressed as "force times speed," so

$$P \sim DV \sim L^2 V^3.$$

But since the power available to the animal $\sim L^2$, it appears that within the limitations of this argument $L^2 \sim L^2 V^3$, implying that on the level the top running speed is independent of size. This suggests that a typical rabbit could run as fast as a typical horse on flat ground, to within a factor between one and two—perhaps there might be a 5 or 10 mph difference, but not a factor of two. As with all fuzzy arguments of this type, there will be counterexamples; thus it is clear that an Olympic athlete sprinting

100 meters in 10 seconds (about 22 mph) could not keep up with a cheetah running at a top speed of 70 mph, but again, that should not be surprising—they are not geometrically similar animals!

Consider next the problem of running uphill at top speed. Now the consequences of doing work against gravity come into play. Indeed, this will be the determining factor, so for simplicity we now neglect air resistance, noting that the top speed attainable will generally be less than on level ground, so the air resistance will be correspondingly smaller. The work done by an object of mass m in moving a vertical distance h is the potential energy gained, mgh. Since the distance l moved up a slope of angle θ is $h \csc \theta$, $l \sim h$, and so the rate of increase of height in time t is a speed

$$v \sim \frac{l}{t} \sim \frac{h}{t}.$$

Now work is done against gravity at a rate

$$mg\frac{h}{t} \sim L^3 v.$$

Recalling that the power available $\sim L^2$ it follows that at top speed

$$L^2 \sim L^3 v \quad \text{or} \quad v \sim L^{-1},$$

so the maximum speed up a hill is inversely proportional to size, perhaps explaining why it is advisable to run away uphill from a bear! Alternatively, if a dog and a pony are rather evenly matched on level ground, the latter may lose the race up a hill.

We have been "running" before "walking," but in order to do justice to a simple model of the latter, the following information is required. The period of oscillation P for small oscillations of a simple pendulum of length L is

$$P = 2\pi \sqrt{\frac{L}{g}},$$

as is shown in many books on applications of ordinary differential equations. Here "small" means that the approximation $\sin \theta \approx \theta$ must be valid, where θ (in radian measure) is the maximum angular displacement of the pendulum from the vertical direction. In practical terms this means that θ may be as large as 0.4 radians, or about 20 degrees of arc. This approximation reduces the governing differential equation for the motion of the pendulum to that of simple harmonic motion, from which the above period is readily obtained. (When θ is too large for the approximation to be valid, the period P is expressible in terms of Legendre's complete elliptic integral of the first kind; this integral will appear in chapter 12 in the context of river meanders.)

The above formula for P is also "explained" quite naturally by the limerick quoted in the book by Rob Eastaway and Jeremy Wyndham:

If a pendulum's swinging quite free
Then it's always a marvel to me
That each tick plus each tock
Of the grandfather clock
Is 2π root L over g.

While passing over this evidence for the superiority of poetry to prose, we can apply this result to the simplest model of gait analysis, namely walking. We do this by considering a swinging leg to be a simple pendulum (it is really more of a compound pendulum, but this will suffice for our purposes). When we run we bend our legs and effectively shorten the lengths of our "pendulums"; from the above formula for T it follows that the frequency of oscillation of the pendulum ($T^{-1} \sim L^{-1/2}$) therefore increases, and our legs oscillate faster.

Consider now two geometrically similar individuals with leg lengths (or any comparable proportions, e.g., height) L_1 and L_2, respectively, but who have the same angle of swing of θ radians at the hip. The ratio of stride lengths is *approximately* equal to the ratio of arc lengths $L_1\theta/L_2\theta = L_1/L_2$ (for small enough angles θ; see the comment below) and the ratio of the periods of swing is $\sqrt{L_1/L_2}$; hence the speed ratio for these individuals is

$$\frac{V_1}{V_2} = \frac{\text{stride ratio}}{\text{time ratio}} = \sqrt{\frac{L_1}{L_2}}.$$

A 6-foot man walking sedately at 3 mph is accompanied, initially at least, by his toddler, who, as it happens is exactly 3 feet tall. Father and child are not geometrically similar, of course, but in the spirit of this chapter let that pass as a second-order correction! Then the ratio of their heights and hence leg lengths is 2, and so the child walks at $2^{-1/2} \approx 0.7$ times the speed of her father, or about 2 mph. This is a rough and ready type of argument, but the principle is straightforward enough. (For those who are concerned about the little girl being left behind, her lung power is sufficient to remind everyone in the vicinity that she has been temporarily but inadvertently abandoned.)

How far down the size scale might this type of argument apply? Without attempting to answer this question directly, we will just proceed and see if the answer makes sense; if it does, then we have a necessary but not sufficient reason for the argument to be valid. Suppose that we notice a cockroach scampering across the floor in a friend's kitchen (certainly not our own). It moves a distance of about a foot, say, in one second. Let us now throw all caution to the winds and use the above reasoning to compare the man mentioned above with his nongeometrically similar counterpart, the cockroach! We will require him to walk a little more briskly now, at 4 mph ($\approx$ 6 ft/sec), to correspond to the cockroach's somewhat brief tour of the kitchen. The ratio of speeds V_1/V_2 is approximately 6, which corresponds to a size ratio of 36, so that the man might walk at the cockroach's pace if he were scaled down to 6/36 feet, or two inches. Note that this must be the *length* of the cockroach, not the height (since its body is perpendicular to its legs!). Notwithstanding the obvious lack of geometric similarity between the two, this is a not unreasonable result unless the pest control bill has been paid.

A comment was made above concerning the approximate equality of arc length and stride for small enough angles of swing θ. The stride length is the length of the chord joining the end points of the arc, and has length $L_c = 2L \sin(\theta/2)$, compared with the arc length $L_a = L\theta$. Even for $\theta = \pi/3$ radians, the ratio $L_c/L_a = 3/\pi \approx 0.95$, and as the angle decreases from $\pi/3$ the approximation gets progressively better.

Next comes *flying*. We will have more to say about this in chapter 13, but we can mention some basic principles at this stage. Once you are up in the air you must stay up, and the power required to sustain flight is proportional to the weight of the object divided by the area of the wings. This ratio is called the *wing loading*. For obvious reasons (by now) this scales as L, the size (length) of the flying object. The ability to stay airborne is dependent on the lift generated on the wings, and the minimum speed necessary to maintain lift is proportional to the square root of the wing loading, $\sqrt{L}$. As a rather extreme example, consider an ostrich whose length is about twenty-five times that of a typical sparrow. The minimum speed for the sparrow is about 20 mph, so according to the above argument the corresponding speed for an ostrich, were it able to fly, would be five times this, or 100 mph! (Be thankful they don't.) Related to this point, we note that the power needed to fly at the minimum speed scales as $L^{3.5}$, as we will show in a later section. This shows that there must be a limit to the size of birds, because a bird that was too large would not have the ability to generate the power needed to fly. The ostrich is in this category. Larger birds do soar as well as flap their wings, but even soaring ability has its limit. We will examine soaring in chapter 13.

We have already mentioned the disappointing fiction of King Kong; unfortunately, many modern science fiction movies persist in illustrating the same ignorance of basic scaling principles. Thus giant insects dominate the planet and brave humans attempt to free themselves from their evil designs. Now in the mind of the producer it may well be that we are dealing with a low-gravity planet on which these creatures can become very large, but it should not be assumed that humans will function the same way as they do on the earth! The moon walks were good examples of this; the lower gravity forced the astronauts to modify their gaits, at least some of the time. It looked great fun, by the way, and I enjoy the special effects of the science fiction movies as well as the next (resident) alien.

So how is structure related to function? As in the examples above, mathematics can help us understand this question within a biological context in that we can quantify these ideas a little. The scale factor α in geometrically similar objects is, as we have already seen, the ratio of corresponding lengths in the figures: $\alpha = L'/L$, for example. Consider a giant ant with scale factor α relative to a normal ant. The two ants are geometrically similar and made of the same material ("ingredients" sounds so gross), so the

$$\text{weight of a giant ant} = \alpha^3 \times \text{weight of a normal ant.}$$

However, we have also seen that the *strength* of any organism depends only on the cross-sectional area of its muscles, so the

$$\text{weight a giant ant can lift} = \alpha^2 \times \text{weight a normal ant can lift.}$$

Now we introduce a related concept: that of *relative strength*. This is defined as the ratio of the weight the animal (or insect) can lift to the body weight of the animal. Thus for the ants the following relationships hold true:

$$\text{relative strength of a giant ant} = \frac{\text{weight a giant ant can lift}}{\text{weight of a giant ant}}$$

$$= \frac{\alpha^2 \times \text{weight a normal ant can lift}}{\alpha^3 \times \text{weight of a normal ant}}$$

$$= \frac{\text{weight a normal ant can lift}}{\alpha \times \text{weight of a normal ant}}$$

$$= \frac{1}{\alpha} \times \text{relative strength of normal ant.}$$

Thus, since $\alpha > 1$, the relative strength of the giant ant is smaller than the relative strength of the normal ant by a factor of α^{-1}.

Let us apply this result to the case of a man and an ant, comparing the man with a giant ant of the same size (72 inches tall or long). It is sometimes inferred that an ant is stronger than a man in the (relative) sense that an ant can lift 3 times its body weight (sometimes more, but we'll go with 3), whereas a typical man can only lift about half his own body weight. In terms of relative strength, that of an ant is 3 and that of a man is 0.5, but to compare their strengths properly their relative size should be incorporated into the "equation," so to speak. If we take a scale factor of

$$\alpha \approx \frac{72 \text{ inches}}{0.5 \text{ inches}} = 144,$$

then the relative strength of the man-sized ant is $\alpha^{-1} \times$ relative strength of normal ant, or $3/144 = 1/48$; this is much less than 0.5, the relative strength of a man. This means that ants are intrinsically weaker than men. From earlier arguments we know that such a giant ant is not a biologically feasible creature (on earth at least, and I suspect elsewhere!); it couldn't even lift its own legs to climb over small obstacles (proving once again that we must suspend our rational faculties if we want to enjoy special effects in movies such as *Return of the Giant Ants*).

All the above applies equally well to bones and other structural material, of course, and goes a long way toward explaining why species of different sizes tend to have such different shapes. We have noted already being very observant, that an elephant is not a scaled-up version of a mouse. An animal the size of an elephant cannot have the shape of a mouse because the ratio of bone strength to body weight would be too small; the bones and muscles of large animals must be disproportionately thicker than the bones and muscles of small animals. Even over a smaller range of sizes this principle generally holds; a big "cat" like a tiger is not merely a scaled-up domestic kitty—look at its leg thickness relative to its torso, for example.

Now let us think about a much smaller biological entity—the cell. A proper balance between the flow of gases (such as O_2 and CO_2) through its surface area and its internal metabolism is possible only if it is not too small and not too large. This is easily appreciated from the surface-area-to-volume arguments used above. Essentially, if the cell is too small, the volume is too small to keep up with the energy requirements of the surface (by diffusion). If the cell is too large, the surface area is too small relative to the metabolism: energy flow and chemical exchange are impeded. Similar considerations for organisms (large collections of cells) show that if a large such organism is to survive this change of scale, it must compensate for its (now) small area-to-volume ratio by selectively increasing its surface area *or* by changing the fundamental mechanisms of energy transport and flow. The typical timescale on which diffusion occurs, given by L^2/D (see chapter 14), where D is the diffusion coefficient, is clearly very sensitive to the size L of the organism. For a small cell, $L \approx 1 \mu$ (1 micron or 10^{-6} m), and for diffusion of oxygen in water, $D \approx 10^{-5}$ cm^2/sec, resulting in a timescale of $\approx 10^{-3}$ sec. For $L \approx 1$ m, the timescale is 10^9 seconds, or ≈ 30 years, so it is a good thing humans do not rely on diffusion at this scale!

Clearly, if diffusion *is* to remain a dominant mechanism in the life of a organism, it must increase its "flatness." Thus flat shapes or long, branched filaments are well suited for organisms that rely exclusively on absorption of nutrients or oxygen directly from their environment, because they can increase in volume without changing thickness and hence the distance through which diffusion must act. Often there is change in shape also; in addition to flattening, surface area is increased via growth of hair, branches, wrinkles, elongations, hollowing-out by means of tube systems, and so on. Nature is very adaptable! Thus diatoms (for which $L \approx 10\text{--}100 \ \mu$) frequently possess needle-like extensions to increase surface area, which in turn both increases their efficiency in nutrient and waste exchange and reduces their tendency to sink. Although some insects increase their surface area by means of hollow tubes (tracheae), this is still only effective over small distances, and for this reason many insects are limited in size to $\approx 1\text{--}2$ cm. Bigger creatures have developed "bulk flow" mechanisms for transporting oxygen, assimilating food, and excreting body wastes. These include the circulatory system, intestinal villi (increasing the absorbing surface area), gills, alveoli in lungs, and lobes in kidneys.

Let us see what bearing these ideas have on the following question— *Why do cells divide when they reach a certain size?* There are important biological details of the mitotic activity of cells which we shall not discuss here, because scaling arguments alone can provide helpful insight into the answer. Following an argument by Cromer, we can define a cell's *viability factor* to be the following ratio:

$$\text{viability factor for a cell} = \frac{\text{maximum amount of available } O_2/\text{unit time}}{\text{amount of } O_2 \text{ required/unit time}}.$$

Then, as before, if two cells are geometrically similar with a scale factor of α between them, it follows using arguments similar to those above that

viability factor for an older cell

$$= \frac{\text{max. amount of available } O_2/\text{unit time for older cell}}{\text{amount of } O_2 \text{ required/unit time for older cell}}$$

$$= \frac{\alpha^2 \times \text{max. amount of available } O_2/\text{unit time for younger cell}}{\alpha^3 \times \text{amount of } O_2 \text{ required/unit time for younger cell}}$$

$$= \alpha^{-1} \times \text{viability factor for younger cell}.$$

Now a young cell must have a viability factor greater than one, so as the cell grows (and ages) the viability factor decreases and eventually approaches the value one. To avoid "death by suffocation" the cell must either stop growing (become dormant) or divide into two smaller cells, each with a larger viability factor. And so the process goes on. . . .

Mention has been made of the importance of shape (or change of shape) if a mechanism-like diffusion is still to be useful as an object grows in size. One interesting measure of volume relative to surface area is the *sphericity index*. Since volume V and surface area A scale respectively as the cube and the square of a linear dimension, it is evident that a suitable *dimensionless* measure of the volume-to-surface-area ratio for an object should involve some power of the ratio $V^{2/3} : A^{-1}$. Although the reciprocal dimensional quantity, namely the surface-area-to-volume ratio, has been used thus far in this chapter, there is no problem with using the sphericity index χ, which is defined as

$$\chi = \frac{4.836 V^{2/3}}{A}.$$

Hmm. What's with the 4.836? After all, it is hardly a household number. It arises by requiring that the sphericity index for a sphere is unity, so if we temporarily replace the above number by α, it follows that for a sphere of radius R

$$\chi = 1 = \frac{\alpha V^{2/3}}{A} = \alpha \frac{(4\pi R^3/3)^{2/3}}{4\pi R^2} = \alpha \frac{(4\pi/3)^{2/3}}{4\pi}$$

from which we obtain $\alpha = (4\pi)^{1/3} \cdot 3^{2/3} \approx 4.836$. Because a sphere has the largest volume to surface area for any closed surface (see chapter 11), it follows that for other shapes, $0 < \chi < 1$.

Let us consider some examples. For a cube it is readily shown that $\chi \approx 0.806$; for two "kissing" spheres (in tangential contact) $\chi = 2^{-1/3} \approx 0.794$. I like the sphericity index because it is always of order one; there is another quantity that is sometimes used in biological contexts, called the flatness index γ, where

$$\gamma = \frac{A^3}{V^2} \propto \chi^{-3};$$

it is generally for that reason larger than χ, being approximately 113 for a sphere and 216 for a cube. The observant, intelligent reader (is there any other kind?) will have noticed immediately for all these examples that (i) the radius of the sphere of side length of the cube has not been specified; and (ii) it is not necessary to do so, because all the dimensional quantities cancel out (by design).

We will mention three more consequences of scale before closing with some general considerations. The first of these is *brain power* and the second is *vision*. The sizes of brain cells change by only a factor of 2 (and hence in volume by a factor of 8) between a mouse and an elephant. The size of the elephant is about one hundred times that of the mouse, which corresponds to a volume increase by a factor of a million. There must, therefore, be correspondingly far fewer brain cells in the brain of a mouse compared with the brain of an elephant. We can put this more prosaically: Tom Thumb wouldn't have the brain power to understand this argument! A similar kind of argument applies to vision, but with surface area the dominating criterion, as opposed to volume. In each human eye there are typically about 1.3×10^8 rods and 7×10^6 cones. The sizes of these light-sensing cells are about the same in mice and humans, but the mice have much smaller eyes and hence fewer rods and cones. This means that the resolving power (the ability to distinguish between two objects) of a mouse eye is far poorer than that of a human eye: thus, as someone said, "at a distance of about six feet the doomed mouse in the nursery rhyme would have trouble distinguishing the angry visage of the farmer's wife from the smiling one of her husband."

The third consequence of scale pertains to hearing. In this highly over-simplified discussion we concentrate on the consequences of basic dimensional considerations. The production and reception of the "speech" of many animals is determined by the vibration of vocal chords and auditory drums, respectively. Here we will consider only the fundamental frequency f_1 for a uniform stretched string of length L under constant tension T, though similar arguments apply to that for a stretched drumskin of uniform thickness. In the study of wave motion along a stretched string or wire, it is customary to work in terms of the mass density μ of the string, which is defined to be the mass per unit length, and which thus scales as $L^3/L \sim L^2$. The frequency of the fundamental mode of vibration is given by

$$f_1 = \frac{1}{2L}\sqrt{\frac{T}{\mu}},$$

so that for constant tension in geometrically similar uniform strings, $f_1 \sim L^{-2}$. The formula for f is a special case of Mersenne's law for the natural frequencies $f_n = n f_1$ ($n = 1, 2, 3, \ldots$), where $n - 1$ is the number of interior nodes of the vibration. When n exceeds one, these vibrations are referred to as harmonics. In what follows only the fundamental frequency f_1 will be used, and the subscript will be recast to refer to the animals "1" and "2," respectively.

If we compare two animals (man and bat) of size L_1 and L_2, respectively, then

$$\frac{f_2}{f_1} = \left(\frac{L_1}{L_2}\right)^2,$$

so that if L_2 refers to the body size of the bat, excluding wing span (which is irrelevant here), then $L_1 \approx 50L_2$ implies that the fundamental frequency for the bat is about 2,500 times that for the man. Since an octave rise in pitch corresponds to a doubling of the frequency, and $2^{11} \approx 2000$, this corresponds to about eleven octaves higher!

Our discussion on scaling has shown us that, while area scales as L^2, volume does so as L^3. Even such simple considerations enable us sometimes to do some detective work. The object of the inquiry is a largish dinosaur called *Dimetrodon*. Apparently members of the Dimetrodon family lived in Texas and Oklahoma and, more important, had a sail on their back. From the currently available data on this creature, the area of the sail grew in proportion to the *volume* of the animal (not varying as the area of the animal, L^2). This implies, if true, that the sail was designed to perform the same function as the elephant's ear (no, not to hear with), namely, to be a temperature-regulating organ that absorbed or radiated heat.

For *any* object, its area is proportional to its length (or width, etc.) squared, but unless such objects are geometrically similar, the constant of proportionality is different for each object. A similar argument applies to volumes. In general, animals, humans in particular, do not grow in a way that maintains geometric similarity. The human arm, which at birth is 1/3 as long as the body, is closer to 2/5 as long by the time adulthood is reached. *Allometric growth* is the name given for growth of one feature at a rate proportional to the power of another. From this definition we see that surface area and volume both grow allometrically with length. Mathematically, we may express an allometric relationship between the quantities x and y as

$$y = bx^a.$$

Ordinary geometric or proportional growth corresponds to $a = 1$. The constant of proportionality is b. As we have already noted, *geometric similarity* corresponds to shape-preserving growth. A striking example of such growth is that of the chambered nautilus. Each new chamber that is added onto the shell is larger than but the same shape as the previous chamber, and the shape of the shell as a whole—an equiangular or logarithmic spiral—remains the same.

However, you have probably noticed that many living things do not preserve exactly the same shape over their lifetimes. For example, an adult human, as noted above, is not simply a scaled-up baby. And a baby, therefore, is not a scaled-down adult! (This may explain why some medieval and Renaissance art involving small children often seems distorted to us today.) Relative to the length of its body, a baby's head is much larger than that of a typical adult. Even the proportions of facial features are different: in a baby the tip of the nose is about halfway down the face, while in an adult the nose is about 2/3 of the way down. That's gravity for you! Although, as we have seen, the body does not scale up uniformly as a whole, different parts do scale up, each with a different scale factor. A baby's eyes grow at one rate to about twice their original (birth) size, while their arms grow at another rate to about four times their original size.

A final point is in order to close this section of the chapter. In describing how "area" and "volume" grow with "size," we have described how certain properties of objects distribute themselves in space. Such considerations also apply to more intangible concepts like energy, be it acoustic, electromagnetic (as in light or radio waves), gravitational, electrical, or magnetic (not to be confused with electromagnetic). These all diminish as the inverse square of the distance from the source (the available energy is distributed over the surface area $4\pi R^2$ of a sphere of radius R; as R increases the energy per unit area decreases in proportion to R^{-2}). Constrained to fewer dimensions (as in a tunnel or a thin flat sheet) the respective energy density (as energy per unit area can be called) is constant or proportional to R^{-1} in dimensions one or two, respectively, neglecting attenuation due to reflections and other dissipation mechanisms.

Even time seems compressed, relatively (but not relativistically) speaking. Heartbeats and wingbeats (where relevant) are faster for smaller creatures, as is the breathing rate. Thus a small creature like a pygmy shrew may have 700 heartbeats/minute, a cat 120, a man 70, an elephant 35, and a whale 20. A midge's wings may beat 1,000 times a second, a bee's 200, a hummingbird's 100, a sparrow's 15. The lifetimes of smaller creatures tend to be shorter, though, not surprisingly, exceptions are fairly common. A butterfly may live for two weeks, a rat a year or two, a woman eighty, and an elephant even longer. And we must definitely not forget sex. Smaller creatures reproduce faster than larger ones; being less complex in general and eating less food (in absolute terms) they can breed their way out of environmental difficulty!

DIMENSIONAL ANALYSIS II

The term dimensional analysis is here intended in its second sense as a formal mathematical technique used to great effect in physics and engineering. It can be used to simplify a problem by reducing the number of variables required to describe the problem to a smallest set of parameter combinations (which may not be unique, however). It is also useful for checking whether one's algebra is correct, for if the dimensions of terms in an equation are inconsistent, the equation is surely so! The units of mass, length, and time (M, L, and T) define the dimensions of the physical quantity of interest (or type of units needed to describe it) and are called *fundamental units*. They are akin to basis vectors for a vector space, such as the unit vectors $\{i, j, k\}$ for the space $\mathcal{R}^3$. Thus the dimension of force is mass $\times$ length/time^2, or MLT^{-2}. The above set of fundamental units is not unique (though the number required is); mass, length and force could have been used just as well.

The central tool of dimensional analysis is the *Buckingham π-theorem*. This theorem indicates how an equation in n variables and m so-called fundamental units (where $m < n$) can be rewritten equivalently in terms of $n - m$ dimensionless parameters. The method does not yield *dimensionless* constants of proportionality, an example being that of the simple pendulum discussed below (and above). In such cases the full physics of the problem must be modeled mathematically. In the case of the simple pendulum, the governing ordinary differential equation must be formulated, simplified where appropriate (which usually means linearization for the case of small oscillations), and solved subject to initial conditions. The resulting sinusoidal solution is periodic in 2π, and it is from this feature that the constant of proportionality (2π) is derived.

The linear algebraic rationale behind the π-theorem is as follows. The space of all possible physical units (e.g., force) can be considered a vector space over the rational numbers, such a vector being represented by the set of exponents (including zero) required to express the fundamental units (e.g., in the case of force, the exponents for $\{M, L, T\}$ are $\{1, 1, -2\}$, respectively). Multiplication or division of physical units is then represented by vector addition within the space. The theorem is stated below:

Buckingham π-theorem. An equation is dimensionally homogeneous if and only if it can be written as

$$\Pi(\pi_1, \pi_2, \pi_3, \ldots, \pi_k) = 0,$$

where Π is some function of the $\pi_i, i = 1, 2, \ldots, k$, which are dimensionless products and quotients of the variables and (dimensional) constants appearing in the original equation.

Sketch proof. Let $\mathscr{L} : \mathscr{R}^n \rightarrow \mathscr{R}^m$ be a linear transformation with range $\mathscr{L} = \mathscr{R}^m$ (so $\mathscr{L}$ is "onto") and $m \times n$ matrix representation $\mathscr{A}$. By a standard theorem in linear algebra (see, for example, the book by Kolman for a proof of this result)

$$\text{rank}(\mathscr{L}) + \text{nullity}(\mathscr{L}) = n$$

or in a common alternative form

$$\dim(\text{image space}) + \dim(\ker \mathscr{L}) = \dim(\text{domain space}) = n.$$

Then the nullity $(\mathscr{L}) = n - m$, n being the number of physical variables and constants, m being the number of fundamental units, and the difference being the number of possible dimensionless groups available (not all of which may be necessary).

As an illustration, we consider the gravitational force F between two point masses m_1 and m_2 in terms of the distance r between them and the universal gravitational constant G (6.7×10^{-11} m^3/(kg s^2)). It is important to note that all the relevant variables and dimensional constants need to be known in advance if the correct functional form is to be determined, and that the functional form is assumed to be algebraic in character, that is, containing no transcendental terms with any of the variables as argument. In this case $n = 5$, $m = 3$, and so there are 2 dimensionless groups. Let

$$\pi = G^a m_1^b m_2^c r^d F^e$$

be a dimensionless combination of the fundamental set $\{M, L, T\}$, where a, b, c, d, e are all arbitrary rational numbers (this could be extended easily to the set of all real numbers). Thus using the known dimensions of the 5 physical variables we can write

$$M^0 L^0 T^0 = (M^{-1} L^3 T^{-2})^a M^b M^c L^d (MLT^{-2})^e$$
$$= M^{-a+b+c+e} L^{3a+d+e} T^{-2a-2e},$$

resulting in 3 simultaneous homogeneous linear algebraic equations in the 5 exponents $a, \ldots, e$. The matrix $\mathscr{A}$ is

$$\mathscr{A} = \begin{pmatrix} -1 & 1 & 1 & 0 & 1 \\ 3 & 0 & 0 & 1 & 1 \\ -2 & 0 & 0 & 0 & -2 \end{pmatrix}.$$

Hence if $\mathscr{Y}$ is the column vector $(a \quad b \quad c \quad d \quad e)^T$ it follows that

$$\mathscr{L}\mathscr{Y} = (0 \quad 0 \quad 0)^T.$$

Since the nullity of $\mathscr{L}$ is 2, let these two "free parameters" be a and b without loss of generality, so that by solving for the remaining three exponents in terms of a and b we find that

$$\mathscr{Y} = a (1 \quad 0 \quad 2 \quad -2 \quad -1)^T + b (0 \quad 1 \quad -1 \quad 0 \quad 0)^T.$$

These two vectors are basis vectors for ker $\mathscr{L}$; that is, ker $\mathscr{L}$, the null space of the exponents, is a 2-dimensional subspace of $\mathscr{R}^5$. Now we can write π in terms of products of powers of the 2 dimensionless groups π_1 and π_2,

$$\pi = \left(\frac{Gm_2^2}{Fr^2}\right)^a \left(\frac{m_1}{m_2}\right)^b = \pi_1^a \pi_2^b.$$

We are at liberty to choose the simplest form consistent with experimental data. In this case setting $a = b = 1$ gives the product

$$\pi_1 \pi_2 = \frac{Gm_1 m_2}{Fr^2},$$

so we obtain from the π-theorem that

$$\Pi(\pi_1, \pi_2) = \pi_1 \pi_2 - 1 = 0,$$

which on solving for F yields the canonical form for the scalar version of the law of gravitation (see Bender, chapter 2).

As another example, consider the problem of the drag force D on a fish (or submarine) of "size" l in a fluid of density ρ (typically water, of course), uniform speed U, and dynamic viscosity μ. As with the preceding example, there are 5 physical and 3 fundamental quantities, so we expect 2 dimensionless groups. We suppress the formal algebra and cheat a little by specifying the variables in each group (because we know in advance which dimensionless groups are the most useful in fluid dynamical problems of this type). The method is well illustrated by this example, nonetheless. Let

$$\pi_1 = \rho^\alpha l^\beta U^\gamma \mu^\delta = (ML^{-3})^\alpha L^\beta (LT^{-1})^\gamma (ML^{-1}T^{-1})^\delta$$

and

$$\pi_2 = D^\epsilon \rho^\eta l^\sigma U^\lambda = (MLT^{-2})^\epsilon (ML^{-3})^\eta L^\sigma (LT^{-1})^\lambda,$$

where the dimensions of the physical variables have been substituted into the expressions for the dimensionless quantities π_1 and π_2. Solving the resulting linear equations for π_1 in terms of α and for π_2 in terms of ϵ we obtain

$$\pi_1 = \left(\frac{\rho l U}{\mu}\right)^\alpha \quad \text{and} \quad \pi_2 = \left(\frac{D}{\rho l^2 U^2}\right)^\epsilon.$$

Again for simplicity choosing the exponents as one and writing the kinematic viscosity $\nu = \mu/\rho$, we have the quantity

$$\pi_1 = \frac{lU}{\nu},$$

which is the *Reynolds number*, and

$$\pi_2 = \frac{D}{\rho l^2 U^2},$$

which is related to the drag coefficient (see chapters 2 and 13).

Our remaining examples involve a rather more direct and informal approach based on the preceding π-theorem. We shall write the physical quantity of interest again as proportional to products of powers of the remaining physical quantities, together with an unknown dimensionless constant of proportionality K. As remarked earlier, K can be determined in general only by a detailed study of the full physical problem, either theoretically, experimentally, or in some combination of the two, and will no longer concern us here. Requiring the governing equation to be dimensionally correct will produce the necessary functional dependence, provided of course that all the relevant physical variables are accounted for in the statement of the problem.

1. Find the volume V of the crater produced by a surface explosion of energy E in soil of density ρ under gravity (g). We proceed as follows. Assume

$$V = KE^a \rho^b g^c.$$

In dimensional terms this is

$$L^3 \equiv M^0 L^3 T^0 = (ML^2 T^2)^a (ML^{-3})^b (LT^{-2})^c.$$

The solution set $\{a, b, c\}$ is $\{3/4, -3/4, -3/4\}$ so

$$V = K\left(\frac{E}{\rho g}\right)^{3/4}.$$

This result is also plausible physically being directly dependent on a positive power of E, and inversely so for the density and gravitational acceleration.

2. Find the radius r of the shock front (expressed as a function of time t) from an atomic explosion of energy E in an atmosphere of undisturbed density ρ. Again, let

$$r = K\rho^a E^b t^c,$$

so that

$$M^0 L^1 T^0 = (ML^{-3})^a (ML^2 T^{-2})^b (T)^c.$$

Now the set $\{a, b, c\}$ is $\{-1/5, 1/5, 2/5\}$, so

$$r = K\left(\frac{Et^2}{\rho}\right)^{1/5}.$$

Since $r(t)$ may be determined approximately from photographic analysis of the shock front, this equation can be written in a form more useful for graphical analysis as

$$\frac{5}{2}\log r = A + \log t,$$

where the constant

$$A = \frac{5}{2}\log\left[K\left(\frac{E}{\rho}\right)^{1/5}\right].$$

If K is known (in fact it is about 1.03), then the energy of the explosion can be determined from the graph of $r(t)$ since ρ is well known (about 1.25×10^{-3} gm/cm^3 for dry air). More details can be found in the book by McMahon and Bonner.

We briefly state several other problems that can be studied in a similar manner (again, see the McMahon and Bonner book for details). There are many more.

3. The pressure P within a soap bubble of radius r and surface tension s:

$$P = Ksr^{-1}.$$

Smaller bubbles burst more noisily than large ones; listen to freshly opened champagne!

4. The (fundamental) frequency of vibration f of a liquid drop of density ρ, radius r, and surface tension s:

$$f = K\left(\frac{s}{\rho r}\right)^{1/2}.$$

The higher the density, the lower the surface tension or the larger the radius, the lower is the frequency; have you ever watched a drop of mercury vibrate?

5. The (fundamental) frequency of vibration f of a self-gravitating fluid (a "star") in terms of its density ρ, radius r and the gravitational constant G:

$$f = K(G\rho)^{1/2}.$$

This is independent of the size of the star! The denser it is, the faster it vibrates. This was an early "liquid drop" model of a variable star; much of the relevant physics is missing, so it is now only of historical interest.

6. The speed υ of gravity waves on the surface of deep water in terms of the wavelength λ and gravity g (with surface tension neglected):

$$V = K(g\lambda)^{1/2}.$$

Long waves outrun shorter ones (and K turns out to be one; see chapter 7).

7. The speed υ of capillarity waves on the surface of a liquid in terms of the wavelength λ and surface tension s (with gravity neglected):

$$V = K\left(s\lambda^{-1}\rho^{-1}\right)^{1/2}.$$

This time short waves outrun long ones (also see chapter 7).

8. The speed of sound c in a gas in terms of its pressure p and density ρ:

$$c = K(p\rho^{-1})^{1/2}.$$

9. The tail-beat frequency f of a fish in terms of its body length l, muscle stress σ (or tension per unit cross-sectional area), and the fluid density ρ:

$$f = Kl^{-1}(\sigma/\rho)^{1/2}.$$

The bigger the fish the slower the tail beats!

10. The period P of a simple pendulum in terms of its mass m, length l, and the gravitational acceleration g:

$$P = K(l/g)^{1/2}.$$

We know from our earlier discussion of the simple pendulum that $K = 2\pi$.

APPENDIX: MODELS BASED ON ELASTIC SIMILARITY

As we have seen already, the simplest version of scaling assumes that the volume (and hence mass M and weight W) of an animal varies as L^3, where L is a characteristic length scale for the animal. Thus on this model, $L \propto M^{1/3}$. Since by the same token the animal's surface area A scales as L^2, it follows that $A \propto M^{2/3} \approx M^{0.67}$. Since the rate of oxygen absorption must vary as the lungs' surface area and the rate of heat loss must vary as the body surface area, it is predicted that the rate at which food energy is used, the metabolic rate, will also vary as $M^{0.67}$. While this simple model is useful, it is not always in exact agreement with what is observed; in other words, sometimes a rather more sophisticated model is needed to explain the observations. In 1932 Max Kleiber found that in the range of mammals from mice to elephants, their rate of heat production was proportional to $M^{0.75}$, not $M^{0.67}$. What is a difference of 12 percent in the exponent between friends? Quite a bit if your friend is an elephant. It is consistently off by too much to be ignored, and, remember, elephants are reputed to have long memories.

Another important "scaling" is that of elastic similarity, discussed in more detail in chapter 12 in connection with the height of trees: the height L and radius R of a cylinder are related by the condition $L \propto R^{2/3}$. Thomas McMahon has suggested that this "model" may be more appropriate in animal models, since most body segments are roughly cylindrical and may be built to withstand buckling. Let us see where this simple change leads us.

The volume of such a cylinder and hence the mass should now be proportional to R^2L, so that

$$M \propto R^2L \propto (L^{3/2})^2L = L^4.$$

Already a difference is evident. Thus, if $L \propto M^{1/4}$ and $L \propto R^{2/3}$, it follows that $R \propto M^{3/8}$; and since the lateral surface area of a cylinder is proportional to the product RL we find that

$$A \propto RL \propto M^{3/8}M^{1/4} = M^{5/8} = M^{0.625} \approx M^{0.63}.$$

It is interesting to note in the light of this result that, while the rate of heat production scales as $M^{0.67}$, the body surface area of animals is observed to scale as $M^{0.63}$—fairly close agreement! Moving right along, however, we can estimate the mass-dependence of metabolic rate by again thinking about the power P used in flexing muscles. Power is the rate at which work is done; for muscle contraction this is the force F exerted multiplied by the velocity v of muscle contraction. The stress $\sigma = F/A_m$ exerted by muscles is about the same for all mammals, where A_m here is the muscle cross-sectional area, so

$$P = Fv = \sigma A_m v.$$

It has been found experimentally that v is also approximately constant in mammals, and since $A_m \propto A$ it follows that

$$P \propto A_m \propto A \propto R^2 \propto (M^{3/8})^2 = M^{3/4} = M^{0.75}.$$

If the power expended and rate of heat production scale in the same way (as is eminently reasonable, since the former implies the latter), then we have our desired result. There are some further implications of this model; in particular the power of the heart muscle should also scale as $M^{0.75}$, as should the lung area, and both these results have been established experimentally. Now since the metabolic rate (and hence oxygen demand of the body) scales as $M^{0.75}$, and is equal to the volume of blood pumped per "beat" times the pulse rate r, it follows that

$$r \propto M^{0.75}/M = M^{-0.25},$$

which again has been verified (and anecdotally we all know that larger animals tend to have lower pulse rates). The $L \propto M^{1/3}$ model predicts that $r \propto M^{-1/3} \propto L^{-1}$, so the former model fits some of the data better than the latter, but even that is a good "first cut," and the $L \propto M^{1/4}$ model is not necessarily the last word either. An excellent account of these types of arguments for a wide variety of mammalian applications can be found in the book by Calder.

CHAPTER FOUR

Meteorological Optics I: Shadows, Crepuscular Rays, and Related Optical Phenomena

> But alas, the daylight is fading, and the shadows of
> the evening grow long.
>
> —Jeremiah 6:4b

APPARENT SIZE OF THE SUN; SHADOWS; TREE PINHOLE CAMERAS; LENGTH OF THE EARTH'S SHADOW

Before we can sensibly discuss shadows cast by the sun, we need to know how large a source of light (in an angular sense) the sun appears to be in the sky; it is about half a degree of arc, as will be established below.

It is well known that the planets move around the sun in ellipses with the sun at one focus; this, after all, is Kepler's first law. The perihelion (= closest) distance r_p of the earth from the sun is about 1.47×10^8 km (or about 9.14×10^7 mi); the aphelion (= farthest) distance r_a is approximately 1.52×10^8 km (9.45×10^7 mi). For now we will work with the *mean* distance r of 1.50×10^8 km (9.30×10^7 mi). As is rather clear from these figures, the earth's orbit is close to circular, being an ellipse with eccentricity e defined by the equation

$$\frac{1-e}{1+e} = \frac{r_p}{r_a} \approx 0.967,$$

from which $e \approx 0.017$ (by comparison, the orbital eccentricities of Mercury and Pluto are, respectively, 0.206 and 0.248; the remaining planets have rather small eccentricities). For future reference note that the mean solar radius $R_s \approx 4.32 \times 10^5$ mi $\ll r$, so the essential geometry of the earth-sun system is well represented by right triangles (see figure 4.1). Our first question is *what is the mean angular diameter θ of the sun as viewed from earth?* From figure 4.1,

$$\tan \frac{\theta}{2} \approx \frac{4.3 \times 10^5}{9.3 \times 10^7} \approx 4.6 \times 10^{-3}$$

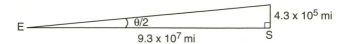

Figure 4.1. The basic (but approximate) geometry to determine the angle θ subtended by the sun (center S) for an observer on earth (E). The radius of the sun is $\approx 4.3 \times 10^5$ miles and the mean distance of the earth from the sun is $\approx 9.3 \times 10^7$ miles.

(to the accuracy required here). This means that

$$\theta \approx 2 \arctan(4.6 \times 10^{-3}) \text{ radians} \approx 9.2 \times 10^{-3} \text{ radians}$$

$$\approx \frac{1}{109} \text{ radians} \approx 0.53°$$

or about half a degree—roughly equivalent to a children's aspirin held at arm's length for a typical adult. This is often quite surprising to people who assume that it and the moon are the size of dinner plates held at arm's length! We will perform this calculation below, using a slightly different value for θ.

Let us recall for inverse trigonometric functions that

$$y = \arctan x \text{ if and only if } x = \tan y,$$

$$-\infty < x < \infty, \quad -\frac{\pi}{2} < y < \frac{\pi}{2}.$$

We are dealing here with small values of x, that is, $|x| \ll 1$, so that the Maclaurin series for $\arctan x$ is

$$\arctan x = x + O(x^3).$$

This is well approximated by the first term, which explains the above result for θ. Returning to the orbital extremes, perihelion and aphelion, it is readily shown that

$$\theta_p \approx 9.4 \times 10^{-3} \text{ radians} \approx \frac{1}{106} \text{ radians}$$

and

$$\theta_a \approx 9.1 \times 10^{-3} \text{ radians} \approx \frac{1}{110} \text{ radians}.$$

We will stick with 1/108 radians as an average value for the angular diameter subtended by the sun at the surface of the earth. Note in passing that, in so doing, we have *not* used the mean of the denominators here; clearly

$$\frac{1}{2}\left(\frac{1}{a} + \frac{1}{b}\right) \neq \left(\frac{a+b}{2}\right)^{-1}, \quad a > 0, \ b > 0, \ a \neq b,$$

but for these numbers it is certainly close.

We are now ready to do some simple geometry by asking the question: *What diameter circular disk held at arm's length subtends 1/108 radians at the observer's eye?* If r is the radius of the disk and a is the distance it is held from the eye (approximately the length of the outstretched arm), then

$$\frac{r}{a} = \frac{1}{216}.$$

It is rather awkward trying to do this with one end of a tape measure between one's teeth, but it can be done. For me, $a \approx 21$ in, so the disk diameter $2r \approx 21/108 \approx 0.19$ in. So a baby aspirin will *easily* "eclipse" the full moon if held at my arm's length, and probably yours as well (the aspirin is about 0.3 inches in diameter). Do *not* attempt this with the sun, however.

At the beginning of their lovely book *Color and Light in Nature*, Lynch and Livingston make the following rather enigmatic statement: *The width of the penumbra (fuzzy outer edge) divided by the distance of the shadow from your hand is the angular diameter of the sun measured in radians.* Naturally, we would like to verify this statement, and use figure 4.2 to do so. In what follows, D is the distance from the extended source (the sun) to the tip of the umbra; the shadow is cast in a flat screen or wall a distance L behind the hand; at the screen the width of the penumbra is $p = p_1 + p_2$ on

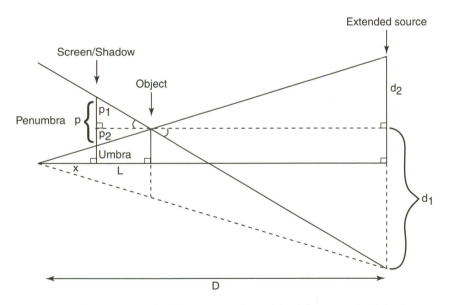

Figure 4.2. The geometry for determining the width of the penumbral shadow cast by an object illuminated by an extended source such as the sun (or at night the moon or a lamp) in terms of the distance of the shadow from the object.

each side of the line of symmetry; the diameter of the sun is $d = d_1 + d_2$; the object casting the shadow (the hand) is a perpendicular distance $D - (L + x)$ from the sun, x being the (generally unobservable) distance from the screen to the theoretical tip of the shadow.

By similar triangles it follows that

$$\frac{p_1}{L} = \frac{d_1}{D - (L + x)}$$

and

$$\frac{p_2}{L} = \frac{d_2}{D - (L + x)}.$$

Since $d/D \approx 1/108$ as shown above, it follows from adding these equations that

$$\frac{p}{L} = \frac{d}{D - (L + x)} = \frac{d/D}{1 - (L + x)/D} \approx \frac{d}{D} \approx \frac{1}{108},$$

as was required.

A little later in their book, Lynch and Livingston discuss the transience of contrail shadows. The word "contrail" is a contraction of "condensation trail," the trail of cloud created by the condensation of moist warm air from jet engine exhaust. They are a beautiful sight and serve as markers of the sometimes complex air motions in the upper atmosphere, revealing wind shear, evaporation rates, and many other features to the practiced eye (see the book *Spacious Skies* by Scorer and Verkaik for more details). We can use the same simple geometry in all these types of problem. In figure 4.3 y is the half-width of a typical contrail shadow (which is like a curtain extending to the surface of the earth) and x, needed for the calculation, is the (unobservable) distance of the shadow tip (were the earth a nonrefracting transparent sphere) from the surface observer. Taking the width of the contrail as about 100 m and its altitude as about 8000 m, we find that

$$\frac{50}{8000 + x} = \frac{1}{216} = \frac{y}{x},$$

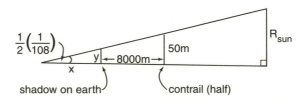

Figure 4.3. The geometry for determining the half-width of a contrail shadow based on the angle of $1/108$ radian subtended by the sun at the surface of the earth.

from which $x \approx 2800$ m and $y \approx 13$ m, so the shadow width is about 25 m; with winds at 8000 m typically being 20–30 m/s this means that the earth-based observer is in the umbra, if at all, for about 1 second at most. This is hard to catch, so be observant!

Using figure 4.4, we may establish a result stated by Minnaert in his delightful book *Light and Colour in the Outdoors* concerning what I call "tree pinhole cameras." The essential idea is that the small spaces between the leaves act as "pinholes," giving rise to the approximately elliptically shaped light patches on the ground. The fact that there is so much shade under a tree, incidentally, is testimony to the effectiveness of the foliage; after all, part of its job description is to intercept light and use it in photosynthesis to provide food for the tree. The effects of "leaf shielding" will be discussed in chapter 12.

Suppose a particular pinhole is at height h above level ground, and d is the horizontal distance from a point directly underneath the pinhole to the center of the light patch produced by the pinhole. Assuming that this patch is an ellipse of major axis b and minor axis k, the latter subtending a *small* angle θ at the pinhole, the angles α and β in the figure are defined implicitly by

$$\sin \alpha \approx \frac{h}{L} \text{ and } \sin \beta \approx \frac{k}{b}.$$

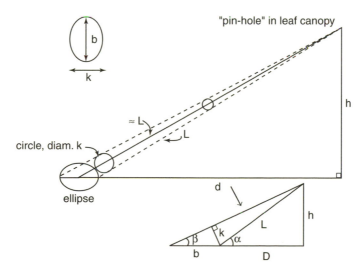

Figure 4.4. Tree pinhole cameras: a small hole in a tree's leaf canopy can produce an elliptical image of the sun on the ground. The major and minor axes are b and k, respectively, and the geometry of the situation is shown reduced to its simplest terms.

Since $\alpha \approx \beta$, we have that $h \approx Lk/b$, but by virtue of earlier calculations we know that $L/k \approx 108$, so

$$h \approx \frac{108k^2}{b}.$$

Walking home from work one day along a path lined with crape myrtle trees, I stopped to measure several such light patches; typical values of b and k were 3 inches and 2 inches, respectively, yielding a value for h of about 12 ft, which looked about right.

Now we ask the question: *How long is the earth's shadow?* A frequent answer to this question, in class at least, is "long." Let us try for a little more precision here. From figure 4.5, triangles *PCO* and *PAE* are (geometrically) similar, so using the notation in the figure, the length of the shadow $l = rd/R$, where $d = PO$. Strictly, this is measured from the center of the earth, but as you may suspect, the radius of the earth is so small compared with that of the sun, this is a very good approximation nonetheless. Since $d = l + s$, it follows that

$$l = \frac{rs}{R - r} \approx \frac{rs}{R},$$

since $R \approx 4.3 \times 10^5$ mi, and $r \approx 4.0 \times 10^3$ mi, so $R \gg r$. The mean earth-sun distance s is 9.3×10^7 mi, so putting in the numbers gives $l \approx 8.7 \times 10^5$ mi, almost a million miles! Being rather delighted with this simple calculation, let us now ask the same thing for the moon: *How long is the moon's shadow?* The numerical differences in this calculation obviously arise because now we are dealing with the moon-earth-sun system, so now $r \approx 1.08 \times 10^3$ mi and s is about the same as for the earth, since the mean earth-moon distance (2.4×10^5 mi) is small compared with s. This time, $l \approx 2.34 \times 10^5$ mi, which is just slightly less than the mean earth-moon distance cited above. This explains why a total eclipse of the sun is not guaranteed to be visible anywhere on earth even if the three bodies are

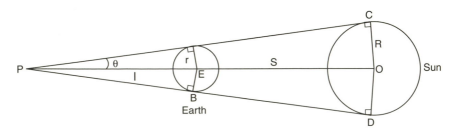

Figure 4.5. A slightly more accurate version of figure 4.1, this time to determine the length of the earth's shadow in space.

collinear (which in general they are not at new and full moon); under these circumstances only an annular eclipse is guaranteed—the moon is often just too far away.

REFLECTIONS FROM A SLIGHTLY RIPPLED SURFACE: GLITTER PATHS AND LIQUID GOLD

Have you ever noticed golden "cylinders of sun" (a "glitter path") on water as the sun sets in the west? I am told that this can happen early in the morning also. This beautiful "liquid gold" effect arises of course, from the reflection of sunlight from the surface of the water, but it is not quite *that* simple. This phenomenon is not reflection from a flat surface in general—the surface is rippled with waves. As Minnaert points out in his book, there are three cases to consider, and we will fill in some of the mathematics for each in turn. The calculations that follow are valid for any light source, including street lights, though in each case we shall ignore the fact that these sources (including the sun) are extended sources. This will not affect the validity of the calculations; it will just modify the effective width of the glitter path. Note from figure 4.6 that for waves with small angles of inclination ($\leq \alpha$) to the horizontal, the angle between the two largest (but opposite) inclinations of adjacent wavelets is 2α. The basic idea is to examine the boundaries of this glitter path or column of light by examining the reflection of light into an observer's eyes from small waves (or wavelets). Minnaert expresses it this way: *If at each location there are a large number of little waves sloping at an angle α, but in all directions of the compass, what is the locus of the waves that will be illuminated?*

Case I: *the observer and the source of light are at equal heights above the water* (see figure 4.7). The observer and source are a horizontal distance

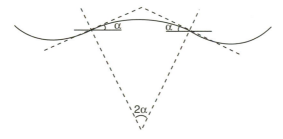

Figure 4.6. The maximum angle between the normals to opposite sides of a wavelet is twice its maximum angle of slope α. This has consequences for the size of observed glitter paths. Redrawn from *Light and Colour in the Open Air* by M.G.J. Minnaert (1954). Dover: New York.

$2l$ apart. Using the symmetry and notation of the figure, it follows that the distances MN and MN' are equal. Were the surface perfectly flat, like a mirror, the source would be symmetrically reflected at the point M into the observer's eye. In terms of the designated angles the following relationships are true:

$$\beta + \alpha = \gamma + \delta, \qquad \beta - \alpha = \epsilon = \delta,$$

so that

$$\gamma = 2\alpha.$$

This means that the angle γ subtended at the eye by the longest axis of the glitter path is equal to the angle between the two largest inclinations of the wavelets. Now we consider the short axis (perpendicular to the long axis); see figure 4.8a, b. From the notation in the diagram, $MP = MP' = h\tan\alpha$, and the width of the light patch is therefore twice this. At the eye, the short axis subtends an angle 2θ, where

$$\tan\theta = \frac{h\tan\alpha}{(l^2 + h^2)^{1/2}}.$$

If α and (hence) θ are small, it follows that the "aspect ratio" a of the glitter path (narrow:long) is

$$a = \frac{2h\tan\alpha}{(l^2 + h^2)^{1/2}} \Big/ 2\alpha = \frac{h\tan\alpha}{\alpha(l^2 + h^2)^{1/2}} \approx \frac{h}{(l^2 + h^2)^{1/2}} \equiv \sin\omega$$

Consequently, if the observer (and source) are high above the water (h large), then $\sin\omega \approx 1$ and the path is close to square. For very oblique observations, $l \gg h$, so the patch is "oblong" in shape.

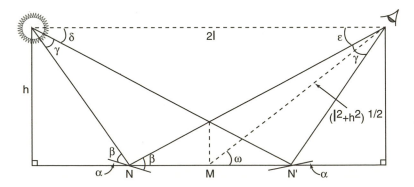

Figure 4.7. The basic geometry for glitter paths produced when the light source (assumed a point source here) and the observer are at the same height h above the water. Redrawn from *Light and Colour in the Open Air* by M.G.J. Minnaert (1954). Dover: New York.

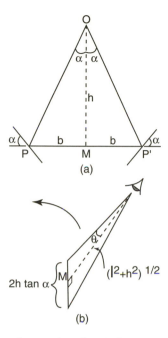

Figure 4.8. (a) The view orthogonal to that in figure 4.7, for a concave up wavelet; (b) an intermediate oblique view. Redrawn from *Light and Colour in the Open Air* by M.G.J. Minnaert (1954). Dover: New York.

Case II is more general and therefore harder to quantify; we shall content ourselves therefore with some broad conclusions (see figure 4.9). Now the height of the source (h') and observer (h) are unequal. Similar arguments to those in case I establish that

$$u + v = 2\alpha = u' + v',$$

i.e.,

$$\gamma + \gamma' = 4\alpha.$$

Note that for a given source position, $\gamma \sim h^{-1}$, that is, the higher the observer (at fixed horizontal position) the smaller the angle subtended by the wavelet reflections. This enables us to make a crude approximation in this more complicated situation, namely by applying this to both source and observer so that

$$\frac{\gamma}{\gamma'} \approx \frac{h'}{h}.$$

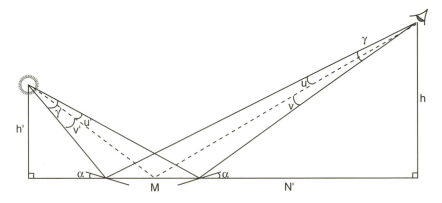

Figure 4.9. The glitter path geometry when the source and observer are at different heights above the water.

Substituting for γ' we have

$$\gamma \approx \frac{4\alpha h'}{h + h'} \to 2\alpha \qquad \text{if } h = h' \text{ as in Case I.}$$

Case III is a limiting one: $h' \to \infty$. This applies to the sun, moon, stars and planets, and other (!) very high lights. Clearly

$$\lim_{h' \to \infty} \gamma = 4\alpha$$

and the distance PP' is now in this limit $2h \tan 2\alpha$, so the ratio of the apparent width to apparent length of the glitter path is approximately $\sin \omega$ as before, but *the dimensions are twice as large*. Go out on a dark night and check all this out for yourself, but please don't fall in the water.

HOW "THICK" IS THE ATMOSPHERE?

The choice of wording may depend on your perspective. In one sense (a mathematical one) the answer could be "indefinitely large" because, although the atmosphere gets thinner as we go higher, the particle number density is never zero (even in deep space). Obviously one could choose an a priori number density as a lower bound, but we will use the concept of *air mass* (AM), as is common in atmospheric science. Simply put, the air mass is the amount of air through which the observer is looking in any given direction. It is a relative distance, though of course it is related to the physical mass of air if the observer is considered to look through an imaginary tube of a given cross section. We consider the simplest case first: that of plane geometry (see figure 4.10). An observer looking directly toward the local zenith point, whether in plane or spherical geometry, is by definition looking through *unit air mass* ($x = 1$). This avoids the issue of how far out the atmosphere actually extends, because the real point at issue is the number of "particles" available for scattering light. If the observer is looking from a plane earth at an angle θ from the vertical direction, then clearly the air mass is just

$$x = \sec\theta, \qquad 0 \leq \theta < \pi/2.$$

For spherical geometry, the situation is more complicated and therefore more interesting (see figure 4.11). Consider first the case in which the observer is looking horizontally, that is, in a direction perpendicular to the zenith. Let R be the radius of the earth; this will be expressed shortly in air mass units also. Recalling that the air mass is unity in the zenithal direction, a judicious use of the Pythagorean theorem yields

$$x = \sqrt{2R + 1}.$$

Obviously R is about 4000 mi, but what is it in air mass units? We perform a "thought experiment" in order to "get a handle" on this problem by imagining the atmosphere to be compressed in such a way that it is of uniform density throughout; this density may be taken to be that at sea level without loss of generality. This will be one air mass since nothing has changed in terms of the amount of air in a vertical direction. Atmospheric pressure at sea level is about 15 lb/in^2 (we do not require any more accuracy than this here), and since 1 cm^3 of air at this pressure weighs about 1.3×10^{-3} gm, it follows that 1 in^3 of such air weighs about $(1.3 \times 10^{-3}) \times (2.54)^3/454$ lb, or approximately 4.7×10^{-5} lb, an atmosphere weighing 15 lb/in^2 must be (in miles) about $[15/(4.7 \times 10^{-5})]/12 \times 5280$ mi or approximately 5 mi. Hence 1 AM $\approx$ 5 miles. With this crucial fact, R may now be expressed as 4000/5 or 800 AM. Note that the fact the earth is not made out of air (!) is

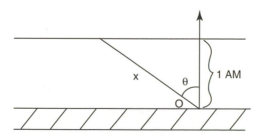

Figure 4.10. The air mass as a function of angle from the zenith (θ) as viewed from a "flat earth." Reprinted from *Color and Light in Nature*, by D. K. Lynch and W. Livingston (1995), with permission of Cambridge University Press.

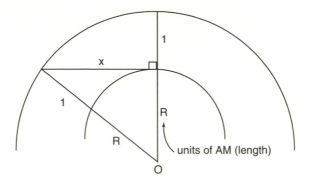

Figure 4.11. The air mass when viewed horizontally on a spherical earth. The radius R of the earth is expressed here in air mass units. Reprinted from *Color and Light in Nature*, by D. K. Lynch and W. Livingston (1995), with permission of Cambridge University Press.

irrelevant here because we need only to know how the two units are related. Substituting this value of R into the radicand above we find that $x \approx 40$ AM. This is both surprising and interesting; the atmosphere is 40 times as "thick" in the horizontal as in the vertical direction. This is certainly a reason for the greater degree of scattering of sunlight in this direction (as a beautiful sunrise or sunset attests).

Now consider the case of an observer looking in an arbitrary direction θ; note that if the observer is above sea level then $\theta > \pi/2$ is possible. By the cosine law (using figure 4.12) it can be seen that x satisfies the quadratic equation

$$x^2 + 2xR\cos\theta - (2R + 1) = 0.$$

Note that the positive root reduces to $x = \sqrt{2R+1}$ when $\theta = \pi/2$. More generally, the positive root is

$$x = R\left(\sqrt{\cos^2\theta + \delta} - \cos\theta\right),$$

where $\delta = (2R + 1)R^{-2}$. Note that δ is a small quantity; if $\delta \ll \cos^2\theta$ then the expression for x may be approximated using the binomial theorem to give

$$x \approx \frac{R\delta}{2\cos\theta} = \left(1 + \frac{1}{2R}\right)\sec\theta \approx \sec\theta,$$

which is the planar result established earlier. For what angles θ is this approximation a good one? Since $R \approx 800$, $\delta \approx 1/400$; if we are willing to accept the strong inequality $\delta \ll \cos^2\theta$ as signifying that $20\delta \lesssim \cos^2\theta$, for example, then this implies that $\cos\theta \gtrsim 0.2$, or $|\theta| \lesssim 78°$. The planar approximation is good for all but very low angles close to the horizon.

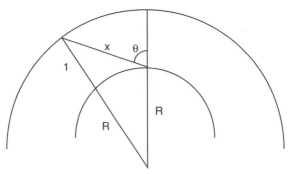

Figure 4.12.   The air mass as a function of zenithal angle θ on a spherical earth.

CREPUSCULAR RAYS AND CLOUD DISTANCES

Crepuscular rays (literally "twilight rays") are formed when clouds block sunlight in such a way that magnificent rays of light interspersed with shadow bands appear to fan out from the direction of the sun. The apparent divergence (and sometimes convergence) is an effect of perspective. The intervening clouds may be many miles away and invisible to the observer, though that is relatively infrequent; usually the blocking clouds are clearly visible. We consider the problem in two parts (figures 4.13 and 4.14). First is the limiting case of a low cloud "resting" at a point C below the observer O's horizon (only fog clouds literally rest on the ground, but low clouds are well described using this particular approach as a bound). As can be seen from the figure, if the sun is at an angle α below the observer's horizon, this low cloud (and any above it, while still below the observer's horizon) can cause crepuscular rays.

In the second part we consider a cloud at height h above the ground at B, but just on the horizon for an observer at O (figure 4.14). Under these circumstances, crepuscular rays may lie at any angle within angle OBD, depending on the cloud location, and on the sun's position (between S_1 and S_2). Outside this range the observer will see no shadows cast by the cloud at B. If h is small compared with the arcs subtending angles α and β, that is, $h \ll \alpha R$, $h \ll \beta R$, the observer-cloud distance d will lie in the range

$$R(\alpha - \beta) \leq d \leq R(\alpha + \beta).$$

Typically, as their name implies, these rays are generally seen at dawn or dusk; as will be shown, this implies that both α and β are small, so this will be assumed for now and justified below. Therefore

$$\cos \beta = \frac{R}{R+h} \approx 1 - \frac{\beta^2}{2},$$

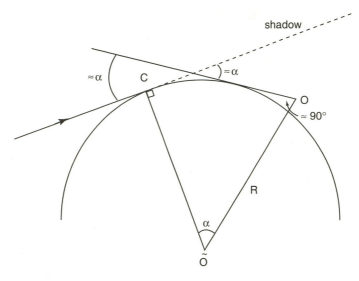

Figure 4.13. Crepuscular rays produced by a low cloud and visible when the sun is at an angle α below the observer's horizon (viewed from O). Note that angle deviation from 90° at O is exaggerated.

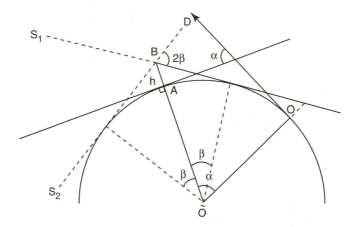

Figure 4.14. A generalization of figure 4.13 for a cloud at height h above the earth's surface. Redrawn from *Light and Colour in the Open Air* by M.G.J. Minnaert, (1954). Dover: New York.

whence

$$\beta^2 \approx \frac{2h}{R+h} \approx \frac{2h}{R},$$

since $h \ll R$ (h being no more than about 5 miles at most, and certainly less for the denser cumulus clouds; typically $h \lesssim 3$ miles.). About half an hour after sunset, at latitude 60°, $\alpha \approx 4°–5°$ (see below), so $\beta \approx \sqrt{6/4000} \approx 0.04$ radians or about 2.2°. If we take $\alpha = 4°$, then $\alpha - \beta \approx 1.8° \approx 0.03$ radians, and $\alpha + \beta \approx 6.2° \approx 0.11$ radians. According to the above bounds on d, this means that 120 mi $\lesssim d \lesssim 440$ mi. Since the distance to the horizon for an observer standing at sea level is approximately 3 miles (as shown below), this explains why crepuscular rays can appear in an apparently cloudless sky! If we redo these calculations for latitude 30°, we find $\alpha \approx 6°–7°$ (we take $\alpha = 6°$), so $\alpha - \beta \approx 3.8° \approx 0.07$ rad, and $\alpha + \beta \approx 8.2° \approx 0.14$ rad. The corresponding bounds on d are 280 mi $\lesssim d \lesssim 560$ mi.

This all hinges of course on (i) whether $h \ll \alpha R$, $h \ll \beta R$ and (ii) whether α is "small." Since $\beta \leq \alpha$, we need concern ourselves only with α in case (ii). In case (i), the first strong inequality is satisfied when $\alpha \gg h/R$, that is, $\alpha \gg 7.5 \times 10^{-4}$. Since a value for α of 0.07 (or 4°) exceeds this value by a factor of about 100, and β is at least one-third α in the latitudes chosen here, the inequalities are well satisfied. For case (ii), the path and "speed" of the rising/setting sun must be considered near the horizon (see figure 4.15). The earth rotates 360° in 24 hours (in fact, it is closer to 23 hours 56 minutes), which corresponds to an angular rate ω of 15°/hr at which the sun migrates across the sky (azimuthally as seen from the poles in their respective long summer "day"). At latitude θ, half an hour before sunrise or after sunset, the angular distance α of the sun below the horizon in each case is $\frac{1}{2} \times \omega \cos\theta$; at a latitude of 60° this is about 3.8° and at 30° latitude it is about 6.5°, which is where the above estimates came from.

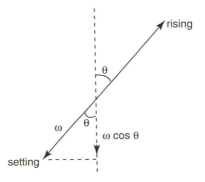

Figure 4.15. The rate at which the sun appears to rise or set in terms of the latitude θ of the observer.

From figure 4.16 we can use similar ideas to discuss the twilight glow produced in particular by an aerosol layer at a height h above the surface of the earth (an aerosol is a suspension of fine solid or liquid particles in the air or other fluid, e.g., smoke, volcanic "cloud" or fog). Suppose again that the sun is an angle α below the observer's horizon, and x is the horizontal distance from the observer O to the aerosol layer. Then, from the diagram,

$$x = \{(R+h)^2 - R^2\}^{1/2} \approx \sqrt{2Rh},$$

since $h \ll R$. Therefore, using similar triangles, we have

$$\tan \frac{\alpha}{2} \approx \left(\frac{2h}{R}\right)^{1/2},$$

or the height of the aerosol layer is

$$h \approx \frac{R}{2} \tan^2 \frac{\alpha}{2}.$$

Incidentally, this same diagram and the corresponding result $x \approx \sqrt{2Rh}$ serve to approximate the distance x from an observer (now at P, a height h above sea level) to the horizon at O. This neglects the effect of atmospheric refraction of course, but since that increases the effective "distance" by about 9 percent (see the article by French) this result represents a lower

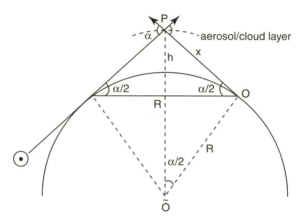

Figure 4.16. The basic geometry for the twilight glow produced by an aerosol layer at a height h above the earth's surface. This figure can also be used to determine the approximate distance to the horizon (neglecting refraction) for an observer, and also the amount by which the moon (or an artificial satellite) falls toward the earth each second. Reprinted from *Sunsets, Twilights and Evening Skies*, by A. Meinel and M. Meinel (1983), with permission of Cambridge University Press.

bound. A useful "rule of thumb" version is easily established if we suppose that h is measured in feet (as is frequently the case) and R is expressed in miles (more accurately than above, we use $R = 3960$ miles here). Then

$$x \approx \sqrt{2 \times 3960 \times \frac{h}{5280}} = \sqrt{1.5h} \text{ mi.}$$

Thus if a tall adult at the seashore looks out to sea, taking $h = 6$ ft, it follows that $x \approx 3$ miles. On a small hill ($h = 150$ ft), $x \approx 15$ miles. The approximation $h \ll R$ is still valid for an observer at the top of Mt. Everest, so taking $h \approx 29,000$ ft, we find $x \approx 200$ miles. There is quite a view from the top, no doubt, although I do not speak from personal experience.

But there is more. The selfsame figure can be utilized yet again, this time to determine how far the moon *falls around* the earth every second, or more generally, how far any given planet falls around the sun in any sufficiently small time interval. We assume the moon's orbit is circular with a radius $R \approx 2.4 \times 10^5$ miles (this assumption is good enough for the present purposes) and a period of 27.3 days. The speed of the moon in its orbit is therefore

$$v = \frac{2\pi \times 2.4 \times 10^5}{27.3 \times 24 \times 60 \times 60} \approx 0.64 \text{ mi/s.}$$

Therefore, if the gravitational influence of the earth were nonexistent at this moment, the moon would travel tangentially a distance of $x = 0.64$ miles in this first second. But using the Pythagorean theorem as before,

$$h = \frac{x^2}{2R} \approx 8 \times 10^{-7} \text{ mi} \approx \frac{1}{20} \text{ inch.}$$

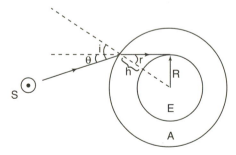

Figure 4.17. Refraction of light from the sun S by the atmosphere A of the earth (E), a simplistic model in which the atmosphere is considered to be uniform and depth h; the sun (or some part of it) is at an angle θ below the horizon but appears to be on the horizon as a result of refraction. By permission of Neil Shea, from "Estimating the Effective Depth of the Atmosphere Using the *Old Farmer's Almanac*," by N. M. Shea and J. Loomis (1997). *The Physics Teacher* 35: 90.

Finally, we close this section with two calculations about the height of the atmosphere (again—but this time using a different approach) and the apparent shape of the sun. In their article about atmospheric thickness, Shea and Loomis point out that at equinox the length of daylight (sunrise to sunset) should be exactly 12 hours at all points on the surface of the earth (except very near the poles, where in practice the day and night are smeared together at these times), but typically almanacs state that daylight will last for 12 hours and 8 minutes. The sun is not a point source of light, and it takes about two minutes (depending on latitude) to completely rise or set. If we take daylight, somewhat arbitrarily, to begin and end when half of the sun's disk is above the horizon, then we still have about 6 minutes to account for and, the authors correctly suggest, this must be due to *refraction* of sunlight by the atmosphere. This enables an estimate to be made for the effective thickness h of the atmosphere (assumed uniform, but with the same cumulative refractive properties as the real atmosphere, which is why the qualifier "effective" is used). From the diagram (4.17) note that if r is the angle of refraction, R the radius of the earth, then

$$r = \arcsin\left(\frac{R}{R+h}\right)$$

and the angle of incidence $i = r + \theta$, where $\theta(t) = \omega t$ is the angular distance of the sun below the horizon. The angular speed ω of the earth about its axis is 7.27×10^{-5} rad/s and t is half the 6-minute discrepancy (the other half occurring at sunset)—180 seconds. This means that $\theta \approx 0.0131$ rad/s. From Snell's law (see chapter 5) it follows that (n being the refractive index of the air)

$$\sin(r + \theta) = n \sin r,$$

from which we find the result

$$\tan r = \frac{\sin \theta}{n - \cos \theta}.$$

Given that $n \approx 1.0003$ for air, it is readily found that $r \approx 1.54$ radians ($\approx 88°$). Since

$$h = R(\csc r - 1),$$

we find on using $R \approx 6.38 \times 10^6$ m that $h \approx 2800$ m. This is about half the previous estimate using air mass, but this is not too bad at all considering the simplicity of each approach. Obviously the atmosphere extends much higher because its density (and hence refractive index) decreases with height, but this provides a useful result if we are interested in the equivalent depth problem for some application or other.

Let us now think about the apparent *shape* of the setting sun—its ellipticity, to be precise. The cause is differential atmospheric refraction: the light from the lower limb of the sun is refracted more than the light from the upper limb, whereas the horizontal diameter is unaffected (both limbs being equally refracted). This "squashing" of the sun can be quite substantial near the horizon—of the order of 10 percent or more depending on the altitude of the sun. A simple mathematical explanation of this phenomenon will make use of some simplifying assumptions; although the air decreases continuously in a direction away from the surface of the earth, we shall consider there to be a discontinuous vacuum-air interface at the "top" of the atmosphere. From Snell's law, $\sin i = n \sin r$, it follows that the differential version is

$$(\cos i)\delta i = (n \cos r)\delta r,$$

where δi is the angular difference in the incident rays from the upper and lower limbs, that is, the apparent diameter of the sun (with no atmospheric distortion), so $\delta i \approx 0.5°$. The angular width of the distorted image is δr, so

$$\delta r = \frac{(\cos i)\delta i}{n \cos r}.$$

Since $i > r$ (or, equivalently, $n > 1$) it follows that $\cos i < \cos r$, so $\delta r < \delta i$, and our result is established, qualitatively at least. To do a full quantitative analysis, we would need to consider the differential equation for continuous refraction, that is, with $n = n(R)$, R being the altitude. The not-unrelated topic of mirages is discussed in chapter 5.

On the topic of the setting sun, it would be remiss not to mention the famous *green flash* (which at the time of writing I have never seen) sometimes observed as the sun sets over water. This has to do with the phenomenon of *dispersion*, which is discussed more fully in the next chapter. Basically, the blue component of the setting sun's image is refracted most and the red component least. The former is therefore highest above the horizon, and since the sun is not a point source of light the intermediate colors overlap except at the top and bottom of the image. It is the top of the sun as it sets that is the last to be seen, of course, leaving a tinge of color as it sinks below the horizon. But shouldn't this correspond to a "blue flash"? Apparently this has been observed, but it appears to be much less common than the green flash. Lynch and Livingston content themselves with the intriguing statement, "Perhaps aerosols play a greater role than we expect or subtle aspects of human vision are involved." *There* is a research topic for the budding atmospheric scientist!

WHY IS THE SKY BLUE?

In a delightful book by Robert Ehrlich entitled *The Cosmological Milkshake* there appears a cartoon showing a child sitting on a bench with her father; the caption reads "The budding urban scientist asks 'Daddy, why is the sky brown?' " Returning to the original question, note that sunlight entering the earth's atmosphere is scattered by the molecules in the air, which are small compared with wavelengths of light. The electric field of the incident sunlight causes electrons in the molecules to oscillate and re-radiate the light; this is what is meant here by the word "scattering." The degree of scattering is inversely proportional to the fourth power of the wavelength of the light; blue light being of shorter wavelength than red, it is scattered the most, and consequently we see blue sky except when we look in the direction of the sun at sunrise or sunset, when the long path through which the light passes depletes the blue light, leaving a predominance of the longer wavelength red light. This phenomenon, coherent scattering, is often referred to as *Rayleigh scattering*, named for Lord Rayleigh, who developed the theoretical basis for this type of scattering. Sunset colors are also determined by the amount of dust (or aerosols in general) in the atmosphere; after major volcanic eruptions they can be really spectacular, as are the sunrises, and can occasionally give rise to so-called "blue moons" (but only once in a blue moon).

By now the wide-awake reader may be thinking: *So why isn't the sky violet, since that has a shorter wavelength than blue light?* The reasons depend on both external and internal factors. First, sunlight is not uniformly intense at all wavelengths (otherwise it would be pure white before entering the atmosphere). It has a peak intensity somewhere in the green part of the visible spectrum, so the entering intensity of violet light is considerably less than that of blue. The other reason is physiological in origin: our eyes are less sensitive to violets than to blues. The scattering of sunlight by molecules and dust is much more complex than described here, of course: there are subtle dependences of color and intensity as a function of angle from the sun and polarization of the light; dust is typically *not* small compared to the the wavelengths of light, and so Rayleigh scattering does not occur; and so on. Excellent readable and informative accounts may be found in the books by Walker and Bohren listed in the references.

Rayleigh scattering applies to more than just light; it works for sound as well under the right circumstances. In their book *Mad about Physics* Jargodzki and Potter state that if one speaks into a large stand of fir trees from a distance, the echo may return raised by an octave! This is based on fascinating discussions in the literature by Crawford and Rinard about such natural "echo chambers." It seems that in 1873 Lord Rayleigh studied this phenomenon; he stated that he heard an echo of a woman's voice that was raised by an octave (perhaps she walked into a tree), but that of a man did not. It transpires that the trees are the scattering centers, analogous to the molecules in the air for sunlight, but this requires that the wavelength of the incident sound should be large compared with the width of the trees. The first harmonic present in the sound (at twice the frequency of the fundamental and hence half the frequency) is scattered $2^4 = 16$ times more than the fundamental, and this may dominate the returning echo. Below we use a dimensional approach (adapted from that in the book by Bohren and Huffman) to try and elucidate the wavelength dependence of Rayleigh scattering.

RAYLEIGH SCATTERING—A DIMENSIONAL ANALYSIS ARGUMENT

Here we hark back to the methods of dimensional analysis discussed in the latter part of chapter 3. Let the intensity I of the radiation scattered from a particle of volume V be described by the expression

$$I = f(V, r, \lambda, n_1, n_2, I_0),$$

where $r, \lambda, n_1, n_2, I_0$ are, respectively, the distance from the scatterer to the observation point, the wavelength of the scattered radiation, the refractive indices of the exterior and interior media, and the intensity of the incident radiation. The refractive indices, being ratios of speeds of light in different media, are dimensionless, and to be specific, we seek the following algebraic functional decomposition:

$$I \propto V^\alpha r^\beta \lambda^\gamma I_0.$$

A particularly simple result follows for the dimensions of both sides of this equation, namely,

$$[L]^{3\alpha}[L]^\beta[L]^\gamma = [L]^0,$$

because the dimensions of intensity "cancel" from both sides. Since the amplitude of scattered radiation is proportional to the number of scatterers, which in turn is proportional to the volume of the (composite) particle, and intensity is equal to the square of the amplitude, we have that $\alpha = 2$. Furthermore, $\beta = -2$ since a dipole (e.g., a molecule in which the positive and negative charges are not coincident) radiates energy in all directions (think of the surface of an expanding sphere of light; the energy per unit area decreases as $(\text{radius})^{-2}$). This means that $6 - 2 + \gamma = 0$, or $\gamma = -4$. Thus, in particular,

$$I_{scatt} \propto \lambda^{-4},$$

which is effectively *Rayleigh's inverse fourth-power law of scattering.* Rayleigh scattering theory applies if the particles are small, $\lesssim 0.1\lambda$; under these circumstances they can be considered as point dipoles. For particles much larger than this size, the light scattered from one part of the particle may be out of phase with that from another part; the resulting interference reduces the intensity of the scattered radiation. There is now a greater intensity in the forward direction, because the cumulative effect of the phase differences is smallest for small scattering angles. *Mie scattering theory* takes into account all these features, and using Maxwell's equations of electromagnetic theory, the scattering intensity can be expressed as an infinite series of so-called *partial waves*; Rayleigh scattering is represented by the first term in this series, so it is a special (or limiting) case of Mie theory. For larger particles, more terms have to be included; indeed, for rainbows (produced by light scattering from raindrops—see the next chapter), the number of terms required is of the order of *five thousand*—the ratio of the drop circumference to the wavelength!

APPENDIX: A WORD ABOUT SOLID ANGLES

At the beginning of this chapter we talked in rather cavalier (but nevertheless appropriate) terms about the angles subtended at the earth by the sun and moon. Because these angles are small (about half a degree of arc) there was really no need to think in terms of areas expressed in an angular sense. However, larger regions of the sky, such as those containing particular constellations, do require to be so expressed on occasion, and for completeness we mention the fundamental idea of *solid angle* here. The solid angle at any point P subtended by a surface S is equal to the area A of the portion of the surface of a sphere of unit radius, with center at P, which is cut by a conical surface with vertex at P and the perimeter of S as a generatrix. The unit of solid angle is called the *steradian*. The total solid angle about a point is equal to 4π steradians. Since one radian equals $(180/\pi)°$ it follows that 4π steradians equal

$$4\pi \times \left(\frac{180}{\pi}\right)^2$$

square degrees, or approximately 41,253 square degrees. The apparent area of the sun (or moon) is $\approx \pi(\frac{1}{4})^2 \approx 0.2$ sq. deg. The bright band of height $\approx 4°$ around an unobscured horizon during the day is approximately $4 \times 360 = 1440$ sq. deg. The bright winter constellation Orion (the Hunter) occupies about 594 sq. deg. (Orion is in fact the 26th largest constellation out of 88). Orion occupies about 1.44 percent of the total sky area; the largest constellation is Hydra (the Water Snake), which covers 1303 sq. deg. (3.16% of the sky); the smallest is Crux (the (Southern) Cross), occupying 68.5 sq. deg. (0.17% of the sky). The famous Ursa Major (the Great Bear) is third in size, with 1280 sq. deg. (3.10% of the sky).

CHAPTER FIVE

Meteorological Optics II: A "Calculus I" Approach to Rainbows, Halos, and Glories

> Like the appearance of a rainbow in the clouds on a rainy day, so was the radiance around him. This was the appearance of the likeness of the glory of the Lord.
>
> —Ezekiel 1: 28a

How much mathematics shall we get into here? Not a great deal, in fact; for those interested in pursuing the mathematics to a much greater level, refer to the author's review and the many references therein. But first some background material on the history and elementary physics of the rainbow is in order. A good overview of the problem can be found in the book by Banks (1999); for more historical and scientific details, those by Boyer and Lee and Fraser are highly recommended.

THE RAINBOW

The rainbow is at one and the same time one of the most beautiful visual displays in nature and, in a sense, an intangible phenomenon. It is illusory in that it is not of course a solid arch, but, like the mirages we will discuss later in this chapter, it is nonetheless real. It can be seen and photographed and described as a phenomenon of mathematical physics, but it cannot be located at a specific place, only in a particular direction. What then *is* a rainbow? It is an *image* of sunlight, displaced by reflection and dispersed by refraction in raindrops, seen by an observer with her back to the sun (under appropriate circumstances). As shown in figure 5.1, the primary rainbow, which is the lowest and brightest of two that may be seen, is formed from two refractions and one reflection in myriads of raindrops. For our purposes we may consider the path of a ray of light through a single drop of rain, for the geometry is the same for all such drops and a given observer. Furthermore, we can appreciate most of the features of the rainbow by using the ray theory of light; the wave theory of light is needed

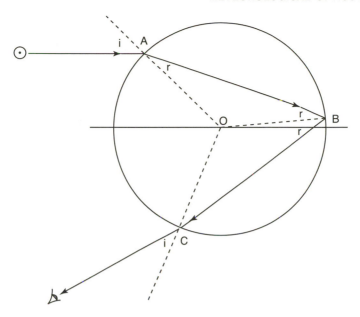

Figure 5.1. The path of a ray of sunlight inside a spherical raindrop that contributes to the formation of the primary rainbow upon exiting the drop. The angles of incidence and refraction are, respectively, i and r; for the primary bow there are two refractions and one internal reflection. The diagram is cylindrically symmetric about the axis of symmetry (the diameter of the drop parallel to the incident ray).

to discuss the finer features such as the supernumerary bows (discussed briefly below).

The first satisfactory explanation for the existence and shape of the rainbow was given by Descartes in 1637 (he was unable to account for the colors, though; it was not until thirty years later that Newton remedied this situation). Descartes used a combination of experiment and theory to deduce that both the primary and secondary bow (larger in angular diameter and fainter than the primary) are caused by refraction and reflection in spherical raindrops. He correctly surmised that he could reproduce these features by passing light through a large water-filled flask—a really big "raindrop." Since the laws of refraction and reflection had been formulated some sixteen years before the publication of Descartes's treatise, by the Dutch scientist Snell, Descartes could calculate and trace the fate of parallel rays from the sun impinging on a spherical raindrop. As can be seen from figure 5.2, such rays exit the drop having been deviated from their original direction by varying but large amounts. The ray along the central axis (#1) will be deviated by exactly 180°, whereas above this point

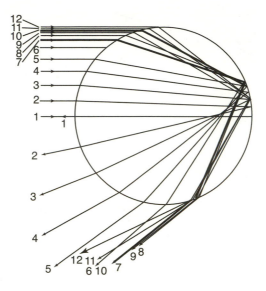

Figure 5.2. The paths of 12 rays incident upon a spherical raindrop. Note that there is a minimum angle of deviation of about 138° corresponding to a direction near that of the emergent path of ray #7. From *Rainbows, Halos and Glories*, reprinted by permission of Robert G. Greenler (now available through *www.blueskyassociates.com*).

of entry the angle of deviation decreases until a minimum value of 138° occurs (for yellow light; other colors have slightly different minimum deviation angles). For rays impinging still higher above the axis the deviation angle increases again. The ray of minimum deviation (#7 in the diagram) is called the rainbow ray. The significant feature of this geometrical system is that the rays leaving the drop are not uniformly spaced: those "near" the minimum deviation angle are concentrated around it, whereas those deviated by larger angles are spaced more widely.

Put differently, in (say) a half degree angle on either side of the rainbow angle (138°) there are more rays emerging than in any other one degree interval. It is this concentration of rays that gives rise to the (primary) rainbow, at least as far as its light intensity is concerned. In this sense it is similar to a caustic formed on the surface of the tea in a cup when appropriately illuminated. The rainbow seen by any given observer, of course, consists of those deviated rays that enter his or her eye. These are those that are deviated by about 138° from their original direction (for the primary rainbow) and into his eye. Thus the rainbow can be seen by looking in any direction that is about 42° away from the line from your eye to the shadow of your head (the antisolar point); the 42° angle is supplementary to the rainbow angle. This criterion defines a circle around the antisolar point, and hence all raindrops at that angle will contribute to your primary rainbow. Of course, at best a semicircular arc will be seen (if the sun is close to setting or has just risen), and usually it will be less than that: full circular rainbows can be seen from time to time at high altitudes on land or from aircraft.

In summary, the primary rainbow is formed (see figure 5.2) by the deflected rays from all the raindrops that lie on the surface of a cone with vertex (or apex) at the eye, axis along the antisolar direction, and semi-vertex angle of 42°. The same statement holds for the secondary rainbow if the semi-vertex angle is about 51° (the supplement of a 129° deviation). These cones will be different for each observer, so each person has his or her own personal rainbow!

Up to this point we have been describing a generic, colorless type of rainbow. Blue and violet light get refracted more than red light: the actual amount depends on the index of refraction of the raindrop, and the calculations thereof vary slightly in the literature because the wavelenghts chosen for "red" and "violet" may differ slightly. Thus for a wavelength of 6563 Å (Angstrom units; 1 Å = 10^{-10} m) the cone semi-angle is about 42.3°, whereas for violet light of 4047 Å wavelength, the cone semi-angle is about 40.6°, about a 1.7° angular spread for the primary bow. A similar spread (dispersion) occurs for the secondary bow, but the additional reflection reverses the sequence of colors, so the red in this bow is on the *inside* of the arc.

In principle more than two internal reflections may take place inside each raindrop, so higher-order rainbows (tertiary, quaternary, etc.) are possible. It is possible to derive the angular size of such a rainbow after any given number of reflections (Newton was the first to do this). Newton's contempory Edmund Halley found that the third rainbow arc should appear as a circle of angular radius about 40° around the sun itself! The fact that the sky background is so bright in this vicinity, coupled with the intrinsic faintness of the bow itself, would make such a bow almost if not entirely impossible, to see (although such claims have been made; see the accounts collected by Lee and Fraser in their book *The Rainbow Bridge*). Jearl Walker has used a laser beam to illuminate a single drop of water and traced rainbows up to the thirteenth order, their positions agreeing closely with predictions. Others have traced up to nineteen rainbows, also under laboratory conditions.

The sky "inside" the primary rainbow is often noticeably brighter than the sky outside it; indeed, the region between the primary and secondary bow is called Alexander's dark band (after Alexander of Aphrodisias, who studied it in connection with Aristotle's theory of the rainbow). Raindrops "scatter" incident sunlight in essentially all directions, but, as we have seen, the rainbow is a consequence of a "caustic" or concentration of such scattered light in a particular region of the sky. The reason the inside of the primary bow (i.e., inside the cone) is bright is that all the raindrops in the interior of the cone reflect light to the eye also, but it is not as intense as the rainbow light, and it is composed of many colors intermixed. *Outside* the secondary bow a similar (but less obvious) effect occurs. Much of the scattered light, then, comes from raindrops through which sunlight is refracted and reflected: these rays do not emerge between the 42° and 51° angle. This dark angular band is not completely dark, of course, because the surfaces of raindrops reflect light into it; the reduction of intensity, however, is certainly noticeable.

Another commonly observed feature of the rainbow is that when the sun is near the horizon the nearly vertical arcs of the rainbow near the ground are often brighter than the upper part of the arc. The reason for this appears to be the presence of drops with varying sizes. Drops smaller than, say 0.2–0.3 mm (about 0.01 in) are spherical: surface tension is quite sufficient to keep the distorting effects of aerodynamical forces at bay. Larger drops become more oblate in shape, maintaining a circular cross section horizontally but not vertically. They can contribute significantly to the intensity of the rainbow because of their size, $\approx$ 1 mm ($\approx$ 0.04 in) or larger, but can only do so when they are "low" on the cone, for the light is scattered in a horizontal plane in the exact way it should to produce a rainbow. These drops do not contribute significantly near the top of the arc because of their noncircular cross section for scattering. Small drops, on the other hand, contribute to all portions of the rainbow.

Drop size, as implied above, can make a considerable difference to the intensity and color of the rainbow. The "best" bows are formed when the drop diameter is $\gtrsim$ 1 mm; as the size decreases the coloration and general definition of the rainbow become poorer. Ultimately, when the drops are about 0.05 mm or smaller in diameter, a broad, faint, white arc called a fogbow occurs. When sunlight passes through these very tiny droplets a phenomenon called diffraction takes place. Essentially, because of its wave nature, the interaction of light with objects comparable in size to a typical wavelength will cause a light beam to spread out. Thus the rainbow colors are broadened and overlap, giving rise in extreme cases to a broad white fogbow (or cloudbow, since droplets in those typically produce such bows). As Greenler points out, these white rainbows may sometimes be noted if one flies above a smooth featureless cloud bank (Lee and Fraser describe a "combo" cloudbow and rainbow observed by J. MacDonald). The rainbow cone intersects the horizontal cloud layer in a hyperbola if the sun's elevation is less than 42° (or an ellipse if it is greater than 42°), as familiarity with the conic sections assures us. This phenomenon can also be seen as a "dewbow" on a lawn when the sun is low in the eastern sky.

Another feature of rainbows produced by smaller drops is also related to the wave nature of light. This time the phenomenon is *interference*, and it produces *supernumerary bows*. These are a series of faint pink or green arcs (2–4, perhaps) just beneath the top of the primary bow, or less commonly just above the top of the secondary bow. They rarely extend around the fully visible arc, for reasons that are again related to drop size. Two rays that enter the drop on either side of the rainbow ray (the ray of minimum deviation) may exit the drop in parallel paths; this *will* happen for appropriately incident rays. By considering the wavefronts (perpendicular to the rays), it is clear that the incident waves are in phase (crests and troughs aligned with crests and troughs). Inside the drop they travel paths of different length. Depending on whether this path difference is an integral number of wavelengths or an odd integral number of half-wavelengths, these waves will reinforce each other (constructive interference) or cancel each other out (destructive interference). Obviously, partially constructive/destructive interference can occur if the path difference does not meet the above criteria. Where waves reinforce one another the intensity of light will be enhanced; conversely, where they annihilate one another the intensity will be reduced. Since these beams of light will exit the raindrop at a smaller angle to the axis than the Descartes ray, the net effect for an observer looking in this general direction will be a series of light and dark bands just inside the primary bow. They are just as much a part of the "rainbow" as the primary bow.

The angular spacing of these bands depends on the size of the droplets producing them. The width of individual bands *and* the spacing between them decrease as the drops get larger. If drops of many different sizes are present, these supernumerary arcs tend to overlap somewhat and smear out what would have been obvious interference bands from droplets of uniform size. This is why these pale blue or pink or green bands are most noticeable near the top of the rainbow: it is the smaller drops that contribute to this part of the bow, and these may represent a rather narrow range of sizes. Nearer the horizon a wide range of drop sizes contribute to the bow, but, as we have seen, at the same time they tend to blur the interference bands.

There are many variations on the theme of rainbows. Greenler has written about such things in some detail in his book *Rainbows, Halos, and Glories*. Reflected-light rainbows, reflected rainbows, lunar rainbows, and also infrared rainbows are mentioned. The magnificently illustrated and scholarly book by Lee and Fraser mentioned above deals with rainbows in art, myth, and science (and also in the advertising industry!).

To begin the elementary mathematical treatment of rainbows, we present a brief (and common) derivation of a generic planar form of Snell's law of refraction, much used in connection with rainbows and halos, among other optical phenomena. What follows is usually stated in calculus texts as a land-and-water problem: an individual on land (say) desires to reach a specific point out to sea in the shortest possible time, perhaps to rescue someone from the encroaching tide, or just because it's there.

However posed, it is the mathematical equivalent of Fermat's principle of least time, or, more correctly, the statement that the path of the ray of light will be such that the total time of travel is constant (in most applications, as here, this will be a minimum). From figure 5.3, the points A and B are fixed, and it is desired to find the point C (the point of contact with the tangent plane in the case of a spherical raindrop) such that the time T taken to travel the path ACB, composed of line segments AC and CB (traversed at speed v_1 and v_2, respectively) is (for our purposes) a minimum. Let T_{AC} and T_{CB} be the times to travel those respective segments, so that with a, b, and d constant

$$T = T_{AC} + T_{CB} = \frac{\sqrt{a^2 + x^2}}{v_1} + \frac{\sqrt{b^2 + (d - x)^2}}{v_2}, \quad 0 < x < d,$$

so that

$$\frac{dT}{dx} = \frac{x}{v_1\sqrt{a^2 + x^2}} - \frac{(d - x)}{v_2\sqrt{b^2 + (d - x)^2}} = \frac{\sin\theta_1}{v_1} - \frac{\sin\theta_2}{v_2},$$

which is of course zero at a critical (or stationary) point. This means that we may state the generic form of Snell's law, which states that

$$\sin\theta_1 = n\sin\theta_2,$$

where

$$n = \frac{v_1}{v_2}$$

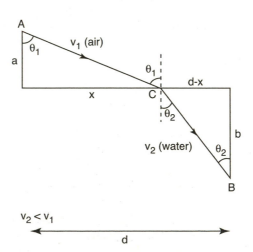

Figure 5.3. The geometry to determine the path ACB of least travel time for a ray of light between the fixed points A and B. This results in the derivation of Snell's law of refraction for light incident on the boundary between two different media.

is what will be referred to below as the refractive index. In the context of meteorological optics, with light rays entering water droplets from air (and exiting into air), the speeds v_1 and v_2 will represent the respective speeds of light in air and in water. Of course, thus far we have not established that T is indeed a minimum time, but, by the first derivative test applied on either side of the critical value x_c for which $T'(x_c) = 0$,

$$x \downarrow \Rightarrow \sin \theta_1 \downarrow \Rightarrow \sin \theta_2 \uparrow \Rightarrow T' < 0$$

and

$$x \uparrow \Rightarrow \sin \theta_1 \uparrow \Rightarrow \sin \theta_2 \downarrow \Rightarrow T' > 0,$$

so this is indeed a minimum time. Now we are ready to examine the primary rainbow. From figure 5.1 note that after two refractions and one reflection the light ray shown contributing to the rainbow has undergone a total deviation of $D(i)$ radians, where

$$D(i) = \pi + 2i - 4r(i)$$

in terms of the angles of incidence i and reflection r, respectively. The latter is a function of the former, this relationship being expressed in terms of Snell's law,

$$\sin i = n \sin r,$$

where n is the relative index of refraction (of water, in this case). This relative index is defined as

$$n = \frac{\text{speed of light in medium I (air)}}{\text{speed of light in medium II (water)}} > 1.$$

Since the speed of light in air is almost that "in vacuo," we will refer to n for simplicity as the refractive index; its generic value for water is $n \approx 4/3$, but it does depend slightly on wavelength (this is the phenomenon of dispersion; without it, as noted earlier, we would only have bright "whitebows").

Now let us examine the behavior of $D(i)$. Note that $D(0) = \pi$. First, a little differentiation is in order. Thus

$$\frac{dD}{di} = 2 - 4\frac{dr}{di},$$

and from Snell's law we find that

$$\cos i = n \cos r \frac{dr}{di},$$

so that

$$\frac{dD}{di} = 2 - \frac{4 \cos i}{n \cos r}.$$

Are there any critical numbers in the domain $i \in [0, \pi/2]$? The condition $D'(i) = 0$ implies that

$$\frac{1}{4} = \frac{\cos^2 i}{n^2 - 1 + \cos^2 i}$$

after some rearrangement and the use of Snell's law again. There are no spurious solutions arising as a result of squaring terms, so at an extremum it follows that

$$\cos i = \left(\frac{n^2 - 1}{3}\right)^{1/2} \equiv \cos i_c.$$

For k internal reflections the corresponding result is

$$\cos i_c = \left[\frac{n^2 - 1}{k(k + 2)}\right]^{1/2},$$

as is readily verified. For general k it can be shown that

$$D_k''(i_c) = \frac{2(k + 1)(n^2 - 1)\tan r}{n^3 \cos^2 r} > 0$$

(see the book by Humphreys), which is positive since $n > 1$ and $0 < r < \pi/2$, except for the special case of normal incidence. Thus the clustering of deviated rays corresponds to a *minimum* deflection, so that the deflection $D(i_c)$ (for $k = 1$ in particular) is a minimum; note, however, that some formulations for $k = 2$ use in effect $-D$ and find a corresponding maximum deflection for this quantity. Note that for $k = 1$, $D(i)$ may be written with sole dependence on i as

$$D(i) = \pi + 2i - 4\arcsin\left(\frac{\sin i}{n}\right).$$

We continue with the case $k = 1$. For a "generic" monochromatic rainbow (the whitebow referred to above), the choice $n = 4/3$ yields

$$i_c \approx \arccos\sqrt{\frac{7}{27}} \approx 59.4°$$

and

$$D(i_c) \approx 138°.$$

The supplement of this angle, $180° - D(i_c) \approx 42°$, is the semi-angle of the rainbow "cone" formed with apex at the observer's eye, the axis being along the line joining the eye to the antisolar point.

Next we mention dispersion, and we can do this in an elementary way or an elegant way. I choose the latter, but before describing it I will point out the essential features of the former. While (as far as I can tell) it is not standardized to the satisfaction of everyone, the visible part of the electromagnetic spectrum extends from the "red" end (7000–6470 Å) to the "violet" end (4240–4000 Å). For red light of wavelength $\lambda = 6563$ Å, the refractive index $n \approx 1.3318$, whereas for violet light of wavelength $\lambda = 4047$ Å, the refractive index $n \approx 1.3435$, a slight but very significant difference! All that has to be done is the calculation of i_c and $D(i_c)$ for these two extremes of the visible spectrum, and the difference computed. Then voilà! We have the angular width of the primary rainbow. In fact, since $D(i_c) \approx 137.75°$ and $139.42°$ for the red and violet ends, respectively, the angular width $\Delta D \approx 1.67° \approx 1.7°$, or about three full moon angular widths.

Now for elegance. In view of the fact that now $D = D(i, n)$ we have that

$$\frac{\partial D}{\partial n} = -4\frac{\partial}{\partial n}\left[\arcsin\left(\frac{\sin i}{n}\right)\right] = \frac{4\sin i}{n\sqrt{n^2 - \sin^2 i}},$$

and so at the rainbow angle ($D \approx 138°$, for which $n^2 - \sin^2 i = 4\cos^2 i$)

$$\frac{\partial D}{\partial n} = 2\frac{\tan i}{n} > 0.$$

This can be used to estimate the angular spread of the rainbow (ΔD) given the variation in n over the visible part of the spectrum (Δn); as noted above ΔD is about 1.7° for the primary bow. In fact, using differentials,

$$\Delta D \approx 2\frac{\tan i}{n}\Delta n,$$

so with a generic value for n of 1.33 and for i_c of 59°, it follows from the above data that

$$\Delta D \approx 2\frac{\tan 59°}{1.33} \times 0.012 \approx 0.03\,\text{rad} \approx 1.7°.$$

Note also that since

$$\frac{dD}{d\lambda} \approx \frac{\partial D}{\partial n}\frac{dn}{d\lambda}$$

for a given value of i (neglecting the small variation of i_c with wavelength λ), and $dn/d\lambda < 0$, it follows that $dD/d\lambda < 0$, and so the red part of the (primary) rainbow emerges at a smaller deviation angle than the violet part, so the latter appears on the underside of the arc with the red outermost.

Now for the secondary rainbow. (For tertiary and higher bows the accounts discussed by Lee and Fraser have been noted above; also see the discussion in Adam 2000a, b; these bows are rarely, if ever seen, so don't look for them; the tertiary is too close to the sun and the higher orders are probably too faint.) In this case there is an additional reflection, so the geometry of figure 5.4 yields a total ray deviation of

$$D(i) = 2\pi + 2i - 6r = 2i - 6r(\text{modulo } 2\pi) = 2i - 6\arcsin\left(\frac{\sin i}{n}\right).$$

Proceeding as before we find that

$$\frac{dD}{di} = 2 - 6\frac{dr}{di} = 2 - 6\frac{\cos i}{n\cos r},$$

so that critical numbers occur when

$$\frac{1}{9} = \frac{\cos^2 i}{n^2(1 - \sin^2 r)} = \frac{\cos^2 i}{n^2 - \sin^2 i}$$

or

$$\cos i = \left(\frac{n^2 - 1}{8}\right)^{1/2} \equiv \cos i_c.$$

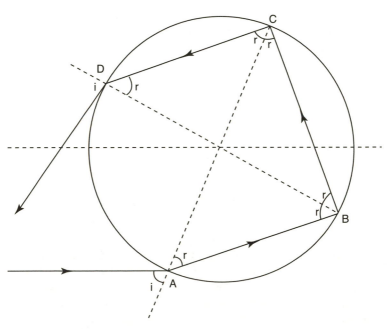

Figure 5.4. The path of light in the formation of the secondary rainbow. There are now two interior reflections and two refractions. As with figure 5.1, the system is cylindrically symmetric about the axis of symmetry.

For our generic n-value of 4/3,

$$\cos i_c = \sqrt{\frac{7}{72}} \approx 0.3118, \text{ so } i_c \approx 71.8°.$$

Then

$$D(i_c) = 2i_c - 6\arcsin(0.7125) = 143.6° - 276.6° = -129°.$$

The negative sign is of no significance here, since the ray geometry is cylindrically symmetric about the central axis, but for this reason sometimes $D(i)$ is defined as the negative of the expression used here, so be careful which one you are using if you calculate $D''(i_c)$. Just as the supplement of $D(i_c)$ for the primary bow is approximately 42°, so for the secondary bow the supplementary angle is 51°; thus the secondary bow is about 9° higher in the sky than the primary bow.

In general, for k internal reflections it can be shown that

$$D(i) = k\pi + 2i - 2(k+1)\arcsin\left(\frac{\sin i}{n}\right),$$

and after an interesting trigonometric calculation, in particular it can be shown that when $k = 1$,

$$D(i_c) = 2\arccos\left[\frac{1}{n^2}\left(\frac{4-n^2}{3}\right)^{3/2}\right],$$

thus expressing the angle of minimum deviation (the "rainbow angle") in terms of the refractive index alone. In principle this can also be done for higher-order rainbows, although the algebra gets messy. Substituting $n = 4/3$ into the above equation gives

$$D(i_c) = 2\arccos\left[\frac{9}{16}\left(\frac{20}{27}\right)^{3/2}\right] = 2\arccos(0.3586) \approx 138°,$$

thus recovering the result for the primary bow. For a detailed review of over two hundred and fifty articles on the mathematical physics of rainbows (and also glories), the author's paper listed in the references should be consulted.

APPENDIX TO RAINBOW THEORY: A LITTLE ABOUT AIRY'S THEORY

Although the historical development of the "mathematical rainbow" has not been emphasized here, it is important to note that the theories of Descartes, Newton and even Young's interference theory all predicted an *abrupt transition* between regions of illumination and shadow (as at the edges of Alexander's dark band when rays only giving rise to the primary and secondary bows are considered). In the wave theory of light such sharp boundaries are softened by diffraction. In 1835 Potter showed that the rainbow ray can be interpreted as a *caustic*, the envelope of the system of rays comprising the rainbow. The word caustic means "burning curve," and caustics are associated with regions of high intensity illumination (geometrical optics—ray theory—predicts an infinite intensity there). When the emerging rays comprising the rainbow are extended backward through the drop, another caustic (a virtual one) is formed, associated with the real caustic on the illuminated side of the drop. Typically, the number of rays differs by two on each side of a caustic at any given point, so the rainbow problem is essentially that of determining the intensity of (scattered) light in the neighborhood of a caustic.

This was exactly what Airy attempted to do several years later. The principle behind Airy's approach was established by Huygens in the seventeenth century: in Huygens's principle one regards every point of a wavefront as a secondary source of waves, which in turn defines a new wavefront and hence determines the subsequent propagation of the wave. Airy reasoned that, if one knew the amplitude distribution of the waves along any complete wavefront in a raindrop, the distribution at any other point could be determined by Huygens's principle. However, the problem is to find the initial amplitude distribution. Airy chose as his starting point a wavefront surface inside the raindrop, the surface being orthogonal to all the rays that comprise the primary bow; this surface has a point of inflection wherever it intersects the ray of minimum deviation—the rainbow ray.

Using the standard assumptions of diffraction theory, Airy formulated the local intensity of scattered light in terms of a "rainbow integral," subsequently renamed the *Airy integral* in his honor (it is related to the now familiar Airy function). He demonstrated that

$$A \propto \int_{-\infty}^{\infty} \cos \frac{\pi}{2} \left(m\xi - \xi^3 \right) d\xi.$$

Here A is the amplitude of the light wave that enters the observer's eye as a function of the angle θ from the direction of the rays of minimum deviation. This integral was first introduced in Airy's 1838 paper "On the Intensity of Light in the Neighbourhood of a Caustic." The significance of the parameter m can be noted from its definition as

$$m = \left(\frac{2ka}{\pi} \right)^{2/3} \left(\frac{\sin^3 \theta}{\cos \theta} \right)^{1/3},$$

which for sufficiently small values of θ (the angle of deviation from the rainbow ray) is proportional to θ. k and a are, respectively, the wavenumber of the light ($= 2\pi/$wavelength) and the drop radius.

Airy's theory enabled him to make predictions about the positions and intensity of the supernumerary bows, which arise from interference between rays exiting the water droplets. However, the theory, while a considerable improvement over geometrical optics, did not adequately explain these interference phenomena in a quantitative manner, but it is certainly sufficient for many practical purposes (as Lee and Fraser emphasize). Further theoretical developments require some of the most powerful tools of mathematical physics; an excellent descriptive account of the historical and mathematical developments in this subject can be found in the article by Nussenzveig.

HALOS

What are halos and what produces them? Two excellent questions! Halos are formed when sunlight is refracted, reflected, or both from ice crystals in the upper atmosphere and enters the eye of a *careful* observer: careful, because even the common types are easily missed. They are usually produced when a thin uniform layer of cirrus or cirrostratus cloud covers large portions of the sky, especially in the vicinity of the sun. Surprisingly, perhaps, they may occur at any time of the year, even during high summer, because above an altitude of about 10 km it is always cold enough for ice crystals to form. In particularly cold climes, of course, such crystals can form at ground level (though we are not thinking here of snow crystals). Very many of these crystals are hexagonal prisms; some are thin flat plates while others are long columns; and sometimes the latter have bullet-like or pencil-like ends. A significant feature of all these crystals is that, while any given type may have a range of sizes, the angles between the faces are the same. Although they do not possess *perfect* hexagonal symmetry, of course, they are sufficiently close to this that simple geometry based on such idealized forms suffices to describe the many different arcs and halos that are associated with them. The halo formation that results from cirrus cloud crystals depends on two major factors: their shape and their orientation as they fall. Their shape is determined to a great extent by their history—the temperature of the regions through which they drift as they are drawn down by gravity and buffeted around by winds and convection currents. Each crystal in all likelihood has a unique path, and in this sense we may make the brash and unproven but highly probable statement (as we will see below) that *no two snowflakes are identical*. If a cloud crystal could be examined for long enough, its shape would allow us in principle to infer the temperature structure along the path it took. In the very simplest terms, plates are more likely to form in two distinct temperature regimes: between

Name	Appearance	Crystal Features	Orientation	Path of Light	Comments
8° and 17° halos	Circular arc around sun or moon	Bullet shaped	Random	Refraction, light through tapering bullet faces	Very rare and hard to see
22° halo	Circular arc around sun or moon	Small columns or plates	Random	Refraction, light through crystal sides	Very common, often brighter on top and bottom
46° halo	Circular arc around sun or moon	Small columns	Random	Refraction, light through top or bottom and one side	Rare
Parhelia to 22° halo	Bright-colored spots of light at same elevation as sun or moon and just outside halo	Medium-sized bullets, capped columns or plates	6 sides, vertical	Refraction, light through crystal sides	Cannot form when sun or moon above about 60°; common
Parhelic circle	Horizontal circle that runs through sun or moon	Medium-sized plates or columns	Some faces, vertical	Refraction, light off vertical faces	Not colored
Tangent arcs to 22° halo (circumscribed halo)	Arcs tangent to 22° halo at top and bottom lying outside halo	Medium-sized columns	Hexagonal, faces vertical	Refraction, light through sides	Shape changes with sun's altitude, merging with halo when sun above 60°
Circumhorizontal arc	Horizontal arc below altitude of sun or moon	Same as for parhelia	6 sides, vertical	Refraction, light in side and out bottom	Cannot form when sun or moon below about 60°
Light pillar	Vertical arc usually above sun or moon	Large-sized plates	6 sides, almost vertical	Reflection, light usually off crystal bottom	Not colored, except when sun is at horizon; usually seen when sun or moon is low in sky
Subsun	Spot of light as far below as sun or moon is above horizon	Medium-sized plates	6 sides, vertical	Reflection, light off crystal top	Must be seen from above—in plane or on mountain

Adapted from Gedzelman (1980).

0 and −4°C and between −10 and −22°C. On the other hand, columns and needles form in the ranges −4 to −10°C and −22 to −50°C. Brrrr!

The above is something of an oversimplification because not only is temperature a determining factor for the shape of the ice crystals, but so is the moisture content, or more accurately, the degree of *water saturation* of the ice. A very useful diagram illustrating this can be found in the 1978 article by Lynch. When the saturation exceeds 108 percent, the temperature alone determines the crystal shape, and, according to Lynch, under these circumstances the various forms are as follows: irregular needles at −3°C, regular needles at −7°, cups or scrolls at −9°, plates at −12°, snowflakes at −15°, plates at −17°, and irregular plates at −23°. Below a saturation level of 108 percent, only plates and columns grow; at levels above 140 percent, the crystals grow so fast that they accumulate *rime*, a type of hoarfrost.

As crystals fall, their shape and the consequent aerodynamic forces acting on them determine their alignment on the way down. Sometimes the orientation is random, or at least distributed in many directions, but frequently as they slowly fall, they are like leaves in the autumn: the large faces are approximately horizontal. This is the case for plate crystals; columnar crystals tend to fall with their hexagonal faces vertical (their long axes nearly horizontal). The stability of their alignment is a problem in classical mechanics: try spinning a book that has been taped shut (or perhaps a brick) about each of three perpendicular axes (two of them in the plane of the cover) as you throw it upward, and see which is the most stable (wobbles the least). We will not pursue the theory here, but the classic book by Goldstein should be consulted (and possibly tossed around) for further details. Whatever the alignment that occurs, each set of similarly drifting ice crystals produces its own family of halos. Details of nine types of halos or related phenomena are listed in table 3, though this is not an exhaustive list. The article by Lynch is very informative and should be consulted for further details. According to the website found at *http://dspace.dial.pipex.com/lc/halo/circular.htm*, halos can be seen on average twice a week in Europe and parts of the United States. The most frequent, according to this site, is the 22° circular halo, followed by parhelia (sundogs) and then the upper tangent arc. In my own rather limited experience, as I walk to work in southeastern Virginia, the sundogs are most common. In order to explain the reasons for the various angles, for instance, 22°, it is necessary to examine the crystal geometry in more detail.

First, we state things a little more formally in terms of theorems (and their proofs) concerning the refraction of light through prisms of apex angle γ and (relative to air) refractive index n (see figure 5.5).

Theorem 1. The deviation or deflection angle for light refracted through a prism is a minimum for symmetric ray paths.

Theorem 2. The minimum deviation angle D_m for a prism satisfies the relation

$$n = \frac{\sin[(\gamma + D_m)/2]}{\sin(\gamma/2)}.$$

From this result we may solve for D_m to obtain

$$D_m = 2\arcsin\left(n\sin\frac{\gamma}{2}\right) - \gamma.$$

Before proceeding to a proof of these theorems we use this result to explain the occurrence of the 22° and 46° halos. As shown in figure 5.6a–c, there are three prism angles in a hexagonal ice crystal prism: 60° (light entering side 1 and exiting side 3); 90° (light entering a top or bottom face and exiting through a side); and 120° (light entering side 1 and being totally internally reflected by side 2). The first two of these create color by dispersion of sunlight; the last contributes nothing directly to a halo. The refractive index for yellow light is $n \approx 1.31$. For the apex angle $\gamma = 60°$,

$$D_m = 2\arcsin(1.31\sin 30°) - 60° = 21.8° \approx 22°$$

and for $\gamma = 90°$,

$$D_m = 2\arcsin(1.31\sin 45°) - 90° = 45.7° \approx 46°.$$

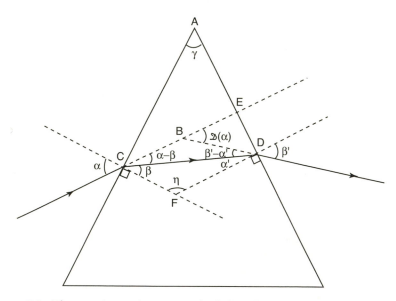

Figure 5.5. The notation and geometry for light refraction through a prism with apex angle γ. $D(\alpha)$ is the deflection angle as a function of the angle of incidence α.

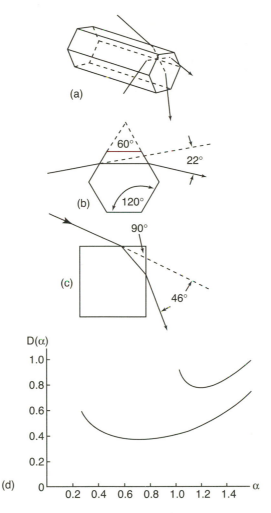

Figure 5.6. (a): A hexagonal ice crystal showing two possible ray paths through the crystal. From *Rainbows, Halos and Glories*, reprinted by permission of Robert G. Greenler (now available through *www.blueskyassociates.com*); (b) the symmetric path through a 60° prism and consequent 22° minimum deviation corresponding to the 22° halo; (c) the symmetric path through a "90° prism" and consequent 46° minimum deviation corresponding to the 46° halo. Reprinted from *Color and Light in Nature*, by D. K. Lynch and W. Livingston (1995), with permission of Cambridge University Press; (d) the deviation angle $D(\alpha)$ for both the 60° prism (lower curve) and the 90° prism (upper curve). Both axes are in radian units (1 *radian* = 180°/π ≈ 57.3°).

Proof of theorem 1. We refer to the diagram showing a nonsymmetric path through the prism (figure 5.5), and here the equations will be numbered to avoid unnecessary confusion. By Snell's law

$$\sin \alpha = n \sin \beta \quad \text{and} \quad \sin \beta' = n \sin \alpha'. \tag{1}$$

From triangle ACD it follows that $\gamma + (90 - \alpha') + (90 - \beta) = 180$, so that

$$\gamma = \alpha' + \beta. \tag{2}$$

Noting from the convex quadrilateral $ACFD$ that $\gamma + 180 + \eta = 360$ we have that

$$\gamma + \eta = 180 \quad \text{and also} \quad D = 180 - \mu, \qquad (\mu = \angle CBD) \tag{3}$$

where D is the deviation angle for the incident ray. From the convex quadrilateral $CBDF$ note that $\alpha + \beta' + \eta + \mu = 360$, so on substituting for η and μ we may write

$$D = \alpha + \beta' - \gamma. \tag{4}$$

From equations (1), (2), and (4),

$$D(\beta) = \arcsin(n \sin \beta) + \arcsin(n \sin[\gamma - \beta]) - \gamma \tag{5}$$

from which, using equation (2),

$$D'(\beta) = \frac{n \cos \beta}{\sqrt{1 - n^2 \sin^2 \beta}} - \frac{n \cos \alpha'}{\sqrt{1 - n^2 \sin^2 \alpha'}}. \tag{6}$$

For an extremum, $D'(\beta) = 0$, i.e.,

$$\frac{\cos \beta}{\sqrt{1 - n^2 \sin^2 \beta}} = \frac{\cos \alpha'}{\sqrt{1 - n^2 \sin^2 \alpha'}} \tag{7}$$

or

$$\cos^2 \beta (1 - n^2 \sin^2 \alpha') = \cos^2 \alpha' (1 - n^2 \sin^2 \beta), \tag{8}$$

$$(1 - \sin^2 \beta)(1 - n^2 \sin^2 \alpha') = (1 - \sin^2 \alpha')(1 - n^2 \sin^2 \beta) \tag{9}$$

or

$$(n^2 - 1)(\sin^2 \beta - \sin^2 \alpha') = 0. \tag{10}$$

This means that an extreme or critical value of D occurs when $\beta = \alpha'$, since $n^2 \neq 1$. Furthermore, this corresponds to a *minimum* value of D, as required since

$$D''(\beta) = n(n^2 - 1)\left[\frac{\sin\beta}{(1 - n^2 \sin^2\beta)^{3/2}} + \frac{\sin\alpha'}{(1 - n^2 \sin^2\alpha')^{3/2}}\right] > 0. \quad (11)$$

Hence

$$\alpha' = \frac{\gamma}{2} \quad \text{and} \quad \alpha = \frac{\gamma + D}{2} \quad (12)$$

and from equations (2) and (4) we find that

$$\beta = \alpha' = \frac{\gamma}{2}$$

and

$$\beta' = \alpha = \frac{\gamma + D}{2}.$$

Note that since all this follows from the condition for an extremum in $D(\beta)$, the value of D in the above equation is D_m. Since $\alpha = \beta'$ it follows that this extremum in the deviation occurs when the ray passes through the prism symmetrically with respect to the bisector plane of the prism angle, and theorem 1 is established. (Reade [2003] calls this result Marriote's equation; for further mathematical and historical details consult his paper, and also Reade [1997].) Returning to equations (5) and (12) it follows that

$$n = \frac{\sin\alpha}{\sin(\gamma - \alpha')} = \frac{\sin[(\gamma + D_m)/2]}{\sin(\gamma/2)},$$

which is theorem 2. Note also that D may be written entirely in terms of the constants of the problem for a particular crystal, n and γ, since from equations (1) and (2), $\beta' = \arcsin(n \sin\alpha')$ and $\alpha' = \gamma - \arcsin(\sin\alpha/n)$ it follows from equation (4) that

$$D = \alpha - \gamma + \arcsin\left\{n \sin\left[\gamma - \arcsin\left(\frac{\sin\alpha}{n}\right)\right]\right\}. \quad (13)$$

All of this goes to show, of course, that not only are ice crystals responsible for some magnificent displays in the sky from time to time, but also they are associated with some rather impressive looking transcendental equations! Sketches of $D(\alpha)$, along with the basic geometry of the corresponding ice crystal, are shown in figure 5.6(d, a–c) for the two apex angles mentioned above: 60° and 90°. Note that the minimum value D_m in each case occurs near the 22° and 46° locations, respectively. As in the case of the rainbow, all possible deviations are present in reality, but it is the "clustering" of deviated rays near the minimum that provides the observed intensity in the halos (unlike the case of the rainbow, no reflection contributes to their formation in these two cases). One way to verify analytically that there is a true minimum is of course to show from equation (13) that when $D'(\alpha) = 0$, $D''(\alpha) > 0$, but this is left as an interesting exercise; it's not as bad as it looks. Note also from the graph that there are restrictions on the angle of incidence α such that outside these, no value of D can be defined for real parameters n and γ. These restrictions arise because of the requirement from equation (13) that, in particular,

$$\left\{n\sin\left[\gamma - \arcsin\left(\frac{\sin\alpha}{n}\right)\right]\right\} \leq 1 \qquad (14)$$

(α lying within the first quadrant). This inequality places a lower bound on α by unfolding expression (13) to obtain

$$\alpha \geq \arcsin\left\{n\sin\left[\gamma - \arcsin\left(\frac{1}{n}\right)\right]\right\}, \qquad (15)$$

where it should be noted that $\sin\theta$ is a one-to-one function in the range $(0, \pi/2)$, so that the arcsin function is monotone increasing in its domain. For $\gamma = 60°$ and $90°$, respectively, this corresponds to $\alpha \geq 13.5°$ and $57.8°$. Regarding the restriction on larger apex angles, it is required from theorem 2 that

$$n\sin\left(\frac{\gamma}{2}\right) \leq 1$$

or

$$\gamma \leq 2\arcsin\left(\frac{1}{n}\right); \qquad (16)$$

for $n = 1.31$ this becomes $\gamma \leq 99.5°$. An apex angle of $120°$ clearly exceeds this; physically this corresponds to total internal reflection within the prism. Arguments of this type enable limits to be placed on the sun's altitude or the latitude of the observer for certain types of halos to be visible (atmospheric conditions permitting). We illustrate this for the circumhorizontal arc (see table 3).

THE CIRCUMHORIZONTAL ARC

The circumhorizontal arc is, as its name implies, an arc parallel to the horizon; it is colored because of dispersion. It is formed as a result of light being refracted through the vertical face of many horizontal hexagonal ice crystal plates, and then out through the lower base. It can only occur for sun elevations greater than $58°$, and this of course imposes limitations on the latitudes in which this particular arc can be seen. We will establish the restriction on solar elevation by examining the limiting condition for total internal reflection within the crystal (see figure 5.7). Let θ_1 be the solar elevation for the observer at O, and let r_2 be the angle of refraction for the ray exiting the base. Clearly it will not exit if $r_2 \geq 90°$, so we will examine the limiting case for total internal reflection, that is, $r_2 = 90°$. We will take the air/ice refractive index to be $n_1 = 1.31$, and the ice/air index is then $n_2 = n_1^{-1}$. By Snell's law applied to the ray AB inside the crystal,

$$\sin \theta_2 = n_2 \sin r_2 = n_2 \approx 0.763.$$

This means that the ray BO will not exit the crystal if

$$\theta_2 \geq \arcsin(0.763) = 49.73°,$$

which implies in turn that

$$r_1 \leq 90° - \theta_2 = 40.27°.$$

For the ray entering the crystal, using Snell's law again we have

$$\sin \theta_1 = n_1 \sin r_1,$$

so total internal reflection will occur if

$$\sin \theta_1 \leq 1.31 \sin(40.27°) \approx 0.847$$

or

$$\theta_1 \leq \arcsin 0.847 \approx 57.86°.$$

Therefore, assuming other atmospheric conditions are appropriate, the circumhorizontal arc will form if $\theta_1 \gtrsim 58°$ as stated. Similar kinds of calculation apply to other halo types to yield the corresponding limitations on

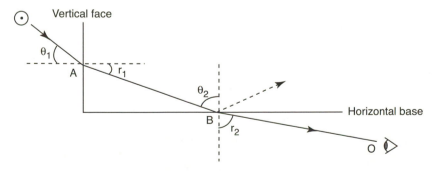

Figure 5.7. The ray path for the circumhorizontal arc; it enters through a vertical face of a hexagonal prism and exits through the bottom base. The dotted ray path inside the crystal indicates total internal reflection. The condition that this does *not* occur places a restriction on the solar elevation θ_1, namely that $\theta_1 \gtrsim 58°$.

solar elevation or perhaps crystal orientation. With regard to the former, let θ_l denote the latitude of an observer in the northern hemisphere, for example, and suppose that it is during the summer, when the sun reaches its highest elevations for those of us north of the equator. Taking into account the 23.5° "tilt" of the earth's imaginary N-S axis to the plane of its orbit, a simple calculation shows that

$$\theta_1 + \theta_l - 23.5° = 90°,$$

from which the above condition for the visibility of the circumzenithal arc (written as $\theta_1 > 58°$) implies that

$$\theta_l < 55.5°,$$

that is, it will not be visible at latitudes above about 56°N. This is the best-case situation, of course; during the other seasons the limiting latitude will be lower because the sun's maximum elevation is correspondingly lower at these times.

Finally, the table on page 94 states properties associated with some of the more well-known halos (and also some of the rarer ones). There are quite a few others to be found if the conditions are right; for example, if you wake up one day at the South Pole, take advantage of being able to see some of the forms not readily discernible in warmer climes.

THE GLORY

Mountaineers and hill-climbers have noticed on occasion that, when they stand with their backs to the low-lying sun and look into a thick mist below them, they may sometimes see a set of colored circular rings (or arcs thereof) surrounding the shadow of their heads. Although an individual may see the shadow of a companion, the observer will see the rings only around his or her head. This is the phenomenon of the glory, initially referred to in the scientific literature as the *anti-corona* (see below). A brief but very useful historical account, along with a summary of the theories, can be found in the book by Nussenzveig. Early one morning in 1735, a small group of people were gathered on top of a mountain in the Peruvian Andes. They were members of a French scientific expedition, sent out to measure a degree of longitude, and were led by two gentlemen named Bouguer and La Condamine; a Spanish captain named Antonio de Ulloa also accompanied them. They saw an amazing sight that morning. According to Bouguer, this was

> a phenomenon which must be as old as the world, but which no one seems to have observed so far.... A cloud that covered us dissolved

itself and let through the rays of the rising sun.... Then each of us saw his shadow projected upon the cloud.... The closeness of the shadow allowed all its parts to be distinguished: arms, legs, the head. What seemed most remarkable to us was the appearance of a halo or glory around the head, consisting of three or four small concentric circles, very brightly colored, each of them with the same colors as the primary rainbow, with red outermost.

Ulloa gave a similar description and also drew a picture. In his account he said:

The most surprising thing was that, of the six or seven people that were present, each one saw the phenomenon only around the shadow of his own head, and saw nothing around other people's heads.

During the nineteenth century, many such observations of the glory were made from the top of the Brocken mountain in central Germany, and it became known as the Brocken bow or the "specter of the Brocken" (being frequently observed on this high peak in the Harz mountains of central Germany). It also became a favorite image among the Romantic writers; it was celebrated by Coleridge in his poem "Constancy to an Ideal Object." Other sightings were made from balloons, the glory appearing around the balloon's shadow on the clouds. Nowadays, while not noted as frequently as the rainbow, it may be seen most commonly from the air, with the glory surrounding the shadow of the airplane. Once an observer has seen the glory, it is readily found on many subsequent flights (provided one is on the shadow side of the aircraft). Some beautiful color photographs have appeared in the scientific literature, a fine example of which can be seen in color plate 8 featuring the photograph by Jim Kaler.

It was mentioned at the beginning of this section that the glory used to be referred to as an anti-corona. It is sensible to ask: What is a *corona*? By this term is meant the atmospheric phenomenon, not the outer atmosphere of the sun (the solar corona) that is visible at the time of a total solar eclipse. The former is a set of diffuse colored rings (with red outermost) similar in appearance to the glory, created by the *diffraction* of light by cloud droplets (usually in altostratus or altocumulus clouds). The size of the rings depends on the drop size; the largest rings are produced by the smallest drops and vice versa. Another related diffraction phenomenon is that of *iridescence*, seen as irregular pale-colored patches of light on clouds that are displaced somewhat from the direction of the sun (or moon). They can be an exquisitely beautiful sight.

Returning to the glory, this phenomenon can be understood in the simplest terms as essentially the result of light backscattered by cloud droplets, the light undergoing some unusual transformations en route to the observer (with a correspondingly complicated mathematical description), transformations that are *not* predictable by standard geometrical optics, unlike the basic description of the rainbow. That this must be the case is easily demonstrated by noting a fallacy present in at least one popular meteorological text. The glory, it is claimed, is formed as a result of a ray of light tangentially incident on a spherical raindrop being refracted into the drop, reflected from the back surface and reemerging from the drop in an exactly antiparallel direction into the eye of the observer (see figure 5.8a). If such a picture is correct, then, since the angle of incidence of the ray is 90°, it follows by Snell's law that the angle of refraction is

$$r = \arcsin\left(\frac{1}{n}\right),$$

where n is the refractive index of the raindrop. For an air/water boundary, $n \approx 4/3$ (ignoring the effects of dispersion here, though this does occur as noted above; note also that for a water/air boundary the reciprocal of n must be used), and so $r \approx 48.6°$. This means (by the law of reflection) that at the back of the drop the ray is deviated by more than a right angle, since $2r \approx 97.2°$; by symmetry the angle of incidence *within* the drop for the exiting ray is also 48.6°, so the total deviation angle (as we saw for the primary rainbow) is

$$D(i) = \pi + 2i - 4\arcsin\left(\frac{\sin i}{n}\right) = -4\arcsin\left(\frac{3}{4}\right) \approx -194.4° \text{ or } +165.6°$$

(modulo 2π) since $i = 90°$. This means that the exiting ray is about 14° short of being "antiparallel." It just won't work as a mechanism for the glory!

There are basically two potential ways out of this. We could ask what value of refractive index n would be necessary for the diagram to be correct; thus $i = 90°$ as before but now $r = 45°$. This means that

$$n = \frac{\sin i}{\sin r} = \sqrt{2} \approx 1.4.$$

Something between water and glass might do it, perhaps plastic! But absent any evidence that clouds are composed of transparent plastic spheres we discard this suggestion. What remains? The other possibility is that somehow the ray travels around the surface (as a surface wave) for part (or parts) of its "trip," the surface portion comprising the missing piece θ, where from figure 5.8(b) (drawn for the symmetric case only)

$$\theta = 180° - 2(180° - 2r) = 4r - 180° \approx 14.4°$$

for $r \approx 48.6°$. The resulting path in the droplet need not be symmetric to account for an antiparallel exiting ray.

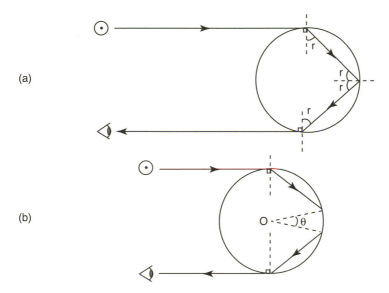

Figure 5.8. (a) *What is wrong with this picture?* An incorrect (but often drawn) ray path in a spherical cloud droplet alleged to contribute to the glory. Although the glory is essentially a backscattering phenomenon, it cannot be produced exactly as shown, because a tangential incident ray will not be returned antiparallel to its original path, as is demonstrated in the text. (b) One possible correct path; it involves the ray traveling as a surface wave around a portion of the droplet surface for a total of about 14° (a symmetric path is shown).

The detailed theory for the glory, using these ideas, has been verified experimentally, though it must be pointed out that a complete theoretical explanation of the glory has been one of the most challenging problems of meteorological optics in the last century (again, see the author's review for details). However, a lovely description of the phenomenon by Pascuzzi has been published, in which he points out the more obvious features seen by an observer. Unlike the coronas sometimes seen around the moon or sun when they are shrouded in clouds, which are caused by diffraction, the glory is a backscattering phenomenon, as indicated earlier. Pascuzzi shows photographs of the glory as the jet descends; these illustrate the fact that the angular diameter of the glory is independent of the observer/cloud distance (unlike the shadow of the aircraft inside the glory rings). It is the diameter of the cloud droplets that determine the angular size of the glory, and these are typically 15–25 microns. Unlike the corona, the light from a glory is strongly polarized.

SNOWFLAKES!

It would be almost criminal to leave this part of the chapter without at least mentioning those most beautiful and diverse of ice crystals known in community form as snowflakes. The individual ice crystals commonly possess complex but regular hexagonal symmetry, growing to at most a few millimeters across. A fascinating website on this topic has been developed by Ken Libbrecht, and it can be found at *www.its.caltech.edu/ ~atomic/snowcrystals/photos/photos.htm*; in particular it is explained there how the sixfold symmetry of an ice crystal at the molecular level determines the symmetry at the macroscopic level. This is understood to occur by means of an instability known as the *Mullins-Sekerka instability*, whereby an initially simple hexagonal ice plate will grow faster at the corners (where the curvature is greatest). This maintains the initial symmetry (now in a more complex form), which then continues since each "arm" is subject to the same mechanism of instability as it grows, and produces what is called a dendritic (tree-like) structure with constantly evolving side branches. Both the effects of diffusion of water molecules and the surface physics of ice need to be incorporated in any theoretical model of snow crystal formation. The growth rates are extremely sensitive to external conditions such as temperature, so it is extremely unlikely that any two crystals will have identical life histories. Mathematically this whole topic is a fascinating example in a class of what are called moving boundary value problems. The above website should be visited for further details.

Libbrecht also addresses what he calls the "no-two-alike" conjecture in answer to the question, is it really true that no two snow crystals are alike? His short answer is "yes," but, as he points out in his longer answer, it all depends on what you mean by "alike." If we mean exactly alike, he notes that all water molecules with two ordinary hydrogen atoms and one ordinary oxygen-16 atom are *identical*, but about one in 5000 water molecules will have an atom of deuterium instead of hydrogen, and one in about 500 will have an oxygen-18 atom instead of its more populous cousin. In practice this means that with these "rogue" molecules randomly scattered throughout each snow crystal, the probability of any two being identical is infinitesimal. However, if by the word alike we just mean "look alike," which is I suspect what most people mean who ask the question, then we might also ask at what level of magnification we are intending to look (an optical microscope can reveal features down to about one micron in size). Libbrecht concludes that any two complex crystals are very unlikely to look similar even at this level of magnification.

MIRAGES

Mirages are fascinating and, by definition, deceptive to the casual observer (for a qualitative description of the relevant physics and historical details on which this introduction is based, see the valuable article by Fraser and Mach). Perhaps no better example of this is afforded by the writings of Robert E. Peary, who in 1906 (en route, as he hoped, to the North Pole), stood on a summit and saw to the northwest at a distance of about 120 miles (as he believed) "snow-clad summits above the ice horizon." This mysterious yet inviting "land" was eventually named "Crocker Land," and in 1913 an expedition, led by Donald B. MacMillan, set out to find and explore it. As they approached its apparent location, he wrote "There could be no doubt about it. Great heavens, what a land! Hills, valleys, snow-capped peaks extending through at least 120 degrees of the horizon." After moving some 30 miles toward this fantastic land, they found nothing. Some might call this a cruel hoax played on them by the "laws" of optics; but it was not an optical illusion, for the image was real enough—it just did not coincide with an object. It was a *mirage*, and that is one convenient way of defining the phenomenon. The mirage of Crocker Land witnessed by Peary and MacMillan was probably an example of what has come to be called the *Fata Morgana*—possibly the most spectacular member of the entire class of mirages. It is the Italian name for the Fairy Morgan, who, as legend has it, was King Arthur's sister, and she possessed, it seems, the ability to create castles in the air (as some people claim mathematicians are want to do, but at least we then inhabit them). The Italian connection comes from a description of such "castles" written by a priest, Father Angelucci, in a letter to a colleague concerning his observations on the morning of August 14, 1643, while looking across the Strait of Messina toward the island of Sicily. He saw what appeared to be a dark mountain range in front of which appeared a variety of different images, including columns, arches, towers, and windows.

Why are there mirages at all and how are they created? The basic answer to each involves recognition of the fact that we "live and move and have our being" inside a huge, locally shapeless lens—the atmosphere itself. Mirages are produced by refraction, but the index of refraction varies in time and space, depending on local climatic and environmental conditions. It is the variations in the refractive index that cause refraction, and these variations are induced by corresponding changes in the density of the air and, to a much smaller extent, its moisture content. The most significant determinator of the refractive index n is the air temperature; since pressure variations are small over the thin layer of air in which mirages are born, the air density is approximately inversely proportional to its temperature. Thus a higher (lower) than normal temperature corresponds to a lower (higher) than normal density, but the spatial gradients of these quantities are also very important. The larger the absolute value of the temperature gradient (its derivative with respect to some direction, usually the vertical), the stronger the corresponding gradient of the refractive index and the greater the degree of refraction that occurs. If the temperature is constant in a region, its gradient is zero and there is no refraction there.

In the article on mirages by Fraser and Mach, it is noted that light rays follow a parabolic path when passing through a region of the atmosphere in which the temperature T varies linearly with height h (or depth z, for that matter—the variable we shall be using later in this section). This means that the temperature gradient dT/dh is constant (rarely achieved in practice in the lowest few meters of the atmosphere). As will be seen from equation (6) below, the curvature of a ray is proportional to the temperature gradient measured perpendicularly to the ray; this is greatest when the ray is traveling parallel to the isotherms—the lines of constant temperature. This curvature causes the image of the object (ship, tree, car, sky, pot of gold) to be displaced away from the position of the object itself. Because (as will be shown below) the ray curves in such a way that the denser, cooler air is "inside" the curved path, the image is always displaced in the direction of the warmer, less dense air. Furthermore, since most mirages involve object-observer distances of at most two or three miles, the curvature of the earth is not a significant factor in mirage production. An excellent qualitative account of the "topology" behind the mirage is provided by Tape (1985).

There is a rather subtle but very interesting point to be made at this stage concerning the relationship between what Fraser and Mach call the "object space" and the "image space." The former is the actual physical space in which the object, observer, and light rays from the object exist. The latter is a deformation of the object space, perceived as such because in general rays are perceived to have traveled in a straight line, so the image is a "back projection" of the rays in the direction from which they enter the observer's eye. This means that the previously flat surfaces of the earth and its atmospheric layers, under this projection or "mapping," become distorted into curved surfaces, some of which may become quite tortuous (depending on the temperature-height profile of the atmosphere; see the Fraser and Mach's article for further details). This mapping will usually involve only the bottom few meters of the atmosphere in practice.

Images may also be magnified or reduced in size if the temperature gradient is variable (if it is constant, there is no magnification). This will generally be the case, because heat from the earth's surface is most efficiently transferred to the atmosphere by convection; as compared with molecular conduction and radiation it requires the smallest temperature gradient to transfer a given quantity of heat. The resulting gradient is a maximum near the surface and decreases with height (as does the temperature), so the temperature profile has nonzero curvature and magnification (or "towering") can occur. Reduction in image size ("stooping") occurs if the temperature gradient decreases as the temperature increases. Fraser and Mach point out that this can happen on sunny afternoons over lakes and sounds as warm air from the land is carried out to the cooler surface of water; this gives rise to mirages displaced upward (superior mirages—see below). The resultant "squashing" of the image is similar to that observed with the setting sun, because the bottom of the object is raised upward more than the top (by virtue of the stronger gradient at the bottom). The reverse kind of phenomenon occurs in towering; the image is inferior because it is displaced downward, the surface being warmer than the air above it, which is duly heated. This means that both the maximum temperature and the maximum temperature gradient are found at the surface, and both of these decrease with height. Such conditions frequently occur over enclosed bodies of water in the early morning or over sun-heated land later in the day. Because of this, the bottom of the object will be displaced downward more than the top, and the image is stretched vertically.

There are many other aspects of mirages that can be considered; we will not do so here, except to note that all the above discussion applies to an atmosphere that is stable (it is *still* in macroscopic terms as opposed to molecular motion). In the presence of wind, for example, surfaces of constant temperature and density are tipped out of the horizontal, with consequent oscillations around their positions of equilibrium. These are gravity waves, and when present they give rise to temperature and density gradients that fluctuate in time and space. If an observer happens to be looking through a region of atmosphere where such fluctuations are occurring, he or she may see many of the finer mirage distortions that rendered Angelucci's fairy castles and Peary's Crocker Land so intriguing and enticing.

We now discuss a little of the mathematics describing the basis of mirages, which is after all the curvature of ray paths of light in the atmosphere. The equations defining a ray path in the x-z plane if the refractive index $n = n(z)$ is a function of depth z only are stated here without proof. The inquisitive reader is invited to consult the text by Officer for the derivation. That book deals with acoustic wave propagation, but a moment's reflection (no pun intended) reminds us that acoustic mirages can also occur, and the same mathematics describes the propagation of any type of wave according to geometrical ray theory—including light and sound. Sound mirages can fool the ears just as optical mirages can mislead the eyes (or more accurately, the brain, when it interprets the data it receives via the senses).

In what follows the arc length along the ray path is denoted by s. The governing equations are

$$n \frac{dx}{ds} = \text{constant} \tag{1}$$

and

$$\frac{d}{ds}\left(n(z)\frac{dz}{ds}\right) = \frac{dn}{dz} \qquad (2)$$

From figure 5.9 note that in terms of the angle θ the ray path makes with the z-direction,

$$\frac{dx}{ds} = \sin\theta = \alpha \qquad (3)$$

and

$$\frac{dz}{ds} = \cos\theta = \beta, \qquad (4)$$

where α and β are the local direction cosines of a point on the curve. If the speed of light in vacuo is c_0, and $c(z)$ is the variable speed of light in the medium (in this case the atmosphere), then the refractive index $n(z) = c_0/c(z)$. From equations (1) and (3) it follows that

$$n\sin\theta = \text{constant}$$

and hence that

$$\frac{\sin\theta}{c(z)} = p \text{ (a constant).} \qquad (5)$$

   This relation states that the sine of the angle of inclination of the ray path to the vertical divided by the speed of light at any depth is always constant; thus if c increases or decreases, so must θ according to this law. The parameter p is constant on any given ray path, but it will differ from path to path. From equation (2)

$$\frac{dn}{dz} = \frac{d}{ds}(n\cos\theta) = -n\sin\theta\frac{d\theta}{ds} + \cos^2\theta\frac{dn}{dz}.$$

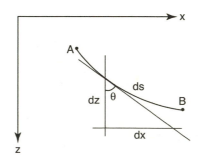

Figure 5.9. The basic geometry for the refraction of light in a plane as a function of depth z in a horizontally temperature-stratified medium.

Solving this we find

$$\frac{d\theta}{ds} = -\frac{\sin\theta}{n}\frac{dn}{dz}$$

and since

$$\frac{dn}{dc} = -\frac{n}{c}$$

we find

$$\frac{d\theta}{ds} = \frac{\sin\theta}{c}\frac{dc}{dz} = p\frac{dc}{dz}. \tag{6}$$

This equation informs us that *the curvature of the ray path is proportional to the velocity gradient of light*. If the speed of light c increases with depth (i.e., decreases with height) then $dc/dz > 0 \Longrightarrow ds/d\theta > 0$ and the rays curve upward. This is the case with the common mirage of "water on the road," in which light rays from the sky direction are concave up from the road and the observer, by projection, sees the resulting effect as water in the distance. The road and the air just above it are hotter than the air somewhat higher up; this means that the air just above the road is less dense than that higher up, and lower density means higher speed of light. The exact relationship between c and the local temperature is sometimes difficult to quantify, but we will examine a relatively simple example below.

Mirages formed under these circumstances are called *inferior* because the image lies below the object. Conversely, if c decreases with depth (or increases with height) then $dc/dz < 0 \Longrightarrow ds/d\theta < 0$ and the rays curve downward. This gives rise to *superior* mirages, so-called because the image lies above the object. I have observed these at the seaside during winter: strange, surreal-looking pillars arising from refraction of light from tanker ships close to or just beyond the horizon. In this case the air just above the ocean surface is colder and hence denser than the air above that level (a temperature inversion has occurred), so the speed of light is less in the lower regions than in the higher ones. It should be pointed out that these differences in density and the speed of light are very small; nevertheless, they are significant enough to produce some quite amazing images, as the examples at the beginning of this section attest.

There are two useful rules of thumb we can apply to the ray paths for both inferior and superior images. First, in each case the concavity of the ray path is always toward the region of denser air. Second, and equivalently, the ray paths always curve toward the region of minimum velocity of light (denser air). Of course, situations may frequently arise when the temperature and density profiles are not monotone, and under these circumstances the ray paths will in general exhibit points of inflection; this can give rise to some interesting mirage forms, as noted in the article by Fraser and Mach.

Continuing with a little more mathematics, let us examine the form of the ray path for the special case of a medium in which the light speed is a linear function of depth,

$$c(z) = c_1 + az, \tag{7}$$

where c_1 is a positive constant and the velocity gradient $dc/dz = a$ can be of either sign. It follows immediately from equation (6) that the curvature κ of the ray path is constant,

$$\kappa = \left| \frac{d\theta}{ds} \right| = |pa|,$$

and so the ray path is an arc of a circle. The radius of curvature R is of course a constant also, namely

$$R = \kappa^{-1} = \left| \left(\frac{d\theta}{ds} \right) \right|^{-1} = \frac{1}{p|a|} = \frac{c_1}{|a| \sin \theta_1}, \tag{8}$$

θ_1 being the angle of inclination to the downward vertical at some point on the path for which z is set to zero, and then $c(0) = c_1$. Note that the bigger the temperature gradient a, the smaller the radius of curvature of the ray path, and the more pronounced the effects of atmospheric distortion. Also, the bigger the initial angle of inclination of the path in the interval $(0, \pi/2)$, again the smaller R will be. If $\theta \in (\pi/2, \pi)$ then the ray is initially upward (recall that z is measured downward here); again the distortion via curvature will be largest when the ray is closest to horizontal in its direction.

While these conclusions have been based on a specific profile for c, they are qualitatively valid for nonlinear profiles also. In this case, however, we can readily compute the equation of the circular ray path. From figure 5.9 and using differentials it follows that

$$x = \int_0^z \tan \theta \, dz = \int_0^z \sin \theta \frac{dz}{\cos \theta},$$

where it is to be understood that the dummy variable is implicit in the integrand, as in those that follow below (so that dz is really dz', etc.). Therefore

$$x = \int_{\theta_1}^{\theta} \frac{\sin \theta}{pc'(z)} \, d\theta, \tag{9}$$

where the derivative with respect to z of equation (5) has been used.

Thus far this expression is valid for all suitable differentiable (nonconstant) profiles $c(z)$. Now we specialize to the case of the linear profile, for which $c' = a$; note also from (5) that, in particular,

$$p = \frac{\sin \theta}{c} = \frac{\sin \theta_1}{c_1}. \tag{5'}$$

Thus

$$x = \frac{c_1}{a \sin \theta_1} \int_{\theta_1}^{\theta} \sin \theta \, d\theta = \frac{c_1}{a} \left(\frac{\cos \theta_1 - \cos \theta}{\sin \theta_1} \right), \qquad (10)$$

and using (7) and (5') it follows that

$$z = \frac{c - c_1}{a} = \frac{c_1}{a} \left(\frac{\sin \theta}{\sin \theta_1} - 1 \right). \qquad (11)$$

A little algebra later (interesting units of time), we find that

$$x^2 + z^2 = 2 \left(\frac{c_1}{a \sin \theta_1} \right)^2 (x \cos \theta_1 - z \sin \theta_1)$$

or

$$(x - R \cos \theta_1)^2 + (z - R \sin \theta_1)^2 = \left(\frac{c_1}{a \sin \theta_1} \right)^2 = R^2, \qquad (12)$$

which is of course the equation of a circle of radius R centered at the arbitrary initial point $(R \cos \theta_1, R \sin \theta_1)$. This was just what we wanted. As a final note, recall the statement that a ray passing through an atmosphere in which the temperature gradient is constant follows a parabolic path. We have not proved this, but will approach it from the other end, so to speak. We have shown that the curvature of a ray is proportional to the temperature gradient measured perpendicularly to the ray, so that this curvature will be least when the ray path is closest to the vertical. Let us arrange our (x, z)-coordinate system so that the origin coincides with the vertex of the parabola $z = ax^2$, where a is a constant of any sign. The curvature κ of the function $z = f(x)$ is known from calculus to be given by

$$\kappa = \frac{|d^2 f/dx^2|}{[1 + (df/dx)^2]^{3/2}}.$$

For the above function,

$$\kappa = \frac{2|a|}{[1 + 4a^2 x^2]^{3/2}},$$

which has a maximum of $2|a|$ at $(0,0)$ and tends monotonically to zero as $|x| \to \infty$, illustrating our point.

IRIDESCENCE: BIRDS, BEETLES AND OTHER BUGS, SOAP FILMS, AND OIL SLICKS

Who has not spent at least a moment or two enjoying the shimmering colors on the surface of a soap bubble as it floats by, the subtle changes of color in the throat plumage of a hummingbird, or the brilliant "eyes" in the tail feathers of a peacock? What about the feathers on the neck of a mallard drake, or the wings of a tropical butterfly? Oil slicks can also exhibit beautiful colors that change in time as the oily film thins. Regardless of the context, all these shifting, shimmering colors arise from the fact that light is a wave, and waves can interact and *interfere* in some very interesting ways. As we will see, such interference is evident when the distance between the reflecting layers (denoted by t below) is comparable with a wavelength. The colors that are observed as a result of interference are very sensitive to the film thickness, which is why we see so many colors on the surface of an ever-changing soap film or on a thinning slick of oil or puddle of gasoline as we slightly change our direction of viewing.

A similar effect is caused by the scales on some butterfly wings. They are made of *chitin*, which is a horny substance also found in the shells of beetles and exoskeletons of other insects. Reflections from these layers of chitin give rise to varying colors as we carefully examine the wings. Blue morpho butterflies, native to Central and South America, have wings that are so bright that on occasion naturalists have reported seeing flashes of blue hundreds of yard away. The layers of chitin in their wings are about a thousand times thinner than a human hair (and my hair is thin), that is, about 110×10^{-9} m (0.11 microns), and this is about one-fourth the wavelength of blue light. The layers are closely packed scales that overlap like shingles or tiles on a roof. Jewel beetles, black-winged damselflies, and peacock tails also exhibit iridescence, as does iron pyrites (a compound of iron and sulfur), which has a thin layer of iron oxide on its surface, permitting light interference to occur. Iridescence is also produced by alternating layers of calcium carbonate and water in abalone shells; these are large, rather flat shells lined with mother-of-pearl.

We will follow here the approach of Cyril Isenberg in his excellent and informative book on soap bubbles. Consider in figure 5.10 light of monochromatic wavelength λ incident at an angle of incidence i on a film of uniform thickness t. Some of the light will be reflected and some transmitted (and subsequently reflected, etc.; see figure). The angle of refraction is θ. The interference pattern will be determined by the two parallel rays B_1C_1 and B_2C_2 emerging from the upper surface B. The refractive index of air will be assumed to be unity, and that of the film to be $n > 1$. We consider the optical path difference Γ between light rays arriving by direct reflection (in direction B_1C_1) and those emerging by refraction and reflection along the path $B_1D_1B_2C_2$. Because the speed of light c, the wavelength λ, and the (fixed) frequency ν are related by the equation $c = \lambda\nu$ it follows that within the soap film (or whatever the refracting medium happens to be) the wavelength is decreased because the speed of light in the medium is decreased. This means that the number of wavelengths per unit length in the film has

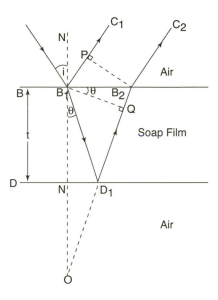

Figure 5.10. Basic notation and geometry for the interference of light in a thin soap film (or layer in the plumage of a bird or a butterfly wing). From *The Science of Soap Films and Soap Bubbles*, by C. Isenberg (1992), reprinted by permission of Dover Publications.

increased, so in anthropomorphic terms the ray "sees" a greater distance to travel within the film than compared with an equal thickness in air. (In his "Bad Science" website, Alastair Fraser has noted that such anthropomorphisms are examples of the so-called "pathetic fallacy." I apologize.) This effective distance is increased by a factor n. Thus from the geometry of the problem

$$\Gamma = n(B_1 D_1 + D_1 B_2) - B_1 P$$

$$= n(OD_1 + D_1 B_2) - B_1 P$$

$$= n(OQ + QB_2) - B_1 P.$$

It is clear that $QB_2 = B_1 B_2 \sin\theta$ and $B_1 P = B_1 B_2 \sin i$ so that

$$n = \frac{\sin i}{\sin\theta} = \frac{B_1 P}{QB_2},$$

or $nQB_2 = B_1 P$, so that the path difference between C_1 and C_2 is given by

$$\Gamma = nOQ = n(2NB_1 \cos\theta)$$

$$= 2nt \cos\theta.$$

This is the fundamental relationship we require, and before proceeding further some comments are in order. At normal incidence ($i = 0$), $\theta = 0$ by Snell's law, and $\Gamma = 2nt$, which is just twice the optical thickness of the film. In addition, there is a phase change of π radians (or half a wavelength) that occurs at an air/medium boundary (meaning that the ray in air then encounters the boundary with the denser medium); this does not occur at a medium/air boundary. Because of this there is an "effective" path difference between the two rays given by

$$\tilde{\Gamma} = 2nt \cos\theta + \frac{\lambda}{2}.$$

The two rays will interfere *constructively* if $\tilde{\Gamma} = N\lambda$, that is,

$$2nt \cos\theta = \left(N - \frac{1}{2}\right)\lambda,$$

where $N = 1, 2, 3, \ldots$, because maxima will coincide with maxima, etc., and so this will correspond to a maximum of the interference pattern. Accordingly, a minimum of the pattern will occur as a result of *destructive* interference when

$$2nt \cos\theta = N\lambda.$$

Since any path difference corresponds to a difference in phase δ, we shall calculate that by considering the rays emerging at C_1 and C_2 to have combined amplitudes $A_r = A(1 + e^{i\delta})$, where we may take A as a real number without loss of generality. The phase difference

$$\delta = \frac{2\pi}{\lambda}\tilde{\Gamma} = \frac{2\pi}{\lambda}\left(2nt \cos\theta + \frac{\lambda}{2}\right).$$

The total emerging intensity associated with these two rays is $I_r = |A_r|^2 = A_r A_r^*$ where the asterisk denotes complex conjugation, so

$$I_r = A^2(1 + e^{i\delta})(1 + e^{-i\delta}) = A^2(2 + e^{i\delta} + e^{-i\delta})$$

$$= 2A^2(1 + \cos\delta) = 4A^2 \cos^2\frac{\delta}{2} = 4A^2 \cos^2\left(\frac{2\pi}{\lambda}nt \cos\theta + \frac{\pi}{2}\right)$$

$$= 4A^2 \sin^2\left(\frac{2\pi}{\lambda}nt \cos\theta\right).$$

Obviously each wavelength will have a corresponding value of I_r. For *extremely* thin films, that is, when $t/\lambda \ll 1$, we have $I_r \approx 0$; this means that the "color" we see emerging is black, because there is zero reflected intensity. If the thickness of the film is increased, then the shorter wavelengths will be the first ones to interfere constructively, and give bright violet interference fringes, followed by blue, green, and so on. An interesting discussion of these colors in a variety of different situations (including umbrella fabric and window ice) is given by Minnaert. There are different *orders* of interference colors associated with the integers N, and in a draining soap film (for example) it is possible to see several at once. Isenberg gives a detailed account of the colors associated with orders up to the eighth.

Let us now return to the morpho butterfly wing, and find out the optimal angle at which to view the wing (assuming interference from a single layer of chitin) in order to see the blue color. There is a slight problem here insofar as I do not know the refractive index of chitin, so in the spirit of Enrico Fermi (see chapter 2 on Fermi problems) I shall attempt to make an educated guess. Chitin is denser than air and I shall assume that it is less dense than glass, which seems reasonable. Its density may be comparable with that of water, but I shall take three values spanning the above range, namely $n = 1.2$, 1.3, and 1.4. I shall also consider only first-order interference ($N = 1$), so that since $\lambda/t = 4$ by hypothesis, the fundamental relationship

$$2nt \cos\theta = \left(N - \frac{1}{2}\right)\lambda$$

reduces to

$$\cos\theta = n^{-1},$$

which for the above three values of n yields $\theta \approx 34°$, $40°$, and $44°$, respectively. A butterfly fluttering by would present its wings at a variety of angles to the observer, so a range of colors would be expected if we include more than one layer of chitin. The color blue appears predominant for this species, however.

Clouds, Sand Dunes, and Hurricanes

> Do you know how the clouds hang poised . . . ?
> Who has the wisdom to count the clouds?
> —Job 37:16a, 38:37a

In this chapter, much of the "scene" is set for the following two chapters on waves and stability, because clouds are wonderful indicators of what kinds of wave motion or atmospheric instabilities are occurring far above us. Although the mathematical descriptions of some of these phenomena (with the exception of the hurricane) are left to chapters 7 and 8, by the time you reach them you will have been exposed to the underlying physics necessary to formulate such descriptions.

CLOUDS

Clouds consist of water (in one form or another) and are high altitude fog; or better, fog is cloud at ground level. Somewhat surprisingly, it doesn't take much water to produce a cloud. A small one with diameter of a few hundred feet contains less water than a full bathtub and would weigh about as much as an adult human. It would probably take water from several thousand Olympic-sized swimming pools to create a typical cumulonimbus thunderhead cloud (don't worry for now about justifying these statements—we'll get to that in a later section).

How do clouds form? From the perspective of an earth-based observer, the "action" starts at the bottom and works its way to the top. Sunlight heats the ground, which in turn heats the air just above it (it is re-radiated by the ground at wavelengths easily absorbed by the moisture in the air). This will cause the air to expand, becoming less dense and hence somewhat buoyant. Generally, the air will be heated unevenly, depending on whether it lies above grass, foliage, water, wood, or concrete, so certain "pockets" or thermals rise faster than others—a degree or two difference from the surrounding air will be sufficient to start the thermal on its way. As the warm air rises into air at lower pressure it expands and cools, eventually causing some of the water vapor present to condense around tiny particles (usually pollen or dust) to form tiny droplets. Myriads of such droplets form a cloud. It is important to note that the level at which a cloud first forms is called the condensation level, and where this is located depends on many factors, the most important of which is the amount of moisture in the atmosphere. At sufficiently high altitudes this can be present in the form of small ice crystals or even snowflakes.

The process just described occurs for isolated thermals that give rise to individual clouds or features such as cloud streets. However, we have all experienced overcast skies, caused by large air masses being forced to rise and cool over regions hundreds of kilometers across. The mechanisms for this may be, for example, the presence of hills or mountain ranges, or one large air mass "wedging" itself under another approaching mass of (warm) air.

There are basically ten different cloud types or classifications, but these are derived from three broad groups of cloud: (i) cumulus ("heap" in Latin), the typical puffy cauliflower-like clouds seen on a summer day; (ii) stratus ("stretched out"), the basically dull gray clouds that cover most if not all of the sky; and (iii) cirrus ("curling"), wispy, high level clouds, sometimes called mare's tails. The ten different cloud types are briefly as follows. At the highest levels (27 km, or $\approx$ 23,000 ft) we find cirrostratus, cirrocumulus, and cirrus. Altocumulus and altostratus are mid-level clouds found between 4 km and 7 km altitude ($\approx$ 13,000–22,000 ft). Low-level clouds ($\leq$ 3 km or $\approx$ 10,000 ft) are stratus, nimbostratus, stratocumulus, and usually cumulus. Cumulonimbus (thunderhead) can be so tall that it extends to the highest altitudes. It has been known to reach 15 km (50,000 ft). The clouds that are most commonly seen to support regular patterns are cirrocumulus ("mackerel sky"), altocumulus (with convection patterns in the form of ripples), and stratocumulus (with rolls and billows). It is important to realize that clouds are not "locked into" any one particular classification; they may change from one type to another or even disappear altogether as temperature, moisture, or wind variations occur.

Now a little more basic science is in order. In meteorology the amount of moisture in the air is often measured in relative terms. Cold arctic air at $-10°C$, containing only 1.8 gm of water vapor for each kg of air, may be saturated. In the tropics at 35°C the corresponding density of water vapor may be 18 gm, ten times higher, but the air is only 50 percent saturated. Thus relative humidity can vary widely at different locations on the earth's surface, from 100 percent (saturated) downward, as we have seen. This is also true at different altitudes. The relative humidity is the percentage of the maximum amount of water vapor the air can support that is in fact present. The stability of the air is also an important factor. When air is stably stratified, the lower surface air has warmer air above, sometimes called an inversion layer. Under such circumstances there is no convection (thermals):

a parcel of air displaced vertically from its equilibrium position tends to return to it. This is easy enough to understand—if displaced upward it is cooler and denser than its surroundings, so it sinks. If moved downward, it is warmer and less dense than its surroundings and so experiences an upward buoyancy force. When the stratification is *unstable*, the forces acting on a displaced air parcel are now in the same direction as the displacement, which tends to continue. This is an example of *convective instability* (of which more in chapter 8) and the layers become mixed as rising and falling of air takes place. Another consideration is the moisture in the air: when a cloud forms, heat is released (the "latent heat of condensation") and this makes that region of air more buoyant and hence more prone to convection. As R. S. Scorer points out, clouds make possible a whole range of motions and patterns that could not occur in a dry atmosphere—even if we could see them.

The ascent of air, then, is a major mechanism of cloud formation, because as we have seen, rising air expands and cools. Above the height at which it has cooled to its *dew point* (100% saturation) the water vapor condenses into tiny droplets that form on condensation nuclei (dust, pollen, or salt particles over sea). If the temperature is cold enough, ice crystals form instead of droplets (called glaciation or freezing). Eventually, as these droplets or crystals grow or accumulate, they may fall as rain, snow, or hail, depending on their history. When air just above the ground is warmed, considerable amounts of water may be evaporated from the ground, either by the vegetation or foliage or by already wet areas. Convection therefore warms the air and also increases its water vapor content. The source of heat, the sun, heats the ground (some of its radiation is absorbed by the ground), and the ground re-radiates it in a longer wavelength that is absorbed primarily by water vapor (this is the basis for the so-called greenhouse effect when the absorbing medium is carbon dioxide). If the air is humid, this results in rapid heating of air about 10 or 20 meters above ground level. In dry air the radiation has to travel much farther before being absorbed by molecules of water vapor, so this layer of warmed air can be 500 meters high (or deep).

Continued warming of course, generates instability and hence thermals by the convective process we have discussed earlier. Thermals get wider as they rise, perhaps by a factor of two or three by the time they reach the condensation level, like an ever-widening cone. Ultimately their ascent will be stopped on encountering a stable layer. Generally speaking, the water vapor content and temperature of air near the ground on a calm clear day are relatively high, but decrease with height. If, however, the temperature decrease (or lapse rate) is greater than the cooling within a rising thermal, the atmosphere is unstable and clouds can grow larger. Each individual cloud may last only fifteen minutes, because formation and evaporative processes (at the surface) cause rapid and continual changes. Between the clouds the air contains less vapor, so the sides, in particular, are being continuously evaporated. The cloud still rises at the top until it reaches a stable layer, so it becomes flattened out and spreads horizontally as does smoke on reaching the ceiling of a room. This process can give rise to anvil clouds.

How high can the cloud "ceiling" be? An upper range can be seen by noting that air achieves *radiative equilibrium* at a temperature of about $-50°$C. This is because the heating from the warm earth below is exactly balanced by the loss of heat to outer space. These circumstances arise in a fairly thick layer between 12 and 30 km (40,000–100,000 ft). This is approximate: the base varies depending on whether we are at the poles or equator or elsewhere. This stable layer (or set of stable layers) is called the *stratosphere*; below this lies the tropopause, which separates the stratosphere from the troposphere, in which most of our "weather" occurs.

SOME COMMON CLOUD PATTERNS

Over the sea the surface temperature during the day does not get as high as over land, and changes more slowly. Under these circumstances convection is mild, and large areas of *stratocumulus cells* of fairly uniform size may occur. Initially they may be 1 or 2 km in width, but they can often grow as large as 100 km without much increase in the depth of the convection layer. If there is a fairly strong wind, the predominant pattern becomes *cloud streets*: the convection cells are lined up along the direction of the wind (more accurately, along the direction of the wind shear). They are readily noticed from the air or from satellite photographs. The width of the streets tends to increase downwind.

The cells themselves can take many different forms: from packed together like a honeycomb with little space in between to "open" reticular (net-like) structures. The former type have strongest up-currents in the middle and sinking motion at the edges. The stable layer at the top in this case is strong, severely curtailing the upward motion of thermals. The open cells are more likely to be associated with rain production. Rain falling into clear air cools it by evaporating into it, producing a downdraft that can render large parts of each cell cloudless. Similar cell-like structures occur at middle levels in altocumulus clouds, sometimes there being no clear distinction between them visually. Altocumulus clouds, however, are more susceptible to cooling from the top by radiating infrared radiation (and heating from below by the reverse mechanism); this process can produce cellular convection cells also.

Billows (about which we will have much to say in later chapters) are formed when a very stably stratified layer of air is tilted for some reason (e.g., airflow over a mountain) creating a large shear as one layer slides over a layer below. This is referred to as the *Kelvin-Helmholtz instability* (see chapter 8), and it may occur without tilting if there is strong wind shear already present in the air. The patterns by which this instability is manifested are the atmospheric equivalent of breaking waves, and they may produce cloud out of clear air if the shear is large enough. When no cloud at all is produced, the instability causes *clear air turbulence*. Billows often occur in the stable layers of the tropopause as well as by flow over hills or mountains. In the latter case the shear is soon canceled out when the air returns to its original level, and so a localized billow cloud pattern may be formed. These clouds are similar to the *lee waves* and *pileus* discussed below. Billows may end up as more or less equally spaced cylindrical rolls lying across the shear, unlike cloud streets, which lie along the shear. There are other significant differences as Scorer points out: streets are the result of heating by a rigid lower boundary (the land or ocean; rigid is a relative term in the case of the ocean). The rolls rotate in opposite directions, still being convectively generated. In billows, on the other hand, the upper and lower flow boundaries are distorted and adjacent rolls do not abut, and they all rotate in a clockwise direction (if the layer is tilted counterclockwise).

Recall that in the absence of wind near ground level, local variations in moisture content and temperature are discrete: each early cumulus cloud that forms is a marker of a slightly more favorable pocket of air originating near the ground below. As we noted above, a moderate wind can change the situation considerably by sweeping away local variations and mixing them downstream. This eventually induces the formation of a uniform condensation level and a somewhat orderly arrangement of cloud features. Winds can change in both speed and direction at different altitudes. Surface pressure systems usually control them in a layer up to about 3 km ($\approx$ 2 miles) above ground in temperate zones; above this the prevailing westerlies dominate. Strong wind shear is manifested by the tops of cumulus clouds being stretched downwind, sometimes producing the well-known "anvil" shape.

Change in shape and other properties of clouds are not only a function of spatial location: age is also an important factor! A newly formed cloud (or one that is continually growing) contains a very large number of tiny droplets close to the cloud surface, which reflect much of the sunlight the cloud receives (such clouds are said to have a high albedo). This is why such clouds appear bright (though the base may appear darker because of the relative absence of light scattered through it). As the cloud or parts of it age, the smallest droplets evaporate, leaving the larger drops (which are fewer in number) to permit more light through the cloud by reflecting less; such clouds generally appear darker. The same principles account for the gray appearance of layer clouds like stratocumulus. They are stably stratified, so little vertical interior mixing occurs, and the cloud evaporates most rapidly at the edges and more gradually in the interior. The cloud turns gray as it ages, a situation common to many of us!

Glider pilots were the first to realize that updrafts in convection-based clouds are stronger than the corresponding downdrafts between them. This sinking stable air is carried down below the condensation level, preventing subsequent thermals from reaching it in those places. This manifests itself as a tendency for subsequent thermals to follow their predecessors through holes in the layer just below the condensation level.

Cloud streets, as previously noted, are rows of cloud elements roughly aligned with the wind direction. A prerequisite for this pattern is a rather uniform depth of convection over a large area (e.g., over the ocean), which gives rise to an orderly arrangement of updrafts and downdrafts susceptible to the influence of the prevailing wind. Over land the spacing between streets is two to three times the thickness of the convection layer, so changes in the spacing indicate corresponding changes in this layer. As atmospheric conditions change, the pattern usually breaks down. Random patches of cloud at the beginning of a day may organize into streets that subsequently disappear as stronger convection occurs with the heat of the day. When convection is vigorous, penetrating the stable layer above, this layer is lifted by the thermal below. In moist air a small cloud cap may condense: this is called pileus cloud. It is a wave cloud formed by the smooth ascent of air into a moist stable layer. The pileus remains in place as the air rises and descends, abruptly vanishing as the wave crest flattens when the air sinks back to its equilibrium level.

What is this "wave" that we mentioned in the preceding paragraph? Bearing in mind that the mathematics of waves will be discussed in chapter 7, we merely note here that there are several different types of wave motion that can occur in a compressible, stably stratified medium like portions of the atmosphere above us (we have already discussed some consequences of unstable stratification). The two major restoring forces of atmospheric motion are compressibility (like "elasticity" or pressure variations in a gas) and buoyancy (due to variations of density acted on by gravity). For low frequencies generated by a disturbance, the waves are essentially *gravity* waves (modified by compressibility); at high frequencies they are *acoustic* (or pressure or sonic) waves (modified by buoyancy). At intermediate frequencies they are dominated by a combination of both restoring forces and referred to as *acoustic-gravity* waves. So what is a gravity wave? We shall discuss this in more detail in the next chapter, but the best example of *surface gravity* waves is the pattern generated when the surface of a pond or puddle is disturbed, say, by a pebble being thrown into it. There *is* another restoring force acting in this case for short wavelengths—surface tension—but we ignore it here because this is most important when the water wavelengths are shorter than about 2 cm, and anyway, it is irrelevant for atmospheric waves.

There is a counterpart to surface gravity waves above about 10 km ($\approx$ 6 miles). It is an acoustic wave in that it is transmitted by air pressure, but the pressure is maintained by gravity acting throughout the depth of the atmosphere at the wave front. From now on, however, we focus on the more commonly noticed *internal gravity* waves that depend on density differences not between water and air (as in the pond), or between the "top" of the atmosphere and outer space, but on differences between adjacent layers within the air. Acoustic waves travel at the speed of sound (≈ 1100 fps or 750 mph in air near the ground, though it is dependent on air temperature). Internal gravity wave speeds are lower, comparable with typical wind speeds. Unlike acoustic waves they are very *dispersive*. This means that waves of different wavelengths travel at different speeds, so a pulse composed of many different wavelengths initially will be smeared out as time progresses. Dispersion is best seen on the surface of our pond—after a few seconds the single pulse or hump of water has spread out into a succession of a dozen or so wave crests (in this case the longer waves have a higher speed than the shorter waves). By contrast, pure acoustic waves are nondispersive: a succession of several distinct notes, for example, is heard as just that, which is just as well for concert-goers!

Returning to internal gravity waves, the property of dispersion ensures that they are readily observable if some mechanism for picking out a particular wavelength is present. Such a mechanism may simply be a mountain with a moderate wind blowing up one side. This can generate *lee waves* in which the particular wavelength observed is the one that travels through the air (on the leeward side of the mountain) with the speed of the wind but in the opposite direction. This mechanism often results in a train of *standing waves* which can be seen when the wave crests are above the saturation level: stationary wave clouds then form.

Sometimes physical barriers like mountains or cliffs will reflect waves impinging on them. These waves may be visible as waves "trapped" by a wind profile that increases speed with height. This is the essential principle behind lee waves; they are trapped within a stable layer low down in the airstream so that their energy is not lost upward. When a layer of low cloud is thin or incomplete, waves trapped within it show gaps in the troughs where the air is forced to descend below its condensation level, whereas the crests contain smooth arch-like wave clouds. *Ship wave patterns* caused by isolated islands or mountains are also possible, though these are best seen in satellite pictures. These patterns will also be discussed in more mathematical detail in chapter 7, but they are the same type of pattern that moves along behind a ship (or a duck on a pond). In this case a mountain is the ship (clearly stationary!) and the air moves past it. These standing waves are visible as lee type waves, though sometimes eddies or vortices are observed to "spin off" the airstream on either side of the mountain (this phenomenon is called vortex shedding, and is well known in certain hydrodynamic situations; we shall discuss one such situation in chapter 12—the humming of wires, the whispering of trees, and the murmur of forests in the wind).

Lee wave patterns are relatively simple, even if produced by complicated ranges of mountains (as opposed to an isolated mountain). Subsequent to encountering such regions, the resulting airflow tends to reflect the first mountain encounter, in terms of the amplitude and wavelength of the lee wave pattern. Often *rotors* will form on the lee side of the mountain: these are eddies, which of course rotate and effectively alter the shape of the ground as far as the airflow above them is concerned. This tends to perpetuate the pattern first set up, since the rotors themselves, which are like waves closed in on themselves, are formed by airflow over the first mountain or range of mountains.

The Size and Weight of a Cloud

So far, we have had a great deal of description for a purported book on applied mathematics. Recall, however, the remarks made in the introduction, where a philosophy of mathematical modeling was set out, namely, to try to understand a given phenomenon at as many complementary levels of description as possible. Since the occurrence, shape, and size of clouds are dependent on so many physical factors, it has been necessary to summarize the essential physics before exploring more about them in mathematical terms. Despair not, however; two small arithmetic calculations in the spirit of chapter 2 follow to close out this section of the chapter!

The first is to determine the *size and weight of a cloud*. Suppose we had the apparatus necessary to measure the dimensions of a cloud shadow on the ground (perhaps by observing it from an aircraft or a balloon). What could we say about the size of the cloud itself from this two-dimensional representation—its shadow? Is the actual cloud larger, smaller, or about the same as far as a typical linear dimension is concerned? It is in fact about the same size as the shadow, at least in the direction perpendicular to the observer-sun direction, because the rays of the light from the sun are essentially parallel (at least as far as this application is concerned). Although the shadow may appear long if the sun is low on the horizon, the width in the perpendicular direction on the ground will give a good indication of its actual size. Obviously, if the sun is high in the sky, both dimensions will be fairly accurate.

What about the weight of a cloud? Margaret LeMone "weighed" a smallish cumulus cloud as it drifted over the plains east of Boulder, Colorado, helped by the fact that from the Appalachians westward the United States is divided into squares one mile on a side, usually marked by roads (see Williams 1997). This particular cloud had a shadow about 0.6 mi (or about 1 km) square, and it was about as high as it was wide. This corresponds to a volume of about 1 km^3 or 10^9 m^3. Typically, each cubic meter of such a cloud contains about half a gram of water in liquid form, so this one had a total of about 5×10^8 gm, equivalent to 500 metric tons. For purposes of comparison, a Boeing 747-400 jetliner with a full load of passengers (up to 524) and fuel has a maximum take-off weight of about 400 metric tons. So even a moderate-sized puffy cloud weighs more than a fully-laden Jumbo! Of course, there would be a bit more room to move around in an aircraft the size of a cloud! This provides the necessary basis for the estimates of size and weight stated at the beginning of the chapter. Thus, prorating this result for a small cloud, say 50 m across, gives a weight of about 60 kg; doing the same thing for our Olympic pool with a volume of about $50 \times 25 \times 2 = 2500$ m^3, we find that it weighs 2500 metric tons, about five times the weight of Margaret's cloud, and, is $5^{1/3} \approx 1.7$ times as large linearly, or about a mile across. And now to another watery question.

Why We Can See Further in Rain than in Fog

We start off with some simple definitions. Suppose that the diameter of a spherical water droplet in centimeters is denoted by L. Further suppose that 1 gm of water (volume: 1 cm^3) is dispersed in the form of a mist or rain shower. This produces N identical droplets, where

$$N = \frac{\text{volume of water}}{\text{volume of a drop}} = \frac{1 \; cm^3}{\pi L^3/6 \; cm^3} \approx \frac{2}{L^3}.$$

Now let V be the *relative volume* of water present in a given unit volume of air (e.g., 1 m^3). The above volume of water yields $V = 10^{-6}$, which is about right for a heavy mist or rain. Now expressing L in meters (essentially multiplying numerator and denominator by 10^{-6}), we have that

$$N \approx \frac{2V}{L^3}$$

and assuming no overlap of drops in the line of sight, each drop will block off an area

$$A = \pi \left(\frac{L}{2}\right)^2 \approx \frac{3}{4}L^2 \; \text{m}^2.$$

The *total area* blocked off is

$$N \times A \approx \frac{2V}{L^3} \times \frac{3}{4}L^2 \approx 1.5\frac{V}{L} \; \text{m}^2,$$

which implies that, for the same relative volume, *smaller drops block off a larger area*; that is, they are cumulatively less transparent than a cloud of larger drops.

Let's put some numerical flesh to this skeleton: consider a linear stack of s one-meter cubes. The total area blocked off is, as an upper bound,

$$A_{total} \approx 1.5 \frac{Vs}{L}.$$

Now suppose that there is only 10 percent visibility, that is, 90 percent of the viewing area is blocked off. How "long" is the fog bank or rain shower? Well,

$$0.9 = 1.5 \frac{Vs}{L},$$

and using our typical value of $V = 10^{-6}$, we find

$$s = 6 \times 10^5 L.$$

For large drops ($L \approx 1$ mm), $s \approx 600$ m; for fine mist ($L \approx 10^{-2}$ mm), $s \approx 6$ m. That seems to accord with my experience; how about yours?

We have discussed cloud formation in some detail and have sketched out some of the wave motion that may be present in the air above us, manifested in various types of cloud. There is a very interesting land-air connection associated with the interaction of cloud street dynamics and, of all things, sand! Like many models, it by no means represents the whole story, and therefore is one among several competing descriptions; see for example the quotation at the end of the next section. In that section we discuss some "induced" waves that are definitely connected with cloud streets, but which can occur a little closer to the ground.

SAND DUNES

First we shall note some general details about dunes, starting with the common *longitudinal sand dunes*. These dunes cover over half the area of the large deserts of the world. Typically, they are aligned in the prevailing wind direction and are spaced about 2 km apart. A possible mechanism for these dunes is that they are formed by large helical eddies in the atmosphere—counter-rotating helical roll vortices aligned along the wind direction with diameters approximately equal to the thickness of the atmospheric boundary layer, and wavelengths about three times this thickness (see the article by Hanna). In other words, they may be a manifestation at the ground of the type of vortices that can cause the development of *cloud streets* in a humid atmospheric environment.

This type of dune was first described in scientific detail by observers in the Libyan desert during the first three decades of the twentieth century. One such observer, W.J.H. King, describes it as follows: "most striking features of the Libyan desert are the curious long, narrow dune belts running across the desert, roughly from north to south, in almost straight lines." The direction of the prevailing wind is from the north. Spacing of the dunes varies between 400 yards and 5 miles, and similar dune patterns have also been observed from airplanes over the Australian, Saudi Arabian, and Algerian deserts. Satellite photographs have revealed the existence of longitudinal dune systems in nearly every large desert of the world. A remarkable feature of these systems is their continuity and geometric regularity. Commonly, smaller-scale dune structures appear on the surface of the larger dunes.

Typically, a longitudinal dune is 20–50 m high, at least 100 m wide, and may range in length from a few hundred meters to several hundred kilometers. The other common type of dune, the *barchan dune*, is by contrast 10–20 m high, about 50 m wide, and crescent-shaped. The concave or hollow side opens to the leeward of the prevailing wind. The leeward or downwind side is called the slip face, and it is steeper than the windward face by virtue of the fact that sand particles are blown up the more gently sloping windward face, and fall down the slip face. The slope of this slip face is equal to the maximum angle of repose (as it is called) under the prevailing weather conditions. It is known that the amount of sand carried by the wind is proportional to $(V - V_t)^3$ (as indicated in the book by Bagnold), where V is the wind speed and V_t is the *threshold* wind speed (when sand grains first begin to be lifted by the wind). Some qualifications to this for grain transportation by water will be noted below. While the direction of the prevailing wind is not necessarily the direction of the net sand transport (storms can move as much sand in one day as the normal prevailing wind moves in a year), this overall assumption is not a bad one in practice.

What of any observational evidence to suggest that atmospheric helical eddies are the culprit in this dune mystery? Leaving aside a body of disputed experimental results from the laboratory (much of which was based on

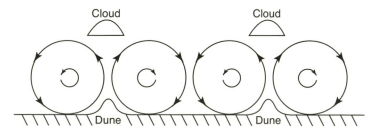

Figure 6.1. The relationship between cloud streets, air circulation, and longitudinal sand dunes in a two-dimensional system. Redrawn from "The Formation of Longitudinal Sand Dunes by Large Helical Eddies in the Atmosphere," by S. R. Hanna (1969), *Journal of Applied Meteorology* 8: 874–883.

laminar or nonturbulent flow; see further comments on this below), there are many direct observations of longitudinal vortices in turbulent air and water. The long parallel cumulus cloud streets that are frequently observed have long been believed to be manifestations of these eddies. The streets form in the upward moving air between every other vertex (see figure 6.1). In bodies of water an analogous phenomenon occurs: there it is known as *Langmuir circulation*. In 1938 Langmuir observed streaks of seaweed spaced 100–200 m apart on the surface of the Atlantic Ocean and oriented in the direction of the surface wind; again, it was suggested that this was the result of alternating left and right helical roll vortices in the topmost ocean layer. Observations of the motion of sheets of paper and *sargassum* bands on the surface have also been made. Returning to the air, the flight of seagulls (under certain atmospheric conditions) has been used to study these longitudinal rolls: above a certain wind speed the gulls soared in straight lines, whereas below that speed they soared in the more familiar circular manner. It has been surmised that they used the circulations associated with the longitudinal vortices to soar in a straight line when it was convenient to do so (see chapter 13 for more on soaring flight).

Returning to cloud streets, it has been found that the average spacing of such features is between 5 and 10 km, which is typically two to three times the thickness of the atmospheric convective layer. According to J. P. Kuettner, the following conditions are conducive to the formation of cloud streets: (i) flat underlying terrain; (ii) little variation of wind direction with height; (iii) wind speed higher than normal; (iv) strong curvature of the wind profile (mathematically, $|V''(z)|$ is "large" in some sense). Typically, the wind increases away from the surface, reaching a maximum at an altitude of about 1 km, and decreases above this height. Finally, (v) the lapse rate of the temperature is unstable near the surface. This means that the temperature decreases with altitude, rendering convection possible.

It appears that these conditions are frequently satisfied in the trade wind regions over tropical oceans, so that cloud streets are a common form of convective activity in these latitudes, and these are where most large deserts are to be found; thus longitudinal vortices are to be expected over these deserts in addition to the oceans. Indeed, satellite photographs show that cloud streets occur at *all* latitudes; their observed spacings average 1–10 km, consistent with the spacings of longitudinal sand dunes.

It is possible to carry out hydrodynamic stability analysis of the equations governing the formation of longitudinal vortices. Such ideas will be developed in more detail in chapter 8; in simplest terms, a small perturbation from the mean flow is assumed to have a form proportional to $exp[-i\alpha t(c_r + ic_i)]$, where α is the vortex wave number, αc_i is the growth rate, and c_r is the phase speed of the disturbance. It is found that the growth rate is a maximum for longitudinal flow vortices along the flow direction having a lateral wavelength of about 2.8 times the fluid depth, according to one model. The lifetime of a typical vortex is short (about 1 hour), so that the effects of the earth's rotation are unlikely to be significant, a conclusion supported by the fact that cloud streets are observed near the equator (where the Coriolis force is negligible).

It has been noted above that the dunes form in the regions of converging air at the flat surface of the underlying terrain below counter-rotating longitudinal roll vortices. How large do these flat regions need to be? A simple calculation helps to answer this: the time period T of the circulations within the observed vortices is about 2000 seconds, and a typical wind speed V is about 10 m/s, so the wind must pass over a flat surface for a distance of *at least* $T \times V \approx 20$ km before longitudinal vortices will develop.

What are the land-based mechanisms for sand dune formation in general? To answer this question, we ask another: how is material transported by water and wind? Significant theoretical work was done by the British mathematician and fluid dynamicist Sir George Gabriel Stokes (1819–1903), who investigated this and concluded that drag experienced by a perfectly round sphere falling *slowly* through a fluid medium is proportional to sphere diameter and speed (i.e., small Reynolds number flow; see chapters 2 and 3, and for more details consult the books by Lamb and Landau and Lifshitz). In Stokes's mathematical model he assumed that the fluid flow was laminar, thus without the eddies that necessarily would increase friction; in practice, this is not the case, not just because sand grains tend to be irregular in shape, but primarily because the environment is quite turbulent! It is known that as water flows over a sand bed at increasing speed, there is (not surprisingly) a threshold speed below which no sand is removed. As the speed increases above this value, individual grains begin to move short distances in an erratic, jerky fashion. This is the process of saltation. Ripples and small dunes can be formed as the speed continues to increase, but eventually a speed is attained when these are swept away and the bottom becomes flat. Beyond these speeds, further dune-like structures may arise. According to research reported by Bascom, the degree of sand transportation (for grains in the size range 0.06 mm to 1 mm) is proportional to very high powers of the water velocity v beyond the threshold—v^{10} for ripples, v^6 for dunes, and v^3 for the transition to a flat bed. Particles that are moved by air follow essentially the same physics, with air being about 800 times lighter than water, so the problem of wind transportation and sedimentation is (in principle) similarly determined, but, as might be expected, things are not quite as straightforward as this.

According to the fascinating account posted on the website of biologist Wayne Armstrong (*http://waynesword.palomar.edu/ww0704.htm*), the origin of sand dunes is very complex, but there are three essential prerequisites for their formation:

1. An abundant supply of loose sand in a region generally devoid of vegetation.
2. A wind energy source sufficient to move the sand grains.
3. A topography whereby the sand particles lose their momentum and settle out (shrubs, rocks, or fence posts—these all can obstruct the forces of wind, causing sand to pile up in drifts and ultimately large dunes).

He also states that

The direction and velocity of winds, in addition to the local supply of sand, result in a variety of dune shapes and sizes. The wind moves individual grains along the inclined windward surface until they reach the crest and cascade down the steep leeward side or "slip face," piling up at the base and slowly encroaching on new territory. Some California dunes with crests only 30 feet high may advance 50 feet a year, posing a serious threat to nearby farms and roads. If the wind direction is fairly uniform over the years, the dunes gradually shift in the direction of the prevailing wind. Vegetation may stabilize a dune, thus preventing its movement with the prevailing wind. Along the Oregon coast, entire forests may cover sand dune areas. Evidence of abrasion on sandblasted surfaces of telephone poles and posts reveals that sand grains seldom travel more than a few feet above the ground. Myriads of sand grains bouncing and rolling up the windward surface of a dune often form a series of ridges and troughs called wind ripples. Bouncing sand grains tend to land on the windward side of each ripple, thus producing a low ridge. Essentially, the spacing of ripples is related to the average distance grains jump. This in turn, is related to the wind velocity and size of the grains.

For centuries, explorers and naturalists throughout the world have described strange sounds emanating from sand dunes. Some of the earliest references about "acoustical" dunes are found in Chinese and Mideastern chronicles dating back more than 1500 years. Marco Polo described weird sounds on a journey through the Gobi Desert, and Charles Darwin mentioned it while traveling through Chile. Some accounts compare the sounds with distant kettle drums, artillery fire, thunder, low-flying propeller aircraft, bass violins, pipe organs and humming telegraph wires. The stationary sand underneath apparently acts as a giant sounding board or amplifier to produce the enormous volume of sound. As noted below, the sand must be very dry for sound production, and under a microscope the grains appear more rounded and finely polished compared with ordinary (silent) sand. Astronomers and geologists have speculated that this remarkable phenomenon may be common in the windy and nearly waterless sand dunes on Mars. . . . Acoustical booming (and seismically responding) dunes are rather widespread, but uncommon, on our planet. Such locations include the Sahara Desert, Middle East, South Africa, Chile, Baja California and the Hawaiian Islands. One of the best places to observe booming dunes in the western United States is Sand Mountain, about 16 miles southeast of Fallon, Nevada. Sand Mountain is composed of two "seif" (sword-shaped) dunes whose summits stand about 390 feet (120 m) above the desert floor. . . . [T]o really appreciate this acoustical phenomenon you must climb to the crest of a dune and then slide down the steep slip face. Going down with an avalanche of sand is sort of like riding down an escalator, ankle deep in sand. As the sand begins to vibrate the sound becomes quite loud, like a low-flying B-29 bomber or squadron of World War II vintage fighter planes.

According to the articles by Lindsay et al. and Criswell et al., there are two types of sand that emit loud, distinct, and often music-like sounds when sheared. The most common is a type of beach sand (found sporadically near oceans, lakes, and rivers) that emits a short note of duration about 1/4 s when poked or stepped on (and, frankly, who wouldn't?). Other descriptions have variously used the adjectives squeaking, singing, whistling, barking, humming, and sometimes screeching and wheezing. The other type of sound is emitted by some desert sand dunes and is of a much lower frequency, sometimes being noted for a frequency and volume approaching that of thunder, which presumably gives rise to the adjective "booming." Such sound recorded at Sand Mountain, Nevada, has been described in terms of its "quizzical charm" when compared with the amplitude trace of an 88-Hz organ tone from the opening stanza of Bach's "Passacaglia and Fugue in C-Minor"! These sounds are "comparable in purity to those from an expertly crafted pipe organ," but we should note the qualifying remarks of Criswell et al., who point out that "the problem is very intriguing and very likely far more complicated physically and mathematically than the simple processes controlling oscillations in an organ pipe."

Despite this complexity, some general features of "the sound of sand" are known. First, booming sand never squeaks and squeaking sand never booms! Second, rain or high humidity will completely eliminate booming, and related to this, warm or hot sand appears to boom best, though heat is not essential. Booming is produced by sand loosely flowing or avalanching down the slip face of a dune. When occurring naturally, the latter can be many meters in extent and several centimeters deep; when produced artificially by pulling fingers (preferably one's own) through the sand, the sounds can occur even though only a few cubic centimeters of sand are displaced. Visually, booming dunes are indistinguishable from their less talkative (even silent) cousins. Sound can accompany so-called "slumping" of meter-sized areas of dune sand that are at or near their natural angle of repose (≈ 30–$35°$; the same phenomenon can be noticed in the classroom, by the way). This can persist from a few seconds to 15 minutes (in both situations) and may produce quite dramatic sounds, sometimes heard (in the case of dunes) at distances in excess of 8–10 km. At distances of 1–2 km, the sound may be perceived as thunder.

A theoretical treatment of these phenomena was carried out by Bagnold, who derived an approximate expression for the frequency f of squeaking sand as

$$f \approx \sqrt{\frac{\lambda K g}{8D}}.$$

Clearly some explanation of the various parameters is necessary. The quantity λ, called the linear concentration, is the ratio of mean grain diameter D to mean separation; g is as usual the local gravitational acceleration, but it is multiplied by the factor K, which represents the increased acceleration induced by the shear stress between slip planes of the displaced material. Prior to the disturbance, $\lambda \gtrsim 17$–22, but during it, λ reduces to about 14; for beach sands, $D \approx 0.2$ cm, $K \approx 25$–40, all of which imply that $f \approx 500$ Hz, in the lower range for squeaking sands. For booming dunes, $K = 1$, $D \approx 0.03$ cm, and $f \approx 240$ Hz, which is in the middle of the observed frequency range observed by one author, but is three to five times higher than that recorded at Sand Mountain, so Criswell et al. suggest that an alternate explanation be sought. That is left as an exercise for the reader.

A final point is in order here. During the review process for this book, an anonymous reviewer made the following helpful comments about the models in this section; they are quoted here (lest the reader should receive the impression that all is "written in stone") to reemphasize the fact that models may be outdated, controversial, and even wrong; "theories for aeolian dunes are under current development, and I don't know if you would get many takers for the helical vortices as an explanation for linear (seif) dunes. My understanding is that these are associated with two prevailing wind directions with an acute angle between, and certainly you wouldn't get a sharp crest from the vortex mechanism. Also, I don't think one generally sees transport laws proportional to such high exponent powers as ascribed to Bascom." So let the buyer beware!

MAYO'S HURRICANE MODEL

This section of the chapter is adapted from a pedagogic article by Ned Mayo. Hurricanes are tropical cyclones with maximum sustained winds of at least 74 mph. In the Pacific Ocean they are called typhoons. Typically they form within bands of latitude 7–20° north and south, where the waters are warm. Waters in higher latitudes are generally too cool to supply enough water vapor to "fuel" the hurricane (the energy source being latent heat from condensing water vapor), and closer to the equator there is insufficient Coriolis acceleration for the tropical disturbance to get organized. Obviously when a hurricane moves over land it weakens because of the absence of water vapor and the consequent loss of (very destructive) energy that is not replenished.

The basic mathematical model is that of a fluid vortex centered on the "eye" of the storm. The eye is characterized by low pressure (relative to the exterior) and descending air; it is surrounded by a vertical annulus of turbulent clouds: the eyewall. The eyewall is a region of high winds and heavy rain, where much latent heat is released, causing the air to rise. As it rises, it is continuously replenished from beneath by an inflow of moist air, spiraling counterclockwise in the northern hemisphere and clockwise in the southern. For the purposes of this model, vertical airflow is assumed to exist only within the eyewall annulus (though, as pointed out by Mayo, the spiraling air, on reaching the eyewall, does spiral upward). In this region the air rises convectively to about 11 km; above this level its upward path is blocked by the temperature inversion of the stratosphere, so the air turns and spirals horizontally outward. This is the exhaust mechanism for the hurricane, but it is not included in this simplified model. The outer eyewall radius is called by Mayo the radius of maximum wind (R_{mw}), representing the smallest radius into which the spiraling air mass can be drawn by reduced eye pressure. This is in fact defined as *cyclostrophic equilibrium* in meteorology: it represents an equilibrium between pressure gradient and Coriolis and centripetal forces; equivalently, the inward pressure gradient provides the necessary centripetal force to maintain the circular motion of the air. The model will be of a "steady-state" hurricane in which wind speeds, dimensions, and pressures are independent of time. Frictional energy losses are therefore assumed to be replenished as they occur by the steady release of latent heat. Furthermore, the hurricane is assumed to be stationary (so that asymmetries in the model are eliminated).

Consider a small element of air of mass dm (kg) located outside the eyewall at a distance r (m) from the origin (located at the center of the eye) and a height z above the sea surface. The central force dF_c needed to keep it in a circular orbit at a tangential speed $w = w(r)$ (m/s) is the sum of the centripetal force and the horizontal component of the Coriolis force, namely,

$$dF_c = \left(\frac{w^2}{r} + 2w\omega\sin\phi\right)dm,$$

where ϕ is the latitude (degrees) and ω is the angular speed of the earth (s^{-1}). To simplify the mathematics we will consider the hurricane to be relatively close to the equator, so that $\sin\phi$ is small, and examine the inner regions where w^2 is large and r is small. This means that the Coriolis term can be neglected in comparison to the centripetal term. Note, however, that the Coriolis force is *essential* for the formation of hurricanes: without it air would flow radially into the low pressure region with no rotation, and so the centripetal force would be zero. Furthermore, as a hurricane moves away from equatorial regions, the Coriolis force becomes more significant and more of the pressure gradient force is required to balance it, leaving less available for the centripetal force. It is to be expected that the horizontal pressure field $p(r)$ would be a continuous function, but, again for simplicity, the present model invokes a piecewise-constant pressure with a step function discontinuity Δp at the outer edge of the eyewall. This fixes the location of the maximum winds there. In view of all the above simplifications,

$$dF_c = \frac{w^2}{r} dm.$$

In cylindrical coordinates,

$$dm = \rho(z)\, dV = \rho_0 e^{-az} r\, dr\, d\theta\, dz,$$

where $\rho(z) = \rho_0 e^{-az}$ is the density of the air, ρ_0 being the density at sea level. According to this model, the density falls off exponentially such that it is half its value at sea level when $z = 7000$ m, thus fixing the constant a to be $ln\, 2/7000$ m$^{-1} \approx 10^{-4}$ m^{-1}. The value of the constant ρ_0 is 1.2 kg/m^3. Thus

$$dF_c = \rho_0 e^{-az} w^2\, dr\, d\theta\, dz.$$

The area over which this force element acts in cylindrical geometry is $dA = R_{mw}\, d\theta\, dz$, so the increment of pressure difference $d(\Delta p)$ across the eyewall needed to keep the airmass dm in circular orbit at speed w is dF_c/dA or

$$d(\Delta p) = \frac{\rho_0 e^{-az} w^2}{R_{mw}}\, dr.$$

The pressure drop Δp is in units of pascals. In an isothermal atmosphere (which ours is not), the pressure field follows an exponential law similar to the density, namely $\Delta p = \Delta p_0 e^{-az}$, where p_0 is the pressure at sea level. This simplification eliminates the exponential terms from the above expression (ignoring a small correction from the differential of e^{-az}) to give

$$d(\Delta p_0) = \frac{\rho_0 w^2}{R_{mw}}\, dr.$$

In order to integrate this expression, the question now arises: how does w depend on r? If angular momentum were conserved, the appropriate relationship would be $wr = $ constant, but at the sea surface friction applies a braking torque to the base of the storm. The most common relation, used by meteorologists and based on the so-called modified Rankine vortex, is that

$$wr^x = \text{constant},$$

where w is the tangential windspeed at radius r. The value of x is determined from observations made in hurricanes; it can be interpreted as a measure of how well angular momentum is conserved, that is, by how close x is to 1 (the value required for conservation of angular momentum). The value $x = -1$ means that w is proportional to r, that is, solid body rotation, and $x = 0$ corresponds to a constant wind (independent of r). Storm data cited by Mayo indicate that at the sea surface $0.4 \lesssim x \lesssim 0.6$. For this (or any other x-value) it follows that

$$W_{\max}(R_{mw})^x = wr^x,$$

where W_{mw} is the tangential windspeed at R_{mw}. Eliminating w from the pressure jump equation yields

$$d(\triangle p_0) = \rho_0 \left[\frac{(W_{\max})^2 (R_{mw})^{2x-1}}{r^{2x}} \right] dr.$$

Integrating, we have

$$\triangle p_0 = \rho_0 (W_{\max})^2 (R_{mw})^{2x-1} \int_{R_{mw}}^{R_0} r^{-2x} \, dr$$

$$= \frac{\rho_0 (W_{\max})^2}{2x - 1} \left[1 - \left(\frac{R_{mw}}{R_0} \right)^{2x-1} \right].$$

Hence

$$W_{\max} = \sqrt{\frac{(2x - 1)\triangle p_0}{\rho_0 [1 - (R_{mw}/R_0)^{2x-1}]}}$$

and

$$w(r) = \left(\frac{R_{mw}}{r} \right)^x \sqrt{\frac{(2x - 1)\triangle p_0}{\rho_0 [1 - (R_{mw}/R_0)^{2x-1}]}}.$$

The integration is carried out from the eyewall to the outer "edge" R_0 of the storm. There is no dependence of w on altitude z in this model because the air density and pressure were taken to decrease at the same rate with altitude. In practice, however, winds reach their maximum speed at an altitude of about 1 km. Below this level friction takes its toll on the wind speed, and at higher altitudes the winds slow, stop, and finally reverse direction as they encounter the stratospheric temperature inversion. To incorporate altitude dependence, the momentum conservation variable x would be a function of z; in all likelihood $x(z)$ would be a monotonically increasing function from the sea surface to $z = 1$ km.

What about the radial component of the wind velocity? It was neglected! This means that the predicted windspeeds will probably be underestimates.

The measured angles of the winds inward from the tangential direction vary from 0° at large radii to 35° near the eyewall. It follows that the ratio of tangential speed to total speed is $\sec\theta$, θ being the above-mentioned angle, so that *at most* the error relative to w is about 22 percent, this occurring at the eyewall. This is offset by the neglect of the Coriolis force (which affords an overestimate of the windspeeds). To illustrate this, the Mayo model predicts for typhoon Tracy (see below) maximum sustained winds of 46 m/s; if the Coriolis force is included, this falls to 41 m/s at a latitude of 12°S—an 11 percent drop.

Mayo tested this model using data from two Pacific typhoons that made landfall in Australia in the mid-1970s: typhoons Tracy and Joan. Tracy was small in diameter and Joan was large. The values of R_0 had to be estimated since, being of little interest at the time, they were not reported. The input values were as shown in the table.

	Tracy	*Joan*
Δp	54 mb	74 mb
R_{mw}	7 km	40 km
R_0	700 km	5500 km
x	0.70	0.57

How were the estimates of R_0 made? Log-log plots of measured winds versus radii were made, and these roughly conformed to straight lines. Extrapolating them to radii at which the wind speed dropped to the arbitrary value of 2 m/s gives the values used for R_0. As pointed out by Mayo, the unrealistically large value of R_0 for Joan is a consequence of neglecting the Coriolis force in a large storm, so only the centripetal force is available to balance the pressure gradient, and this requires a sufficiently large air mass to be set in motion. If that force is *not* neglected, the value of R_0 for Joan is reduced to 650 km. However, the present model does not significantly affect wind predictions near the core of the typhoon; it does cause an overstatement of its radial extent and the strength of the winds in the outer reaches of the storm.

THE KINETIC ENERGY OF THE STORM

The kinetic energy (KE) of the rotating air mass dm will be denoted by dE, where dm is defined as above. Hence

$$dE = \frac{1}{2}w^2\, dm = \frac{1}{2}w^2\rho_0 e^{-az}r\, dr\, d\theta\, dz$$

$$= \frac{1}{2}(W_{\text{max}})^2(R_{mw})^{2x}\rho_0 e^{-az}r^{1-2x}\,dr\,d\theta\,dz.$$

The total KE is therefore given by the triple integral

$$E = \frac{1}{2}(W_{\text{max}})^2(R_{mw})^{2x}\rho_0 \int_0^H \int_0^{2\pi} \int_{R_{mw}}^{R_0} e^{-az}r^{1-2x}\,dr\,d\theta\,dz,$$

where H is the height of the top of the storm. Hence after some simplification we find

$$E = \pi\rho_0(W_{\text{max}})^2 \frac{1-e^{-aH}}{2a(1-x)}\left[\left(\frac{R_{mw}}{R_0}\right)^{2x}(R_0)^2 - (R_{mw})^2\right],$$

where

$$(W_{\text{max}})^2 = \frac{(2x-1)\triangle p_0}{\rho_0[1-(R_{mw}/R_0)^{2x-1}]}.$$

For typhoon Tracy, with $H = 1$ km, $W_{\text{max}} = 46.2$ m/s, and $E = 9.31 \times 10^{15}$ joules (J); since one megaton of TNT is equivalent to the release of 4.2×10^{15} J, this corresponds to approximately 2.2 megatons of TNT, or about a hundred times that of the atomic bomb dropped at Hiroshima or Nagasaki in 1945. If Tracy's KE were released as an earthquake, it would measure about 7 on the Richter scale.

Mayo discusses other aspects of his model, including a sensitivity analysis (which is designed to investigate the "robustness" of the model) in the paper listed in the bibliography.

CHAPTER SEVEN

(Linear) Waves of All Kinds

> ... the currents swirled about me; all your waves and
> breakers swept over me.
>
> —Jonah 2:3b

DESCRIPTIVE ASPECTS OF WAVE MOTION

Wave motion can occur in a wide variety of situations in the natural world, and it is something with which we are all familiar. We can observe waves on the surfaces of oceans and lakes or when a pebble is dropped into a pond; waves are generated and propagated when a musical instrument is played (correctly or otherwise), when a radio station transmits programs, or when a solar flare occurs. In the previous chapter we noted that clouds can often be indicators of wave motion, especially in the presence of wind. Violent disturbances on or below the earth's surface, such as explosions or earthquakes, give rise to wave motion that can subsequently be detected experimentally. These examples of wave motion all have two very important characteristics in common:

(i) energy is propagated from points near the source of waves to points which are distant from it, and

(ii) the disturbances travel through the medium (whatever that may be) without giving the medium as a whole any permanent displacement. (A caveat: obviously one characteristic of earthquakes is that such displacement *does* occur at the surface.)

In the example of the pond, the ripples spread outward over the surface carrying energy with them, but if we watch, say, a cork on the surface we see that it (and hence the water) does not move with the waves, but bobs up and down periodically. It is found that whatever the nature of the medium through which the waves are transmitted, whether it be air, a stretched string, a liquid, an electric cable, or deep space, these two properties are common to all types of wave motion and enable them to be described.

Mathematically, they are all governed by a certain partial differential equation called the wave equation, and the solution of this equation when

interpreted appropriately is the mathematical description of the corresponding idealized physical phenomenon of wave motion. It must be emphasized
that in this chapter we are dealing with what is called *linear theory*. This
means that the wave amplitudes (or particle displacements) are sufficiently
small (technically they are infinitesimal) that all their squares and higher
powers can be neglected in the subsequent mathematical analysis. This is
a severe approximation with great limitations, to be sure, but nonetheless
linear theory has provided some very powerful and useful results because
the mathematics is generally more tractable than for nonlinear theory, and
therefore amenable to analytical techniques. Some examples dealing with
nonlinear waves will be discussed in chapter 9.

Waves can be separated basically into two distinct types: transverse
waves and longitudinal waves. Transverse waves may be conveniently illustrated by means of our pond example; we have already noted that, whereas
the waves or ripples gradually move outward in ever-increasing circles
across the surface of the pond, the cork, initially at rest, bobs up and down
several times, eventually coming to rest again as the waves pass on. The
cork merely serves as a convenient indicator of the behavior of the fluid
beneath it, that is, the (nearly) vertical motion of the water particles. It is
evident that when the ripples reach the point where the cork is located, the
water particles suddenly acquire energy that they retain for only a short time
and then relinquish to horizontally neighboring particles as the waves pass
on. It is therefore the energy that travels in a horizontal direction with the
ripples, not the water itself, although for a time the water particles acquire
a certain amount of kinetic energy in the vertical direction.

Waves in which the motion of the particles of the medium is at right
angles to the direction of wave propagation are called transverse. Although
water waves have been used to introduce this concept, it should be noted
that such waves are not *strictly* transverse, because the particle motions in
general have horizontal components as well as vertical ones. They therefore
tend to move in elliptical or even circular paths depending on the depth of
the water and the wavelength (defined as the distance between successive
crests, or successive troughs of the waves) of the waves.

What of the second characteristic of wave motion, namely, that the
medium suffers no permanent displacement? A pebble may initiate the wave
motion and obviously sinks to the bottom of the pond, but the cork bobs
up and down. The medium must in loose terms have an inherent elasticity
in order to support this wave motion, for when the particles are displaced
there must be a "restoring force" (or forces) that will stop their motion in
one direction and accelerate them back toward their equilibrium or initial
position. For transverse waves to exist, the particles must be coupled to one
another by forces such that each can do work on the next one in order to
transfer energy to it. In our example, due to the presence of unbalanced

forces on the molecules that find their way into the surface of a liquid, that surface acts like a stretched membrane, and these surface tension forces furnish most of the necessary elasticity for the propagation of the ripples. In larger waves, however, gravity is the dominant restoring force, inasmuch as larger quantities of water are lifted up and brought below their equilibrium position.

There are many other examples of transverse waves, such as waves that occur on a taut string when it is plucked. This phenomenon in fact is more representative of pure transverse waves than are water waves, for the reasons discussed above. For the other basic type of wave, the longitudinal wave motion we shall concentrate on is that of sound waves in gases. This conveniently illustrates the underlying physical concepts involved in the propagation of longitudinal waves, but such motions are not restricted to gases alone. A metal rod struck at one end with a blow aimed along its length, for example, can support such waves.

Gases lack the necessary elastic properties for the propagation of transverse waves because the only way a molecule can exert a force on another is by colliding with it. This gives rise to an elasticity of "compression": a fundamentally different kind of restoring force from that discussed for water waves. Gases therefore transmit variations in pressure in the form of longitudinal waves in much the same way that the free end of a compressed spring will oscillate if released with one end fixed. Between certain limits such waves, in air, for example, can be detected by the human ear: these are of course sound waves. Longitudinal waves may be further visualized by considering a medium consisting of long strings of beads separated by light springs. If a bead is displaced so as to vibrate back and forth along the direction of the string, the springs connecting it with adjacent beads are alternately compressed and extended, causing these beads to vibrate back and forth also. This transfer of vibrational energy occurs gradually along the string in each direction (if our initial bead was not on the end of the string), with any particular bead merely vibrating back and forth about its equilibrium, but a procession of compressions and extensions is actually propagated along the string.

These examples have been used to illustrate the fundamental types of wave that can exist in various media, namely, transverse and longitudinal waves. Basically any system, gas, liquid, or solid, can support waves of one kind or another (or even possessing characteristics of both), provided there is some type of force present that tends to restore displaced particles toward their equilibrium position. This restoring force generally tends to "overshoot," thus necessitating the force to act in the opposite direction and giving rise to one cycle of what we know as wave motion. The restoring forces due to surface tension, gravity, and elasticity or compression have already been mentioned, but there exist other kinds of forces tending to restore equilibrium. Magnetic fields, for example, have associated with them a "magnetic tension" analogous to tension in a taut string. As a consequence, waves can propagate in transverse fashion along magnetic field lines, in the earth's upper atmosphere in laboratory situations, or even in the "atmospheres" of the sun and other stars. The word atmospheres has been written inside quotation marks because they are generally not electrically neutral gases but *plasmas* consisting of positively and negatively charged atomic particles—ions and electrons. This situation in itself can give rise to a variety of plasma waves resulting from *electrostatic* restoring forces between different particles.

INTRODUCTORY THEORETICAL ASPECTS

The section above gave a descriptive introduction to wave motion. Let us now start to fill out this subject with some mathematics. We need to note the following basic characteristics of most kinds of wave motion, following from (i) and (ii) above:

(iii) The time and space evolution of wave-associated quantities are described by a partial differential equation—a wave equation.

(iv) Associated with the waves is a velocity of propagation, v, broadly defined by

$$v = \left(\frac{\text{elastic factor}}{\text{inertia factor}} \right)^{1/2}. \tag{1}$$

(v) The waves can in general undergo reflection, refraction, diffraction, interference, and so forth (they can under appropriate circumstances, exhibit all the properties that are normally associated with light waves).

Some obvious but fundamental definitions related to wave motion are (a) the wave frequency is the reciprocal of the wave period, or

$$v = T^{-1}, \tag{2}$$

where the period is simply the time interval between the passage of successive wave crests (or troughs), and (b) the wave speed = frequency × wavelength, or

$$v = v\lambda \tag{3}$$

We have already noted the distinction between transverse and longitudinal waves. Before we concentrate our attention on specific, readily observable kinds of waves, let us recall that the elasticity factor or measure of restoring force in equation (1) can arise in connection with a wide variety of situations; some such forces are associated with tension in strings or magnetic fields, compressibility, surface tension, gravity, rotation, electric charge, elasticity, and curvature of space-time; indeed, waves can exist almost anywhere! We shall concentrate on those arising in puddles, ponds, rivers, lakes, and seas. But first we shall discuss a simple yet general equation called the *wave equation*. Whence does it come?

Let us sketch out the method for derivation of the wave equation using an old undergraduate friend, the stretched string. Let the string lie along the x-axis in equilibrium, under uniform tension T. Then consider a small displacement in the y-direction and examine the forces acting on a *small* element Δx on the string. By using Newton's second law (force = rate of change of momentum = mass × acceleration) and using the appropriate limiting procedures (à la differential calculus, as many textbooks show), we arrive at the wave equation for the string displacement $y(x, t)$:

$$\frac{\partial^2 y}{\partial x^2} = \frac{1}{v^2}\frac{\partial^2 y}{\partial t^2}, \tag{4}$$

where the velocity

$$v = \left(\frac{\text{tension in string}}{\text{density of string}}\right)^{1/2} = \sqrt{\frac{T}{\rho}}. \tag{5}$$

The assumption of small displacement is equivalent to requiring that the slope $|\partial y/\partial x|$ of the string at all points x is small compared to unity. As noted in the previous section, this assumption (linearity) is made for mathematical tractability and within its limitations is physically very useful. It is in principle possible to obtain a bound on the error induced by this assumption; such a bound for approximating a function by a Taylor polynomial of degree n is established in most calculus texts, for example. However, most of the *really* interesting wave phenomena occur when $|\partial y/\partial x|$ is *not* small in this sense, and this provides mathematicians with some headaches!

Gravity-Capillarity Waves

Not all wave equations are as simple-looking as that expressed by equation (4), but the basic principles underlying their derivation are similar. Accordingly we focus attention on the expression on the speed of waves propagating under the combined effects of gravity and surface tension—the so-called gravity-capillarity waves. As mentioned above, these waves, or related ones, are readily observable in nature; in fact the town in which I lived in Northern Ireland (Portstewart) is probably among the best places in Europe for observing surface gravity waves, that is, long waves on the surface of the ocean. For the combined effects of both forces

$$v^2 = \left(\frac{g\lambda}{2\pi} + \frac{2\pi\gamma}{\lambda\rho}\right)\tanh\left(\frac{2\pi h}{\lambda}\right), \tag{6}$$

where, as before, λ = wavelength, ρ = fluid density, and h and γ are the fluid (water) depth and coefficient of surface tension, respectively. The gravitational acceleration has magnitude g. Recall, by the way, that

$$\tanh x = \frac{\sinh x}{\cosh x} = \frac{e^x - e^{-x}}{e^x + e^{-x}}.$$

It is interesting to consider some special cases of (6).

Deep Water Waves

Although we shall qualify this condition a little later, "deep" here means that the wavelength is small compared with the depth of the water,

$$\frac{h}{\lambda} \gg 1,$$

which means that

$$\tanh\frac{2\pi h}{\lambda} \approx 1.$$

Under these circumstances

$$v^2 \approx \frac{g\lambda}{2\pi} + \frac{2\pi\gamma}{\rho\lambda}, \tag{7}$$

which represents the (wave velocity)^2 for disturbances that "feel" the effects of gravity and surface tension, but do not "feel" the bottom of the channel, reservoir, or other body of water. Furthermore, for "long" waves in this category, where

$$\frac{g\lambda}{2\pi} \gg \frac{2\pi\gamma}{\rho\lambda},$$

it follows that

$$v^2 \approx \frac{g\lambda}{2\pi}. \tag{8}$$

Thus

$$v \propto \sqrt{\lambda},$$

which is the correct relationship between velocity and wave length for ocean waves, which are completely dominated by gravity. Crudely—the longer the wavelength, the faster the wave moves. At the other extreme, for "short" waves, where

$$\frac{g\lambda}{2\pi} \ll \frac{2\pi\gamma}{\rho\lambda},$$

we find that

$$v^2 \approx \frac{2\pi\gamma}{\rho\lambda}. \tag{9}$$

Thus

$$v \propto \frac{1}{\sqrt{\lambda}} = \lambda^{-1/2}.$$

These waves (ripples) are completely dominated by surface tension, and the shorter they are the faster they move.

Shallow Water Waves

For shallow water waves, the depth of water is small compared with the wavelength,

$$\frac{h}{\lambda} \ll 1.$$

Under these circumstances it may be shown from a Taylor series expansion about the origin (a Maclaurin series) that

$$\tanh \frac{2\pi h}{\lambda} \approx \frac{2\pi h}{\lambda}.$$

These waves do "feel" the bottom. For most problems of interest in this wave situation, the second term in (6) is negligible, so that the following result is valid for gravity waves in shallow water:

$$v^2 \approx gh. \tag{10}$$

This is an important result: it means that the wave speed is *independent of wavelength*. This implies that all the waves travel with the same speed, and any complex initial wave configuration may retain an identifiable shape for quite some time afterward. Actually, the strong inequalities we have employed to distinguish between deep water waves and their shallow water counterparts do not need to be enforced so strictly; sometimes it is sufficient to demand that $h < \lambda/2$ for shallow water waves, for example. It all depends on the context.

Let us return to equation (7) for deep water capillarity-gravity waves, for there is quite a bit more information we can extract. In the extreme cases given by equations (8) and (9), we have seen that the square of the speed behaves in a linear and a rectangular hyperbolic fashion, respectively, as functions of wavelength. In the intermediate region, where the terms $g\lambda/2\pi$ and $2\pi\gamma/\rho\lambda$ are comparable, both restoring forces are comparable, and the respective graphs of $\upsilon(\lambda)$ must intersect. This and other aspects are illustrated in figure 7.1a–d.

The speed υ is clearly a minimum since $\upsilon''(\lambda) > 0$ at the critical wavelength λ_c given by the solution of

$$\upsilon'(\lambda) = 0,$$

which is

$$\lambda_c = 2\pi\sqrt{\frac{\gamma}{g\rho}}. \tag{11}$$

In cgs units $\lambda_c \approx 1.72$ cm for water. For wavelengths less than or greater than this, the dominant restoring force tends to be, respectively, surface tension or gravity. If we substitute (11) back into (7) we find the corresponding minimum velocity to be

$$\upsilon_{\min} = \sqrt[4]{\frac{4g\gamma}{\rho}} \approx 23 \text{ cm/s} = 0.23 \text{ m/s}.$$

This means that any breeze or gust of wind with speed less than 0.23 m/s will not generate any propagating waves, other than a transient disturbance. Wind speeds above this minimum value will in principle generate two sets of waves with wavelengths on each side of λ_c, one set with $\lambda < \lambda_c$ (capillarity waves) and one set with $\lambda > \lambda_c$ (gravity waves). Note that these results may be derived without the use of calculus: the arithmetic-geometric inequality gives the required result. This inequality tells us, in particular, that if $a > 0$ and $b > 0$ then

$$\frac{a+b}{2} \geq \sqrt{ab},$$

with equality occurring if and only if $a = b$. This result, which tells us that the arithmetic mean is never less than the geometric mean, is easily established by considering the inequality $\left(\sqrt{a} - \sqrt{b}\right)^2 \geq 0$, and can be generalized to a set of n positive numbers, but we need only two here. We do this by writing equation (7) for brevity as

$$\upsilon^2 = \alpha\lambda + \frac{\beta}{\lambda}, \quad \alpha > 0, \beta > 0.$$

Then it follows by the above inequality that the sum of these two terms is never less than $2\sqrt{\alpha\beta} = 2\sqrt{g\gamma/\rho}$. Since the minimum of υ^2 occurs when the minimum of υ does, the corresponding result stated above for $\upsilon_{\min}$ is established.

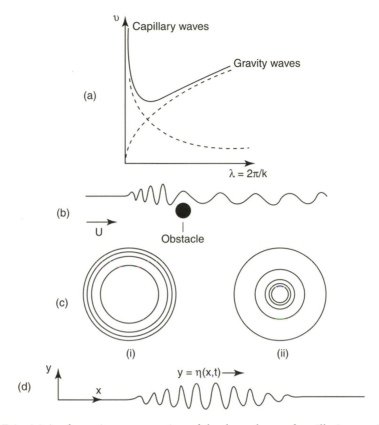

Figure 7.1. (a) A schematic representation of the dependence of capillarity-gravity wave speed on wavelength. For sufficiently short (long) wavelengths, the waves are referred to as capillary waves (gravity waves). The wavenumber $k = 2\pi/\lambda$. (b) Side view of stationary waves generated by a uniform flow U past a small partially submerged object (such as a rock or twig). Note that the small wavelength capillary waves are ahead of the obstacle, and the longer gravity waves are behind it. (c) A "snapshot" of two wave patterns produced on the surface of a pond, one produced by a large pebble and the other produced by a raindrop. In the light of the material discussed in this chapter, determine which pattern corresponds to each of these situations. (d) A wave group (note that the envelope of peaks and troughs is sinusoidal in linear theory). The quantity $\eta(x, t)$ represents the surface elevation as a function of one spatial variable and time. From *Elementary Fluid Dynamics,* by David Acheson (1990), Clarendon Press: Oxford, by permission of David Acheson.

It is appropriate at this point to add a comment on wave refraction: we now have a simple (not to say simplistic) model explaining why ocean waves always/usually/sometimes (delete whichever does not accord with your experience) line up parallel to the beach, even if far out to sea they are approaching it obliquely. Fix the wavelength of any particular wave you are interested in. Far out, the wave is in deep water ($\lambda \ll h_{deep}$) and so $\upsilon \propto \sqrt{\lambda}$. Nearer in, the wave is in shallow water ($\lambda \gg h_{shallow}$) and so $\upsilon \propto \sqrt{h_{shallow}}$, which is of course smaller than $\sqrt{\lambda}$. So that part of the wavefront nearest the beach slows down compared to that farther out, and the whole thing tends to "slew" around.

Dispersion Relations

Thus far we have examined the relationship between speed and wavelength for the small but very important class of waves known as gravity-capillarity waves. These relationships have been based on equation (6) above, and the question naturally arises as to how it is obtained. The governing differential equations for small amplitude motion on the surface of a fluid are satisfied, in particular, by what are called *plane wave solutions*, and the relationship between the frequency ν of the wave and its wavelength λ (or a related quantity called the wavenumber, $k = 2\pi/\lambda$) determined by these solutions is referred to as a *dispersion relation*. Equation (6) is just such a dispersion relation, where from equation (3) the wave speed $\upsilon = \nu\lambda$.

This is therefore the bottom line: the "quick and dirty" way of obtaining this relationship between the frequency and wavelength of a wave-like solution to a partial differential equation is to seek one in the form of a complex exponential function of space and time. This relationship, the dispersion relation, will be different, of course, for different types of wave. We illustrate the concept with a simple example concerning the omnidirectional propagation of sound in a medium. The Cartesian form of the wave equation governing the space and time evolution of small-amplitude sound waves of speed c in a uniform three-dimensional medium is given by

$$\frac{\partial^2 u}{\partial t^2} = c^2 \nabla^2 u \equiv c^2 \left(\frac{\partial^2 u}{\partial x^2} + \frac{\partial^2 u}{\partial y^2} + \frac{\partial^2 u}{\partial z^2} \right). \tag{12}$$

The dependent variable $u(x, y, z, t)$ can represent *small* pressure deviations from the ambient pressure as a sound wave traverses the medium; it may also represent a velocity component or displacement in this case. (Notice the similarity with equation (4) when $u = u(x, t)$ only.) It is worth reminding ourselves that this emphasis on small-amplitude (theoretically, infinitesimal) waves is necessary because the form of the solutions we shall be seeking is based on linear theory.

The speed of the sound waves (assumed constant here) and of waves other than water waves will generally be denoted by c, and a particular point in space is defined by the radius vector $\mathbf{r} = \langle x, y, z \rangle$. A *plane wave solution* of this equation is one of the form

$$u(x, y, z, t) = \text{Re}\,(Ae^{i(\mathbf{k} \cdot \mathbf{r} - \omega t)}) = \text{Re}\,(Ae^{i(kx + ly + mz - \omega t)}), \qquad (13)$$

where k, l, m, and ω are real quantities; A is a (possibly complex) constant, the *wavevector* $\mathbf{k} = \langle k, l, m \rangle$, and ω is the angular frequency. In the chapter 8 we shall allow ω to be a complex quantity. The wavenumber $\kappa = |\mathbf{k}| = \sqrt{k^2 + l^2 + m^2}$, and the wavelength $\lambda = 2\pi/\kappa$. By substituting the complex solution $Ae^{i(\mathbf{k} \cdot \mathbf{r} - \omega t)}$ into equation (12) an algebraic equation results, namely, the dispersion relation $\omega = \omega(\kappa)$, where

$$\omega^2 - c^2 \kappa^2 = 0 \qquad (14)$$

or

$$\frac{\omega}{\kappa} = \pm c.$$

This expression represents the wave or phase speed of waves propagating parallel and antiparallel to the wave vector. A plane wave is one component of an integral superposition of such exponential forms if the wave equation is defined in an infinite domain. There are many different types of waves that can be supported by the earth's land, sea, and atmosphere, often named after the scientists who made major contributions to the topic. Some of the waves that can exist are called acoustic, atmospheric, baroclinic, Eady, gravity, internal, Kelvin, lee, Love, planetary, Rayleigh, and Rossby. We will not state the governing differential equations, but for some of these waves we will briefly discuss their dispersion relations and some of the consequences that follow from them. There are also combinations of various types, such as acoustic-gravity waves (mentioned in chapter 6), which, as their name implies, are supported by the restoring forces of both buoyancy and compressibility in the atmosphere. In terms of the quantities present in equation (13), the acoustic-gravity wave dispersion relation is

$$m^2 = k^2 \left(\frac{N^2}{\omega^2} - 1 \right) + \left(\frac{\omega^2 - \omega_a^2}{c^2} \right), \qquad (15)$$

which is valid for an isothermal atmosphere (one in which the sound speed c is constant). There are two naturally occurring frequencies in this model. The first is the so-called Brunt-Vaisala frequency N; it is the frequency at which a small "parcel" of air will oscillate if displaced vertically from its equilibrium position in a stably stratified atmosphere. It is the highest frequency at which gravity waves may propagate (see figure 7.2). The quantity ω_a ($> N$) is called the acoustic cut-off frequency; it is the lowest frequency at which gravity-modified acoustic waves may propagate. They are defined as follows for an isothermal atmosphere:

$$N = \sqrt{\frac{(\gamma - 1)g}{\gamma H}} \text{ and } \omega_a = \frac{1}{2}\sqrt{\frac{\gamma g}{H}} = \frac{c}{2H},$$

where H is density scale-height, defined by

$$H = -\left(\frac{1}{\rho}\frac{d\rho}{dz}\right)^{-1}$$

($\rho(z)$ being the atmospheric density profile). Note that $m^2 < 0$ when $N^2 < \omega^2 < \omega_a^2$; this corresponds to horizontally propagating waves that are attenuated exponentially in the vertical direction. They are sometimes referred to as evanescent or surface waves.

We now restrict ourselves to waves for which the first term in relation (15) is much larger than the second. For the earth's atmosphere, this corresponds to compressibility-modified gravity waves with wavelengths of the order of a few kilometers (at most). Furthermore, we impose a uniform

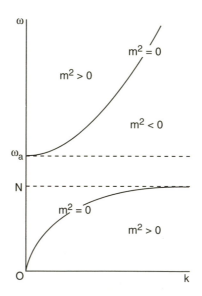

Figure 7.2. A schematic representation of the $\omega(k, m)$ dispersion relation (15) for acoustic gravity waves. When $m^2 > 0$ the waves can propagate vertically; otherwise they cannot.

wind (sometimes called a zonal wind) U flowing in the x-direction. The determining dispersion relation then takes the form

$$m^2 = k^2 \left(\frac{N^2}{\Omega^2} - 1 \right), \tag{16}$$

where $\Omega = \omega - kU$ is the Doppler-shifted frequency. Equation (16) may be rearranged to give

$$\frac{\Omega^2}{N^2} = \frac{k^2}{m^2 + k^2}. \tag{17}$$

The most easily observed example of this type of wave is in the lee of mountains: these are lee waves originating as a result of air being forced to flow over a mountain under stable atmospheric conditions. The air is set into motion in the form of gravity waves as it moves downstream from the mountain. If their amplitude is large enough, and the temperature and humidity are in the right ranges, clouds will form in the crests of these waves (see chapter 6). They will appear stationary with respect to the mountain as the horizontal component of wave velocity v relative to the surface ($v = \omega/k$) is zero. If $k \ll m$, then equation (17) implies that $m \approx N/U$, and since the wind supplying their energy is near the ground, the majority of the energy associated with them (and traveling with what is called the *group velocity*) is propagated upward. This brings up an interesting and seemingly paradoxical property of gravity waves: for $k \ll m$ the vertical component of their group velocity is antiparallel to the vertical component of the wave velocity. We demonstrate this for the case of $U = 0$, but first it is necessary to define this new term "group velocity" $\boldsymbol{v}_g$ (or group *speed* v_g in one spatial dimension and direction). It will be developed at length below in connection with gravity-capillarity waves, so for now all we need to know is that in two spatial dimensions, as here, $\boldsymbol{v}_g$ is the vector

$$\boldsymbol{v}_g = \left(\frac{\partial \omega}{\partial k}, \frac{\partial \omega}{\partial m} \right).$$

It is readily shown from (17) that the vertical components of group and wave velocity are, respectively,

$$\frac{\partial \omega}{\partial m} = -\frac{kmN}{(m^2 + k^2)^{3/2}} \approx -\frac{kN}{m^2} \quad \text{and} \quad \frac{\omega}{m} = \frac{kN}{m(m^2 + k^2)^{1/2}} \approx \frac{kN}{m^2}.$$

Gravity waves are not dominant in the overall energy budget in the lower atmosphere, but above an altitude of about 75 km they are very significant. Note that on a planetary scale tidal motions are a special case of gravity waves in water with long time and spatial scales. On these large scales in the atmosphere, there are wave motions generally known as *planetary waves*. In their simplest form they arise as a result of the variation of the

Coriolis parameter with latitude, and are known as *Rossby waves* (named after the meteorologist Carl Rossby who discovered their existence in the 1930s). They are manifested as mid-latitude westerly winds flowing around the earth in the troposphere (where the "weather" occurs—the lowest 4–10 miles of the atmosphere, depending on latitude, being thickest at the equator and thinnest at the poles) and, to a lesser extent, in the stratosphere immediately above the troposphere. Their pattern is variable; their number, amplitude, and wavelength are changeable; and they influence the movement of air masses from the polar and mid-latitude regions, the balance of heat exchanged between these air masses, and also, to some extent, the development and track of storms.

In terms of the mean flow U the governing dispersion relation is

$$\omega = k\left[\frac{\beta}{k^2 + l^2 + m^2 f_0^2/N^2} - U\right], \tag{18}$$

where in the so-called β-plane approximation, the Coriolis parameter $f = 2\Omega \sin\phi$ is written $f = f_0 + \beta y$ (ϕ being the latitude and the quantity Ω now representing the angular velocity of the rigidly rotating earth). It turns out that $f_0^2 \ll N^2$, and hence the vertical wavelengths are the order of 1 percent of the horizontal wavelengths. Sometimes these waves are forced by surface features, such as mountains or large land masses, in which case they are stationary with respect to the surface (like lee waves and the water waves associated with sticks in streams). Thus, when $\omega/k = 0$, equation (18) implies that

$$m^2 = \frac{N^2}{f_0^2}\left[\frac{\beta}{U} - (k^2 + l^2)\right].$$

Thus $m^2 > 0$ when

$$0 < U < \frac{\beta}{(k^2 + l^2)}. \tag{19}$$

Now U is generally a westerly flow, so it is clear that if the flow is either easterly ($U < 0$) or large and westerly ($U > \beta/(k^2 + l^2)$) no vertical propagation occurs ($m^2 < 0$). This result is in fact confirmed by the observed lack of planetary wave activity during the summer months when the mean stratospheric flow is easterly. From the inequality (19) it is apparent that, for a given value of U, waves of large horizontal wavelength are most readily propagated in these atmospheric regions.

WAVE SPEED AND GROUP SPEED (AGAIN)

Having surveyed a rather wide variety of waves, large and small, we are now in a position to distinguish between two fundamental types of *speed* associated with waves, concentrating for the present on the water waves discussed in an earlier section of this chapter. In terms of the speed v of a wave with frequency v and wavelength λ, it follows from equation (3) that the following relationship is the equivalent of the constant-speed formula "distance equals speed times time":

$$v = v\lambda = 2\pi v \frac{\lambda}{2\pi} = \frac{\omega}{k},$$

where $\omega = 2\pi v$ defines the *angular* frequency of the wave and $k = 2\pi \lambda^{-1}$ is defined as the *wavenumber*. This speed is known as the wave speed or phase speed; it is the speed of an individual wave as it propagates through the system. This speed (or velocity—a vector quantity—in more than one dimension) will in general depend on the wavelength and hence the wavenumber k. If this is the case, waves with different wavelengths travel at different speeds, and so an initial "clump" of such waves will tend to spread out over time. This phenomenon is known as *dispersion* (by analogy with a similar phenomenon in optics); hence the terminology "dispersion relation." The dispersion relation for an unbounded system that can support waves is the relationship (in one spatial dimension) $\omega = \omega(k)$. We shall reexamine equation (6) in terms of this more commonly used relationship. In particular, we can see that gravity-capillarity waves in a channel of depth h satisfy the dispersion relation (in quadratic form)

$$\omega^2 = v^2 k^2 = \left(gk + \frac{\gamma}{\rho} k^3 \right) \tanh kh,$$

where g and γ are the constants defined above. Before reexamining some limiting cases of this equation, we point out another fundamental speed associated with a group of waves: the *group speed* v_g (or group velocity $\mathbf{v}_g$ if the wave propagates in more than one spatial dimension). This is the speed at which such a group moves through the system, and it may not be the same as the wave speed. If these speeds are different, as is frequently the case, a group will not consist of the same waves as it moves along. In one dimension,

$$v_g = \frac{d\omega}{dk},$$

while in more than one (e.g., three as illustrated here), the wavevector $\mathbf{k} = (k, l, m)$ and

$$\mathbf{v}_g = \nabla_k \omega \equiv \left(\frac{\partial \omega}{\partial k}, \frac{\partial \omega}{\partial l}, \frac{\partial \omega}{\partial m} \right).$$

Returning to the one-dimensional wave system defined above, we shall calculate v and v_g for some limiting cases. Although the equations for the wave (or phase) speed v have been determined above, we shall compare it directly with the group speed in each case. First, consider again the so-called deep-water case for which the wavelength is small compared to the fluid (usually water, of course) depth, i.e., $kh \gg 1$. Then $\tanh kh \approx 1$, and in this limit

$$\omega^2 = \left(gk + \frac{\gamma}{\rho}k^3\right).$$

Furthermore, if we restrict waves to those for which the second term is negligible compared with the first (gravity waves), then we write

$$\omega^2 = gk,$$

and it follows from their respective definitions that

$$v = \left(\frac{g}{k}\right)^{1/2} \propto \lambda^{1/2}$$

and

$$v_g = \frac{1}{2}\left(\frac{g}{k}\right)^{1/2} = \frac{1}{2}v.$$

So for gravity waves (including ocean swell) the group speed is less than the wave speed.

The other extreme for deep water waves is that of very short waves (capillarity waves, dominated by the forces of surface tension), for which the first term in the above equation is negligible compared with the second. Then

$$\omega^2 = \frac{\gamma}{\rho}k^3,$$

and so

$$v = \left(\frac{\gamma}{\rho}k\right)^{1/2} \propto \lambda^{-1/2}$$

and

$$v_g = \frac{3}{2}\left(\frac{\gamma}{\rho}k\right)^{1/2} = \frac{3}{2}v.$$

In this case the group speed exceeds the wave speed.

What follows is an interesting but related interlude before addressing the topic of shallow water waves; at least I think it is interesting. I am fortunate to be able to walk to my place of work, about a mile each way, and most of the time I don't get caught in the rain. Occasionally, of course, I do, and then the most interesting part of my walk arises after the rain has all but passed and before the trees have had a chance to dry off. Then puddles abound, and drops of water falling from the trees into these puddles create circular ripples that move outward in a well-defined manner. Nothing surprising so far, you may say, but wait. I have noticed that behind the waves there is an expanding circular central region of calm water. How fast does this region expand? This requires an interesting calculation concerning the group speed of gravity-capillarity waves in deep water; shallow water is not relevant as will be demonstrated below.

First though, what do we *really* mean by "deep" and "shallow" water? We have used (and will continue to do so) the strong inequalities $kh \ll 1$ (or $\lambda \gg h$) for shallow water, and $kh \gg 1$ (or $\lambda \ll h$) for deep water, but how large is "large"? Recall that these inequalities arise from considering the extremes of the function $\tanh kh$, so if we require that $\tanh kh > 0.95$, say, for our waves to be considered "deep" water waves, then this means that $kh \gtrsim 2$ or $\lambda \lesssim 3h$. Bascom uses the criteria $h > \lambda/2$ for deep water waves and $h < \lambda/20$ for shallow water waves, noting that the full hyperbolic tangent formula must be used for the intermediate regime $\lambda/20 < h < \lambda/2$.

Now let us assume for the moment that my puddle is deep in the sense that its depth exceeds half that of the shortest wavelength that I observe on its surface. Then we write

$$\omega^2 = \left(gk + \frac{\gamma}{\rho} k^3 \right) = gk + bk^3,$$

where $b = \gamma/\rho$. Differentiating this expression implicitly, we find that the group velocity of deep water gravity-capillarity waves is

$$v_g(k) = \frac{d\omega}{dk} = \frac{g + 3bk^2}{2(gk + bk^3)^{1/2}}.$$

If this group velocity has a minimum value for some wavenumber k, then the last wave group to move out from the center of the disturbance will travel with this speed, and since the energy also travels with the group speed, the region

$$r < (v_g)_{\min} t$$

is calm. A further differentiation shows that

$$\frac{dv_g}{dk} = 0 \quad \text{when} \quad k = \left(\left[\frac{2}{\sqrt{3}} - 1 \right] \frac{g\rho}{\gamma} \right)^{1/2} = k^*,$$

and furthermore it is readily shown by the first derivative test that this critical number defines a *minimum* of v_g. This means that there *is* a region of calm water that expands with speed

$$(v_g)_{\min} = v_g(k^*) = \frac{\sqrt{3}-1}{(2/\sqrt{3})^{1/2}(2/\sqrt{3}-1)^{1/4}}\left(\frac{g\gamma}{\rho}\right)^{1/4} \approx 1.09\left(\frac{g\gamma}{\rho}\right)^{1/4}.$$

Now it's time to put in the numbers. We use $g = 9.81$ m/s^2, $\gamma = 0.074$ N/m, $\rho = 10^3$ kg/m^3 to find that $(v_g)_{\min} \approx 18$ cm/s (about 0.4 mph). The wavelength corresponding to this minimum speed is just $\lambda^* = 2\pi/k^* \approx 4.4$ cm. This means that for the deep water approximation to be valid, $h \gtrsim 2$ cm, which may well have been the case (I did not measure it at the time).

A word of warning: the algebra is a little tricky, and if like me you have a tendency to be careless, then go slowly. Initially I found a speed of about 1.5 cm/s and went back out next time it rained to find this was far too slow. So either the deep water approximation was invalid or I had made a mistake (or both!). I checked my calculations and found a mistake. Of course, it may still be the case that the puddle was not deep enough for me to neglect the hyperbolic tangent in my calculations, but that is an even more tricky procedure, so we can take the easy way out and graph v_g in this most general case. In general a minimum in v_g is seen to occur as before, but if we then examine the other limiting extreme of shallow water waves, for which $kh \ll 1$, we find that

$$\omega^2 = ak^2 + bk^4; a = gh, b = \frac{\gamma h}{\rho}.$$

Then proceeding as above

$$v_g(k) = \frac{d\omega}{dk} = \frac{a + 2bk^2}{(a+bk^2)^{1/2}},$$

from which it follows that

$$\frac{dv_g}{dk} > 0;$$

that is, *no* minimum value of v_g occurs for nonzero k, so the calm water phenomenon cannot occur in strictly shallow water. Taking into account the fact that a minimum still exists for intermediate depths down to about one-twentieth of a wavelength, this places the depth of the puddle $h \gtrsim 0.2$ cm, which is undoubtedly true for "my" puddle! Now go and answer the question posed implicitly in figure 7.1c!

We stay with shallow water waves, for which, as you must know by now, $kh \ll 1$ (the waves are long as compared with depth—they now "feel" the bottom), we know that when this condition is satisfied, $\tanh kh \approx kh$. Then the dispersion relation reduces to

$$\omega^2 = k^2 \left(g + \frac{\gamma}{\rho} k^2 \right) h.$$

Again, for "gravity-dominated" waves, the second term is negligible, and in this limit

$$\omega^2 = k^2 g h,$$

from which

$$\upsilon = (gh)^{1/2},$$

that is, independent of wavelength, and so nondispersive, as previously noted. This means that the result

$$\upsilon_g = \upsilon$$

is not very surprising. For surface tension as the dominant restoring force, the dispersion relation becomes

$$\omega^2 = \frac{\gamma}{\rho} k^4 h,$$

so that

$$\upsilon = \left(\frac{\gamma}{\rho} h \right)^{1/2} k \propto \lambda^{-1}$$

and

$$\upsilon_g = 2(\gamma h)^{1/2} k = 2\upsilon.$$

While these results pertain to a specific (but very common) system of waves, we can derive an important relationship between the two speeds discussed here, one which applies to all wave systems for which the wave speed is a differentiable function of wavelength. Recalling that in general $\omega = \omega(k)$, and hence that $\upsilon = \omega(k)/k$, we obtain

$$\upsilon_g = \frac{d\omega(k)}{dk} = \frac{d(\upsilon k)}{dk} = \upsilon + k \frac{d\upsilon}{dk} = \upsilon + k \frac{d\upsilon}{d\lambda} \cdot \frac{d\lambda}{dk}$$

$$= \upsilon + k \frac{d\upsilon}{d\lambda} \left(-\frac{2\pi}{k^2} \right) = \upsilon - \lambda \frac{d\upsilon}{d\lambda}.$$

This result informs us, amongst other things, that when the wave speed is an increasing (decreasing) function of wavelength, the group speed is less than (greater than) the wave speed. We now implement these ideas concerning wave and group speeds in a rather unusual context: the loves, life, and times of water striders.

Jearl Walker has written a fascinating and highly recommended account of the biomechanics and physics associated with these little creatures. Typically they live on the surface of water in a slowly moving stream and can move at speeds of up to a meter per second, especially if disturbed by intruders into their domain. They don't swim; they push against the surface of the water like a sprinter against starting blocks, which enables them to glide smoothly over the surface, generating waves as they go. They may also use waves to great advantage for detection of objects or potential mates!

When the insect "kick-starts" its motion, it sets up a wave pattern or group associated with about five observable individual wave crests separated by about a millimeter; these are short wavelength capillary waves, surface tension being the dominant restoring force. As we have seen, for such waves the group speed v_g exceeds the wave speed v, so a crest in an expanding wave group will appear at the front and grow in amplitude as the group envelope overtakes it, finally diminishing in amplitude and ultimately disappearing as it exits from the rear of the group.

We have earlier noted that the minimum speed of disturbance required to set up a set of traveling waves (as opposed to a transient group of waves initiated by the initial "kick") is 0.23 m/s. If the speed of the water glider relative to the water is less than this, no waves are generated. Traveling faster than 0.23 m/s will generate two sets of waves: the short capillary waves ahead of them and the longer gravity waves behind them, both in a downward parabola-like shape with the vertex close to the insect's head. It is very similar to the pattern produced by a fishing line in a moving stream. However, these are not necessarily deep water waves, at least as far as the gravity waves are concerned; if so, the "angle" of the gravity wave pattern will depend on the speed of the glider, being narrower if the gliding speed is higher (assuming the speed exceeds the minimum of 0.23 m/s). For gravity waves in shallow water, we will show below that the angle of the "wedge" is

$$\psi = 2 \arcsin \frac{\sqrt{gh}}{v_s},$$

where h is the depth of the water and v_s is the speed of the strider. If the water is deep in the sense already described, then there is no such easy or explicit formula as this for capillary waves, but the angle between the asymptotes of the parabolic crest shapes has the same qualitative behavior: the faster the strider, the narrower the angle.

Another interesting feature of water striders, or any insects on the surface of water, is the shadows they cast. This is unrelated to waves, of course, having to do with optics. The shadows can be quite fascinating; they are often cartoon-like caricatures reminiscent of Disney or Far Side characters—blobs representing eyes, torso, hands, and feet joined by thin neck, arms, and legs! A stationary strider rests its weight on all six legs, each one creating a shallow depression on the water surface. If the water surface were completely flat, the shadow of the insect cast by the sun would be a pretty accurate representation of the size and shape of the strider (unless the sun is low, when some elongation would occur). However, the effect of the depressions in the surface is to refract the sunlight out and away from beneath the insect, thus increasing the size of the shadow areas associated with its legs on the surface. The body, being raised above a flat region of the surface, casts a normal shadow, so the overall effect is an apparent enlargement of the extremeties!

Hilaire Belloc wrote a poem in *Cautionary Verses* about water beetles, and it is close enough in spirit to be "valid" for striders also; it is also quoted by Hildebrandt and Tromba:

> The WATER BEETLE here shall teach
> A sermon far beyond your reach:
> He flabbergasts the Human Race
> By gliding on the water's face
> With ease, celerity and grace;
> But if he ever stopped to think
> Of how he did it, he would sink.

We now leave these little creatures to their own devices, and start to think about the bigger picture, so to speak. Ocean waves are frequently classified according to their periods, and their energy is distributed accordingly. The following list is provided by Bascom. Those waves with the shortest periods, less than one second, are ripples; "wind chop" has periods from one to four seconds; fully developed seas, five to twelve seconds; swell, six to twenty-two seconds; surf beat, about one to three minutes; tsunamis, ten to twenty minutes; and tides with periods of twelve or twenty-four hours. All except the ripples are gravity waves.

Waves in deep water have a fairly uniform, undulating shape. Certain features of these waves are well established, at least observationally (but the theory is very complex and we will not pursue it here). If a wave is too steep, it breaks. Wave steepness is defined as the ratio of the height of the wave (from crest to trough) to its length: when this ratio exceeds about 1:7 the wave breaks, forming whitecaps. Another criterion used to study breaking waves is the angle at a wave crest: if this is rather less than about 120° then the wave breaks. This is qualitatively understandable if we imagine a wave moving up an increasingly shallow beach region; as the depth decreases the wave height increases, which in turn effectively shortens its length while steepening its sides. By now the crest is moving faster than the water below, and it falls forward and breaks, dissipating most of its energy at the shoreline. Many fascinating patterns can be observed at the shoreline as a result of this interaction between the waves and the shore.

In contrast to deep water waves, ocean waves moving into shallow water display a variety of shapes. Jearl Walker mentions four basic types: spilling, plunging, collapsing, and surging. Without defining them precisely, it is sufficient for our purposes to recognize that they are consecutive phases in the life of a wave as it approaches the shore and is ultimately dissipated. A detailed account of many patterns associated with these phases can be found in Walker's article and also the book by Bascom. Sand waves are particularly interesting; they can be found on the damp area of a beach left by the receding tide and also parts still being washed by waves. They can have wavelengths of several centimeters, and involve sand deposition by the surf as it interacts with the shore. Their shape depends on the speed of the surf; if they are large enough they can interfere significantly with the water flow and modify it for the next upsurge (a type of feedback mechanism). Obviously sand erosion and deposition in sufficiently large quantities are of interest to engineers concerned with the changing topography of beaches, especially in heavily populated regions.

Along many shorelines, strange *cusps* (the Latin word for spear points) may be seen, set out with remarkable periodicity, particularly at high tide. They occur when the wind blows at an angle to the beach, and are symmetrically placed, crescent-shaped depressions concave toward the sea. The beach is lowest at a cusp and highest in between. Each cell produces a circulating current pattern. Along the beach, the cusps are synchronized by an *edge wave* (see below) running along the beach in the breaker zone. This wave dips at each cusp and rises in between. Their diameters can range from less than a meter along the shore of a lake to 100 km or more for major shoreline features along an ocean. The latter are sometimes called giant cusps (not surprisingly) or shoreline rhythms. According to Bascom, their origin is still something of a puzzle, though some information on their generation and formation is available. The pattern appears to be maintained by the interaction of two waves that arrive consecutively on the shore, with the source of periodicity being the above-mentioned edge waves, created along the shoreline by waves coming in from deep water. These deep water waves are gradually refracted by the shallowing shoreline, and as they are subsequently reflected, part of their energy becomes trapped close to the shore and moves along the beach.

What are edge waves, then? Interestingly, like many other fluid dynamical phenomena, edge waves can be studied in the privacy of one's own home;

indeed, the famous experimental physicist Michael Faraday did just this in 1831. A detailed account of his experiments and observations can be found in the article by Walker. The "home-grown" edge waves are manifested as spoke-like patterns arising when the surface of a liquid responds to external vibrations. A simple way to do this is with a wine glass; we are all familiar with the fact that such a glass may be made to "sing" by stroking the rim with a wet finger (many bored spouses at dinner parties have been known to perform this experiment quite absentmindedly, at least until a well-aimed kick under the table jolts them from their reverie).

The physics behind the singing phenomenon is that the moistened finger tends to slip and stick periodically on the rim, and this produces vibrations in the glass, wherein opposite sides (say at 12 o'clock and 6 o'clock) expand and contract radially, while the sides at 3 o'clock and 9 o'clock contract and expand out of phase, respectively. This occurs when a wave is produced that has the same natural frequency of oscillation as the wineglass; this phenomenon is called *resonance*, and it is relatively easy to produce in an empty wineglass. If the glass is nearly full, however, it is sometimes possible to generate and see edge waves, because the upper part of the glass vibrates more strongly than the base, naturally enough (however, a glass with wine in it, being more massive, has a lower resonant frequency than if the glass is empty). The pattern of edge waves is strong-weak-strong and so on around the glass, moving along with the finger. Apparently, this is an example of an instability in the surface of the wine (it works for water also!) induced by the oscillations of the glass. Normally this instability is "relieved," so to speak, by the generation and propagation of capillary waves, but the creation of a stationary pattern of edge waves is another way of doing this. This is an example of parametric resonance, and it generates edge waves with a frequency exactly half that of the driving oscillation. For more information about parametric resonance and its characteristics, the papers by Walker and Lanzara et al. should be consulted.

SHIP WAVES AND WAKES

For those who live near ponds, rivers, or lakes these are perhaps the most commonly visible and yet most overlooked form of wave patterns. The name is a little deceptive, because ducks and swans (in particular) do just as good a job in producing them, indeed perhaps *better*, because the waves and wakes they produce are small enough in scope that they can be appreciated in their totality by the observer. However, we shall continue to refer to the moving source of waves as a ship, even though we know that ducks and swans qualify! The analysis that follows is adapted from the treatment by Tricker in his classic 1964 book on the subject of waves and wakes. In

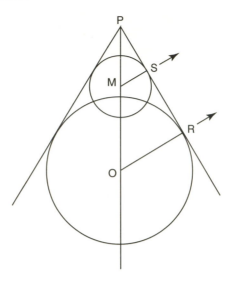

Figure 7.3. A ship at P generates waves that expand in a circular fashion; two are shown. Redrawn from *Bores, Breakers, Waves and Wakes*, by R.A.R. Tricker (1964), M. & B. Elsevier: New York.

figure 7.3, suppose that a ship (considered here as a point P—a ship with very few passengers!) travels with uniform speed V from O to P. Let this be our basic unit of time (we measure time in multiples of the time to travel distance OP). Then the distance OP in the figure is the speed of the ship in these units.

Suppose that the ship, while at O, generated a wave of such a wavelength that its (wave) speed is represented by OR. Then by the time the ship has reached P the wave has spread out into a circle of radius OR. Similarly, waves generated elsewhere, such as at M, would have spread out into circles, the radii of which are proportional to the time it takes for the ship to move from M to P. Clearly, they will all define and reinforce the common tangent line PR (see figure 7.4). Obviously, a ship may be expected to generate a wide range of wavelengths (and hence speeds) as it moves through the water, just as a stone does when thrown into water. However, the wake of a ship is formed by the waves that are able to keep up with the ship, so those waves will correspond to a particular wavelength out of the spectrum of possible waves generated by the ship's passage through the water. It is the waves that are of exactly the right speed and in the right position to reinforce each other that will contribute to the visible wake. Since PR is perpendicular to OR, PRO must lie in a semicircle with PO as diameter. PR is the wavefront of the waves generated as the ship passes along OP, and these keep pace with it as they travel in the direction OR. The energy

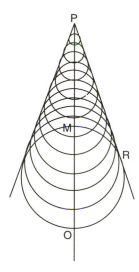

Figure 7.4. Many such waves (as in figure 7.3) reinforce a common tangent line *PR*. Redrawn from *Bores, Breakers, Waves and Wakes*, by R.A.R. Tricker (1964), M. & B. Elsevier: New York.

these contain, however, will travel at the group velocity, which for surface gravity waves is half the wave velocity, and will therefore be found along the line *PS* (where $OS = SR$; see figure 7.5).

Exactly the same arguments apply to waves of appropriate speed in other directions, *OR'*, which have their energy at the halfway point *S'* (see figure 7.6). As we consider more such directions, it becomes apparent that the energy of such waves lies on the smaller circle with *MO* as diameter (this follows from the fact that the equation $r = a\cos\theta$ defines a circle of diameter a in polar coordinates; and $OS' = OM\cos\theta$). The energy of all the waves in the wake, then, will be concentrated along the envelope of these small circles, drawn for all points along the path of the ship. *It is these common tangent lines that are noticed when we see the oblique part of the wake.*

From figure 7.7, note that $MO = \frac{1}{2}OP$. Thus if $CM = r = CT$, $CP = 3r$, and so the semi-apex angle of the wake is

$$\theta = \arcsin\left(\frac{1}{3}\right),$$

that is, $\theta \approx 19.47°$ or $19° 28'$, so the angle between the two "arms" of the wake is 2θ or approximately $39°$. Provided the water is deep, that is, the ratio of wavelength to depth is small, this angle is independent of the speed of the ship. An alternative proof of this result is provided at the end of this section. The wavefronts of the waves in the side arms of the wake

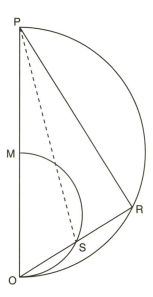

Figure 7.5. A geometric representation of a wavefront traveling in the direction *OR*, and the corresponding location of its wave energy. Redrawn from *Bores, Breakers, Waves and Wakes*, by R.A.R. Tricker (1964), M. & B. Elsevier: New York.

are not parallel to the arms themselves, that is, they are not parallel to *PT* in figure 7.8. The waves at *T* were generated when the ship was at *O*, and so the wavefronts will be perpendicular to *OT* and thus parallel to *MT*. Had the waves not fallen back because of the effect of the group velocity, they would have reached the position *PR*. Suppose that they cross the line of the wake at an angle ϕ; the geometry on this accompanying figure shows that $\phi \approx 35° \, 16'$.

The remainder of the wake is composed of waves that move behind and in the same direction as the ship; their wavelengths are such that they travel at the speed of the ship. Another geometrical argument can be used to determine this wavelength in terms of the wavelength of the waves in the lateral arms of the wakes. In figure 7.9 the velocity of the ship (and the waves astern) is denoted by V_s and velocity of the side arm waves is denoted by V_w. Clearly,

$$V_w = V_s \cos \phi,$$

and in terms of the corresponding (gravity-dominated) wavelengths, in deep water,

$$V_w = \sqrt{\frac{g \lambda_w}{2\pi}}$$

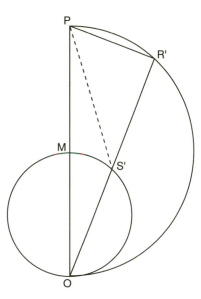

Figure 7.6. A representation similar to figure 7.5 for a wavefront traveling in the direction *OR'*. Redrawn from *Bores, Breakers, Waves and Wakes*, by R.A.R. Tricker (1964), M. & B. Elsevier: New York.

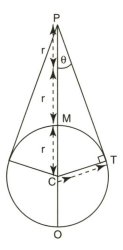

Figure 7.7. Geometry for the angle between the "arms" of the ship wake. Redrawn from *Bores, Breakers, Waves and Wakes*, by R.A.R. Tricker (1964), M. & B. Elsevier: New York.

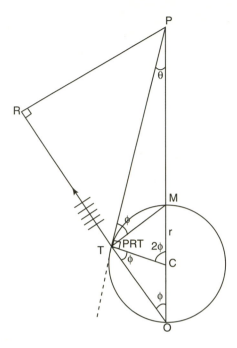

Figure 7.8. Geometry for the "side arms" of the ship wake. Redrawn from *Bores, Breakers, Waves and Wakes*, by R.A.R. Tricker (1964), M. & B. Elsevier: New York.

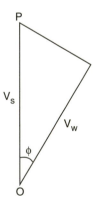

Figure 7.9. The relationship between the ship velocity, the side wave velocity, and the angle between their respective directions. Redrawn from *Bores, Breakers, Waves and Wakes*, by R.A.R. Tricker (1964), M. & B. Elsevier: New York.

and

$$V_s = \sqrt{\frac{g\lambda_s}{2\pi}},$$

so that

$$\frac{\lambda_w}{\lambda_s} = \frac{V_w^2}{V_s^2} = \cos^2\phi = \frac{1 + \cos 2\phi}{2} = \frac{1 + \sin\theta}{2} = \frac{2}{3}.$$

This ratio may be determined directly from photographs of ship wakes (see figure 7.10).

Note that this theory is approximate in the sense that some things have been neglected. The length, width, and shape of the hull have been ignored, along with the waves generated by various other parts of the hull. Certainly a similar pattern of waves will be generated by the stern, and these two wave trains will interfere. . . .

A further point of interest is that the generation of ship waves requires energy, and this limits the maximum speed that the ship can attain (over and above the loss of energy due to friction with the water). If two geometrically similar ships have dimensions L_1 and L_2, respectively, and if traveling at

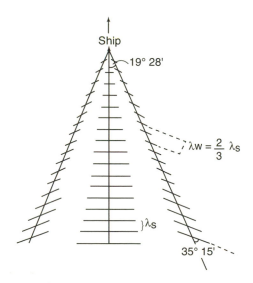

Figure 7.10. Summary of the geometric relationships between fundamental angles and wavelengths for the ship wave problem. Redrawn from *Bores, Breakers, Waves and Wakes*, by R.A.R. Tricker (1964), M. & B. Elsevier: New York.

speeds V_1 and V_2 they generate waves of wavelength λ_1 and λ_2, respectively, then on deep water:

$$\frac{V_1^2}{V_2^2} = \frac{g\lambda_1}{g\lambda_2}.$$

If, as is reasonable to expect, the wavelengths are proportional to the linear dimensions of the ships, then

$$\frac{V_1^2}{L_1} = \frac{V_2^2}{L_2},$$

which is called *Froude's law* of corresponding speeds.

SHIP WAVES IN SHALLOW WATER

All waves travel with the same speed $\sqrt{gD}$ in shallow water of depth D. The group velocity is (therefore) the same as the wave velocity. Waves generated as the the vessel moves through various points of its path will spread out with this speed regardless of wavelength, and will therefore all lie on circles having a common tangent through P, the present position of the ship. Clearly, from figure 7.11,

$$\sin\theta = \frac{\sqrt{gD}}{V_s},$$

so that as the speed increases, the wake becomes narrower. In contrast to the deep water case, the wavefronts of the wake lie along PT so the angle ϕ is now zero. Obviously for this theory to make sense, the ship speed must exceed $\sqrt{gD}$ ($\sin\theta < 1$ for a wake to occur), so there is no wave capable of "keeping up" with the ship. *But what happens if $V_s < \sqrt{gD}$?* Speed-boats moving in shallow harbors or estuaries are affected by this "critical speed." The boat starts off with no problem, but when this critical speed is neared, the waves created by the vessel build up into a large wave just ahead of it, so in a sense it is going uphill all the time and doing so with difficulty! It is like trying to break the equivalent of the sound barrier! (It can be done if the boat is capable of a sudden burst of power that carries it quickly beyond the critical speed before the bow wave has had time to form.)

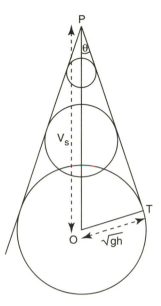

Figure 7.11. Geometry for ship waves in shallow water. Redrawn from *Bores, Breakers, Waves and Wakes*, by R.A.R. Tricker (1964), M. & B. Elsevier: New York.

APPENDIX: MORE MATHEMATICS OF SHIP WAVES

We have seen that the wave speed V of gravity waves on deep water satisfies

$$V^2 = \frac{g}{k}.$$

It is these waves, stationary relative to the ship, that are produced as it moves across the water. They are analogous to the mountain lee waves that have been discussed in chapters 6 and 7, except that the mountain is not moving and the ship is. This does not change the mathematics one iota, so let us consider the water to move past a stationary ship and examine the geometry of figure 7.12, in which a ridge-like disturbance is inclined at an angle $90° - \phi$ to the stream moving in the x-direction, where $\phi \in (-\pi/2, \pi/2)$. The component $V \cos \phi$ across it produces waves, and for the point P with coordinates (r, θ) on the crest (say) of a wave in the total wave pattern, it follows from the diagram that

$$X = r\cos(\theta + \phi). \tag{1}$$

Also the wavenumber for a particular lee wave satisfies the relationship

$$k = \frac{g}{V^2 \cos^2 \phi}. \tag{2}$$

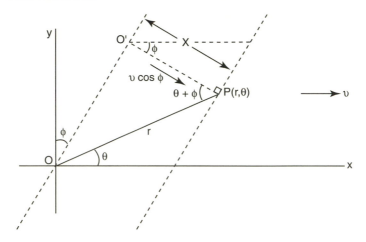

Figure 7.12. The notation and geometry used in the derivation of the wedge angle and parametric system (10) for ship waves in deep water. From *Dynamics of Meteorology and Climate*, by Richard Scorer, Praxis Publishing Ltd., copyright 1997, reprinted by permission of the publisher and Richard Scorer.

The total pattern behind the disturbance is the sum of all these lee waves, and the locus of P is the envelope of all the individual lee wave crest lines on which

$$kX = N, \tag{3}$$

where N is the phase corresponding to crests (this equation "converts" the distance X to a phase). The condition for constructive interference is that this phase is independent of the angle of inclination ϕ, or

$$\frac{\partial}{\partial \phi}(kX) = 0. \tag{4}$$

These four equations imply that

$$r \sec^2 \phi \cos(\theta + \phi) = \frac{NV^2}{g} \equiv a \tag{5}$$

and

$$2 \tan \phi = \tan(\theta + \phi), \tag{6}$$

from which equation we find that

$$\tan \theta = \frac{y}{x} = \frac{\tan \phi}{1 + 2 \tan^2 \phi}. \tag{7}$$

This means that θ is zero when $\phi = 0$ and as $\phi \to \pm \pi/2$; straightforward differentiation then shows that the maximum value of $\theta(\phi)$ occurs when $\tan^2 \phi = 1/2$, or

$$\tan \theta_{max} = 2^{-3/2} \text{ so } |\theta_{max}| = 19°28'.$$

This means that the wave pattern is confined in a wedge of angle $2\theta_{max} \approx 39°$. The parametric equations for the crest lines can be found as follows. Clearly in polar coordinates,

$$x = r \cos \theta = a \cos^2 \phi \sec(\theta + \phi) \cos \theta$$

from equation (5). After a little algebraic manipulation it may be shown that

$$x = a(2 - \cos^2 \phi) \cos \phi \tag{8}$$

and similarly that

$$y = r \sin \theta = a \cos^2 \phi \sec(\theta + \phi) \sin \theta,$$

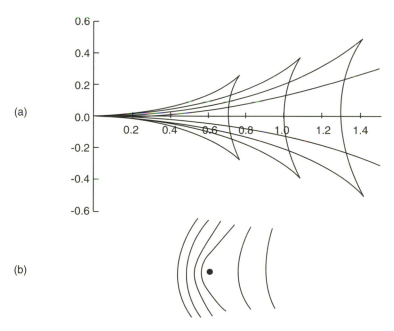

(a)

(b)

Figure 7.13. (a) the locus of ship wave crests in deep water for several values of a (0.7, 1, 1.3, 2) according to the system (10); (b) the corresponding locus for capillary waves ahead of an obstacle in a moving stream (e.g., a twig) or an object moving on the surface of water (e.g., a water strider). From *Elementary Fluid Dynamics*, by David Acheson (1990), Clarendon Press: Oxford, by permission of David Acheson.

from which we obtain

$$y = a \sin \phi \cos^2 \phi. \tag{9}$$

Thus the locus of the crest line in the total pattern in polar coordinates is

$$a((2 - \cos^2 \phi) \cos \phi, \sin \phi \cos^2 \phi), \tag{10}$$

which is illustrated in figure 7.13a for several values of a; the corresponding pattern for capillarity waves ahead of an obstacle is shown for comparison in figure 7.13b.

Plate 1. A plant exhibiting a "spherical" leaf arrangement that possesses a very large surface area to volume ratio (see chapter 3).

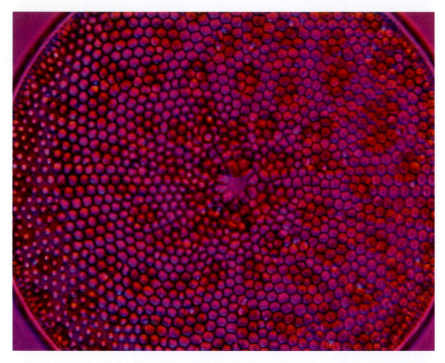

Plate 2. A diatom (*Coscinodiscus* sp.) exhibiting hexagonal spaces (aerolae) (see chapter 3). By permission of Lisa Drake, Old Dominion University.

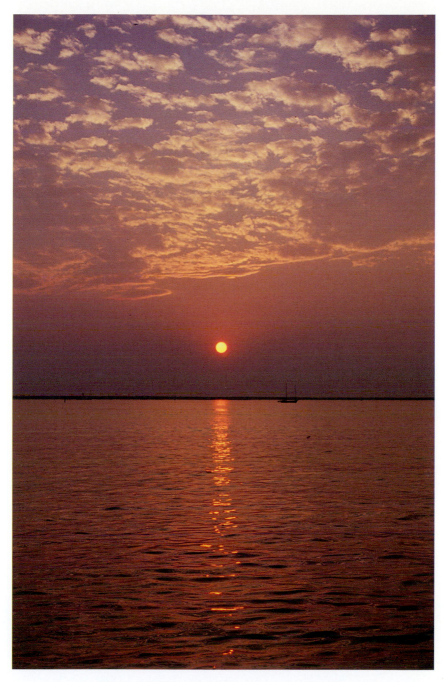

Plate 3.   Sunset and glitter path (see chapter 4).

Plate 4.   Clouds at twilight (see chapter 4).

Plate 5. The author's elongated shadow by Bras d'Or lake (Nova Scotia) at sunset (see chapter 4).

Plate 6. Primary and secondary rainbow with Alexander's dark band and supernumerary bows (Kootenay Lake, B.C.) (see chapter 5). By permission of Alistair B. Fraser, from Lee, R. L., and Fraser, A. B. (2001), *The Rainbow Bridge: Rainbows in Art, Myth and Science* (Pennsylvania State University Press: University Park).

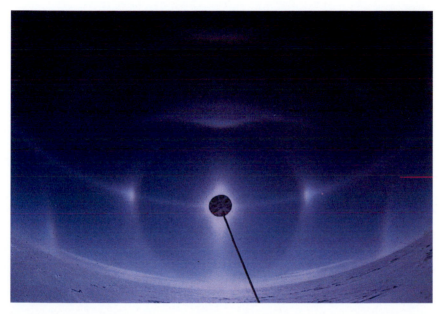

Plate 7. Halo display at the South Pole, Jan. 2, 1990 (see chapter 5). By permission of Walter Tape, from *Atmospheric Halos* (1994), American Geophysical Union: Washington D.C.

Plate 8.   A glory (see chapter 5). By permission of J. B. Kaler.

Plate 9.   Sand dunes: slowly moving desert waves (see chapter 6).

Plate 10.   Ship waves (see chapter 7).

Plate 11.   A partially breaking wave (see chapter 9).

Plate 12.   Billow clouds viewed nearly edge on (see chapters 6, 8).

Plate 13.   Scattered clouds reflecting light from the setting sun (see chapter 4)

Plate 14. A daisy exhibiting spiral seed pattern (see chapter 10).

Plate 15. A petunia exhibiting characteristic pentagonal symmetry (see chapter 10).

Plate 16. Leaf arrangement exhibiting phyllotaxis (see chapter 10).

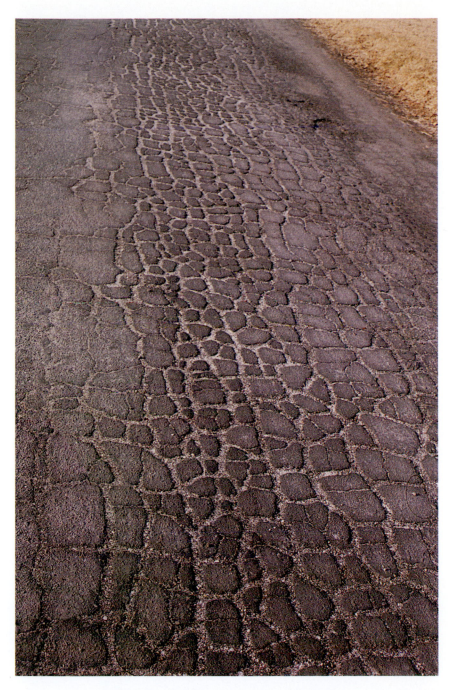

Plate 17.   Cracks in tarmac (see chapter 11).

Plate 18. Cloud streets (see chapter 6).

Plate 19.   Sunset with a long cloud pattern (see chapter 4).

Plate 20.   A Brahminy Kite (see chapter 13). Copyright Peter Basterfield, ARPS.

Plate 21. A cecropia silkmoth (*Hyalophora cecropia*) with a penchant for mathematics textbooks visiting the author's office (see chapter 14).

Plate 22. An Owl butterfly (*Caligo memnon memnon*) on a lunch break (see chapter 14).

Plate 23.   A cheetah; notice the striped tail (see chapter 14).

Plate 24. A thirsty reticulated giraffe exhibiting cell-like coat patterns, and also bilateral symmetry as a bonus (see chapter 14). Note the waves in the foreground! Copyright Don Getty.

Stability

Unstable as water, thou shall not excel. . .
—Genesis 49:4a (KJV)

"The problem with linear theory is that it is not nonlinear."

The second quotation above is from the very first professional presentation I made in fear and trepidation as a first-year graduate student. I remember the occasion as if it were yesterday; I was talking about a particular class of waves that may exist in the atmosphere of the sun and other stars (or perhaps just in my mind), and the above line brought a laugh. I needed it, and so did the audience. Now, about thirty years later, I had intended to use it in the next chapter, which introduces the topic of nonlinear waves, but since this present chapter deals, in one sense, with the *transition* to nonlinearity, I decided to retain it here.

As noted earlier, linear theory deals with waves that are, strictly speaking, infinitesimal. In reality, of course, this is impossible to achieve: all waves are nonlinear according to this definition. It is a fortunate practical reality that many of the predictions of linear theory still work for waves of small amplitude, but ultimately, if conditions are right, the system will become nonlinear. This can occur when the forces driving the wave motion are such that the amplitude of a wave increases, ultimately vitiating the assumption of small amplitude. Sometimes as this happens, mechanisms tending to oppose wave growth, such as dispersion or dissipation, may become more important and actually balance the growth tendency in such a way that the wave maintains its shape for a considerable distance. Under these circumstances it is called a *solitary wave*, about which more will be said in the next chapter. In general, the increase in wave amplitude will lead to the ubiquitous but poorly understood phenomenon of turbulence. What is of interest here is the transition to nonlinearity as predicted by linear theory. In a sense this represents the "other side of the coin" from the previous chapter, where plane wave solutions with time harmonic dependence were sought, that is, waves with time-dependence of the form $exp(-i\omega t)$, ω being a *real* number (see equation (13) in chapter 7). The fundamental idea in linear stability theory is that ω is permitted to be a complex quantity, $\omega = \omega_r + i\omega_i$, where of course the respective real and imaginary parts ω_r and ω_i are real. This is a significant difference because now the temporal dependence will be oscillatory with exponential growth (if $\omega_i > 0$ and $\omega_r \neq 0$) or oscillatory with exponential decay ($\omega_i < 0$ and $\omega_r \neq 0$), because

$$e^{-i\omega t} = e^{-i\omega_r t} e^{\omega_i t}.$$

If $\omega_r = 0$ then the wave just grows or decays exponentially with no oscillatory behavior whatsoever; sometimes the case $\omega_i > 0$ and $\omega_r \neq 0$ is referred to as *overstability*. If there is no dissipation in the system (which is unlikely in practice) then both signs for ω_i arise from the equations; if $\omega_i < 0$ for each wave then any disturbance to the equilibrium state decays away in time and so the system supporting it is said to be linearly stable. This means that the equilibrium is returned. This being so, without loss of generality we will focus on linear instability, wherein at least one wave exists for which $\omega_i > 0$. We will consider this for a variety of situations that have bearing on previously discussed topics such as clouds and gravity waves. A more general mathematical definition of stability (and by implication instability) will be provided when we address convection.

Let us examine first the following physical situation in some mathematical detail. A deep layer of inviscid fluid of density ρ_2 flows with uniform speed U over another deep layer of density $\rho_1 > \rho_2$ which is at rest. The interfacial layer has coefficient of surface tension T. Here the reason for the adjective "deep" is that then we may safely neglect the effects of any boundaries, because they are too far away to influence the physics significantly (this is a gross oversimplification, of course). As noted in chapter 7, the governing dispersion relation between angular frequency ω and wavenumber k (here $k > 0$ without loss of generality) is obtained by seeking solutions to the governing differential equation for the dependent variable $\eta(x, t)$ in the form

$$\eta(x, t) = A e^{i(kx - \omega t)}, \tag{1}$$

where A is a (possibly complex) constant amplitude. For the interfacial stability problem stated above, η (or more accurately, Re η) represents the (small) perturbation of the interface along the interface (x-direction) and in time. The gravitational acceleration is vertically downward, naturally enough. By the principle of superposition (not defined here, but don't worry about it), any reasonable linear perturbation of the equilibrium state (a flat interface) can be expressed as an integral superposition of complex exponential functions of the form above; this has the added benefit that we can, in effect replace a differential equation for the problem by an algebraic one, just as we did to obtain a dispersion relation. An algebraic equation is generally easier to solve, and for our purposes it is quite sufficient to illustrate the essential ideas at work in this system. The dispersion relation is quadratic in ω and can be written in the form

$$(\rho_1 + \rho_2)\omega^2 - 2\rho_2 Uk\omega + \rho_2 U^2 k^2 - k[k^2 T + g(\rho_1 - \rho_2)] = 0. \quad (2)$$

Note that, given the form (1) in which k is a positive real number and ω may be a complex one, the system will be stable (not growing in time) *if and only if ω is real*. Therefore we are interested in the conditions under which this will (or will not) occur, and those can be determined by examining the discriminant of the quadratic equation (2), which we temporarily write in the form

$$A\omega^2 - B\omega + C = 0$$

(the definitions of $A, B,$ and C being obvious). Clearly a necessary and sufficient condition for stability is that "$B^2 \geq 4AC$":

$$(\rho_1 + \rho_2)[k^2 T + g(\rho_1 - \rho_2)] \geq \rho_1 \rho_2 U^2 k \quad (3)$$

after a little algebra. Values of wavenumbers k for which (3) is satisfied correspond to wavelengths for which the system is stable. However, generally disturbances consist of a *range* (usually infinite, in principle) of wavenumbers, so that to be stable *for all values* of $k > 0$ it is necessary that

$$\rho_1 \rho_2 U^2 \leq \min_k\{(\rho_1 + \rho_2)[kT + gk^{-1}(\rho_1 - \rho_2)]\}. \quad (4)$$

This minimum can be determined by the standard techniques of calculus, but it is much less cumbersome (and therefore more elegant) to employ the arithmetic-geometric mean inequality in the form

$$a + b \geq 2\sqrt{ab}, \quad a \geq 0, b \geq 0,$$

which as noted above is easily established. Thus

$$kT + gk^{-1}(\rho_1 - \rho_2) \geq 2\sqrt{(\rho_1 - \rho_2)gT},$$

so that from (4) we find

$$U^2 \leq 2(\rho_1^{-1} + \rho_2^{-1})\sqrt{(\rho_1 - \rho_2)gT} \quad (5)$$

for stability to all disturbances for which $k > 0$. (Incidentally, these results are similar for $k < 0$, as may be seen by changing the sign of x in (1) and taking account of any corresponding sign changes in spatial derivatives; negative values of k just means that the disturbance propagates in the opposite direction.) Note some consequences of all this algebra. From equation (5) it is apparent that if either T or g is zero, instability must occur, no matter how small the relative speed U; the presence of both surface tension and gravity is needed to prevent the instability. However, note from inequality

(3) that either T or g (but not both) being zero will guarantee stability for a given speed U for *some range of k* (and hence λ). This is illustrated in figure 8.1, where the expression "$B^2 - 4AC$" is seen to become negative for a range of k. For sufficiently short waves (large enough k values) or sufficiently long waves (small enough k values) the system is stable, but the system is generally unstable for an intermediate range of k.

Some interesting results can also be obtained if we systematically "turn off" various parameters in the problem as originally stated. Thus, if we set each of ρ_2, U, and T equal to zero in equation (2) we obtain the very simple dispersion relation

$$\omega^2 = gk \qquad (\text{or in general } \omega^2 = g|k|), \qquad (6)$$

which is exactly that for surface gravity waves! Since $k > 0$ the system is stable, and we can justifiably regard such waves as special cases of hydrodynamic stability. If only U and T are set equal to zero, then a slight generalization of equation (6) results, namely

$$\omega^2 = gk\left(\frac{\rho_1 - \rho_2}{\rho_1 + \rho_2}\right). \qquad (7)$$

In this case the result represents the dispersion relation for a special case of *internal gravity waves* propagating along the interface between two adjacent layers of fluid (the less dense fluid (ρ_2) overlying the denser one (ρ_1)). These waves can propagate in the interior of a stratified fluid. They can be observed, for example, in estuaries between layers of fresh and salt water; as pointed out in the book by Drazin and Reid, the fresh water upper surface may be very smooth while strong internal gravity waves occur at the salt/fresh water interface about a meter or two below. In this situation, the factor $(\rho_1 - \rho_2)/(\rho_1 + \rho_2) \approx 10^{-3}$, so for the same value of k (or wavelength) in (6) and (7), the internal gravity wave frequency ω is lower than the surface gravity wave frequency by about a factor of 30. We shall have more to say about this below.

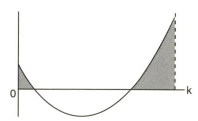

Figure 8.1. A schematic representation of the inequality (3) governing the linear Kelvin-Helmholtz stability of waves under the influence of gravity and surface tension. For wavenumbers *below* the curve (shaded regions) the system is stable; for the intermediate wavenumbers lying above the curve, linear instability occurs.

Notice, furthermore, that if the denser fluid is placed above the less dense one ($\rho_1 < \rho_2$ now) the system is unstable for all values of k; this is called the Rayleigh-Taylor instability and is readily noticed when you have a glass of water in your watch-hand and someone asks you for the time! Now we examine the case of surface capillarity-gravity waves by setting $\rho_2 = 0$ (and letting $\rho_1 = \rho$) along with $U = 0$. The resulting dispersion relation is

$$\omega^2 = gk + T\frac{k^3}{\rho}, \tag{8}$$

which is a generalization of the result (6). The final special case that we consider here occurs when $\rho_1 = \rho_2 = \rho$ and $T = 0$ in equation (2), resulting in a special case of the Kelvin-Helmholtz instability (due here to shear in the form of a vortex sheet, a discontinuity in the speed U; this is of course present in the problem as originally posed). Under these circumstances there are two complex conjugate roots of the dispersion relation,

$$\omega = \frac{1}{2}Uk(1 \pm i),$$

the positive root guaranteeing instability for all values of k. This instability comes about when, simply put, the destabilizing effects of changes of wind speed with altitude are sufficient to overcome the stabilizing effects of buoyancy when the denser fluid is at the bottom.

The study has an interesting history; in 1871 Lord Kelvin used this type of analysis to model wind-generated ocean waves. Somewhat later, in 1890, Helmholtz applied the theory to billow clouds, which are atmospheric indicators of the presence of strongly sheared winds. A magnificent photograph of such clouds, viewed approximately perpendicular to the direction of the "wave" propagation, was made by Paul Branstine near Denver, Colorado (see the note by Colson). Details of the prevailing atmospheric conditions are provided in the article; apparently a pronounced jet stream was located over the area at the time and with the same orientation. The paucity of information notwithstanding, a worker at the Denver branch of the U.S. Weather Bureau is quoted as saying that "it appears that the combination of rather marked temperature advection, sufficient moisture content for condensation, and a rather sharp wind shear was responsible for this unusual set of breaking atmospheric waves." A simple linear approximation to the data supplied in the article for that day confirms this; the wind shear is approximately 200 knots in 30,000 feet and the temperature dropped about 60°C over the same range of altitude.

Let us insert some numbers into inequality (3) (with $T = 0$) and try to "guesstimate" some typical critical wavelengths for instability to occur. It must be emphasized that these "back of the envelope" calculations are just that; I have not consulted any meteorological data, but rely upon reasonable estimates for typical wind speeds and so forth. From (3) we find that for instability to occur (a precursor to billow waves), the wavelength λ must satisfy

$$\lambda < \lambda_c = \frac{2\pi \rho_1 \rho_2 U^2}{g(\rho_1^2 - \rho_2^2)} = \frac{2\pi (\rho_1/\rho_2) U^2}{g(\rho_1^2/\rho_2^2 - 1)}. \tag{9}$$

This latter form has the advantage that we only require the relative density with respect to the upper layer, and we can try to pick reasonable values for this. Now $g = 9.8 \approx 10 \text{ ms}^{-2}$, and mixing our units, a 23 mph wind is about 10 ms^{-1}, so all that remains is to estimate the ratio ρ_1/ρ_2; the density differences will not be large (but they may well be larger than the differences in water, in relative terms) so let us be conservative and use $\rho_1/\rho_2 = 1.01$. Then for these numbers $\lambda < \lambda_c \approx 3000 \text{ m} = 3 \text{ km}$. According to these figures, any disturbance with wavelength less than about 3 km will be unstable. Note that this critical wavelength λ_c is proportional to the square of the wind speed, so that any increase in U will induce a significant increase in λ_c: a doubling of U to 46 mph increases λ_c by a factor of four to approximately 12 km.

But since the density ratio is the weakest point in this number game, how does λ_c vary with $x = \rho_1/\rho_2$? Since we require physically (for stable stratification) that $x > 1$, the domain of the function

$$f(x) = \frac{x}{x^2 - 1}$$

is $(1, \infty)$ and the range is $(0, \infty)$. It decreases monotonically over the domain, but in practice, barring bizarre atmospheric conditions, x is unlikely ever to exceed 1.2, so the effective range may well be from about $f(1.01) \approx 50$ to $f(1.2) \approx 3$. This of course can make a significant difference to the value of λ_c; uncertainties in this ratio will render such estimates crude at best. However, the point of these calculations is to show that they are relatively easy to perform, and, given sufficiently accurate data, can be used to determine a reasonable upper limit to the scale of unstable atmospheric disturbances. Observationally it is known that, while shear-induced billow clouds vary considerably in wavelength, they are typically less than 1 km. This is well below the upper bounds determined above! Bizarre atmospheric conditions might occur for snow mixed in with the air, by the way; the resultant density can increase while the mixture still behaves as a fluid.

INTERNAL GRAVITY WAVES AND WAVE ENERGY

Let us return to the simple situation of a layer of water of density ρ_2 overlying a denser layer with density ρ_1, and this time out to sea but near the shore. If the wind blows, then near the shore the upper layer can pile up relative to the lower layer (similar to what occurs in those executive toys filled with two different oils, when tipped one way or the other, but in that case the denser fluid piles up at the low end since gravity is doing the pulling). In other words, it pushes down the denser layer near the shore, but not farther out to sea; furthermore, we will consider a steady state to have been achieved after the wind has been blowing uniformly for some time. Consider two points on the same horizontal level, one out to sea (P_s) and one in the shallows actually on the land (P_ℓ) see figure 8.2. If the circulation under the surface is slow, the situation is almost hydrostatic and the pressures at the two points are approximately equal. Suppose that the wind had raised the surface of the water near the shore by d m relative to the level out to sea; by how much has the level of denser water been depressed? We call this amount H and follow the approach due to Barber.

Suppose that above the point P_s the thickness of less dense water is h_1 m and the thickness of denser water is h_2 m. The pressure (relative to atmospheric) at this point is therefore

$$p_s = g(\rho_2 h_1 + \rho_1 h_2).$$

At the point P_l the pressure is

$$p_l = g\big(\rho_2[h_1 + d + H] + \rho_1[h_2 - H]\big)$$

because the water of lower density now has a depth $h_1 + d + H$ above P_l and the depth of the denser water above the same point is reduced by H. Equating these two pressures in accordance with our assumptions we find that

$$\rho_2(d + H) = \rho_1 H$$

or

$$H = \frac{\rho_2 d}{\rho_1 - \rho_2}.$$

Figure 8.2. The initiation of internal waves. A layer of water lies above a slightly denser layer (dotted region), and the wind pushes the surface layers inland toward the beach (shaded region), displacing the denser region as shown (points P_s and P_l are on the same horizontal level). Subsequent oscillations occur; these are internal waves.

The lower layer of water may be colder or saltier (or both) than the upper layer, but the density difference will in general be small, perhaps one part in a thousand. Suppose (not unrealistically) that the densities are $\rho_2 = 1.030 \times 10^3$ kg m^{-3} and $\rho_1 = 1.031 \times 10^3$ kg m^{-3}. Then $H = 1030d$ m, and for $d = 0.03$ m (3 cm) and 0.05 m (5 cm), respectively, $H \approx 31$ m and 52 m. In retrospect, this is eminently reasonable: since the density difference is one part in a thousand, the denser water would need to be lower by about $1000d$ m to maintain the same pressure at P_l. Perhaps it is reasonable to state that, depending on circumstances, the density difference may vary between one part in ten thousand and one part in one hundred, with corresponding wave amplitudes H.

When the wind stops blowing, the levels tend to revert to their previous positions and the denser water slides back up, but it may overshoot its equilibrium position and slowly oscillate about it for some time. This is an internal wave in action. Another cause of these waves is relative motion between two streams of fluid, just as in the case of billow clouds in the atmosphere—both being examples of the Kelvin-Helmholtz instability. In the water, the amplitudes will be comparable to H, which as we have seen may be at least 30 m. Because the oscillations are so slow (see below) there is little energy associated with them; submarines sometimes "ride" on them inadvertently. As we have seen, for wavenumber k such waves have phase speeds given by

$$c = \frac{\omega}{k} = \sqrt{\frac{g\lambda}{2\pi}\left(\frac{\rho_1 - \rho_2}{\rho_1 + \rho_2}\right)} \approx \varepsilon\sqrt{\frac{g\lambda}{2\pi}},$$

where

$$\varepsilon = \sqrt{\left(\frac{\rho_1 - \rho_2}{\rho_1 + \rho_2}\right)}, \quad 7 \times 10^{-3} \lesssim \varepsilon \lesssim 7 \times 10^{-2}$$

for a range of density differences from 10^{-4} to 10^{-2}. Thus these internal gravity waves travel much more slowly than their cousins on the surface for which $\varepsilon \equiv 1$. Their periods T are increased by the factor ε^{-1}, and from the equation for c are given by

$$T = \varepsilon^{-1}\sqrt{\frac{2\pi\lambda}{g}} \approx 0.8\varepsilon^{-1}\lambda^{1/2} \text{ s}$$

if λ is measured in meters. Thus if $\lambda = 1$ km, for example, then for a density difference of 10^{-3}, $\varepsilon \approx 0.02$ and

$$T \approx 25\varepsilon^{-1} \text{ s} \approx \frac{1}{3} \text{ hr.}$$

Speaking of wave energy, we can perform some calculations on this very topic. How much energy do they have? We will focus first on the deep water linear surface gravity waves to simplify the mathematics, and then modify the result for internal gravity waves. There is a quick, dirty, and valuable physical way to obtain this result, and also a longer but more mathematically satisfying procedure; it would be nice to adopt a bipartisan approach and do both, but in the interests of space we will retain the more physical approach.

Imagine a thin column of water in the wave a height z above the mean level (were there no wave), with width δx and length y parallel to the crest lines (out of the page as you read this! Don't get wet). It has gained potential energy by an amount equal to mass × gravity × height, or $\rho z y \delta x \times g \times z$; to represent the whole crest it must be integrated over the whole length $\lambda/2$. Thus if the wave profile is sinusoidal (as linear theory demands), then $z = A \sin kx$, A being the wave amplitude, which we will take as real for now, then the potential energy in one wavelength is

$$\int_0^{\lambda/2} \rho g y A^2 \sin^2 kx \, dx = \frac{1}{2}\rho g y A^2 \int_0^{\lambda/2} (1 - \cos 2kx)dx = \frac{1}{4}\rho g y A^2 \lambda.$$

But wait a moment—what about the rest of the wavelength? Haven't we forgotten the trough, and since that is *below* the mean surface doesn't the total potential energy cancel out to zero? That is a question worth thinking about. Since in linear theory the troughs and crests are symmetrical, we can think of each wavelength as having been formed by scooping out the water from the flat surface (making a trough) and piling it up to make a crest, so it is not necessary to do any more calculations for the potential energy of the wave. This isn't quite what happens of course, but I did say this was quick and dirty.

So now all that remains is for the kinetic energy of the wave to be calculated. That's easy; just double the above result! But wait another moment—what is going on now? Nothing more than a little mathematical sleight of hand, perfectly acceptable, but it needs a little more clarification. It is well known that two equal linear wavetrains are equivalent to a standing wave, because, for example,

$$A \cos(kx - \omega t) + A \cos(kx + \omega t) = 2A \cos kx \cos \omega t,$$

so that at each location x the surface rises and falls in time. Less familiar is the fact that two suitably chosen standing waves are equivalent to a propagating (or progressive) wave. This follows, for example, with two waves of equal amplitude one-fourth of a wavelength out of phase and one-fourth of a period out of phase, because

$$A \cos kx \cos \omega t + A \sin kx \sin \omega t = A \cos(kx - \omega t).$$

When one of these is at maximum amplitude, the other is momentarily "flat," with zero net potential energy and all its energy in the form of particle motion. But one-fourth of a period later, these profiles are interchanged, so there must be "equipartition" between the averaged potential and kinetic energies. Hence the total energy is $\frac{1}{2}\rho g y A^2 \lambda$; per unit surface area (dividing by $y\lambda$) it is

$$E = \frac{1}{2}\rho g A^2,$$

which is independent of the wavelength or frequency of the wave. Nowhere has the depth entered in, other than through the assumption of "deep" water. This result is standard in one sense for all waves: energy is proportional to the square of the amplitude. Continuing the calculation, for an amplitude A of 1 meter, and density of water of 1000 $\mathrm{kg\,m}^{-3}$, the energy per square meter for such a wave is

$$E = \frac{1}{2} \times 10^3 \times 9.8 \times 1 = 4900 \text{ joules m}^{-2}$$

or about one-tenth of the energy of a steel breaking-ball when it smashes into a building. This energy is delivered at the group speed, which is half the wave speed, $c/2$. For a wavelength of, say, 100 m, the wave period T is 8 s and the group speed is about 6 m/s, so the energy per meter length of wave in the y-direction is about 30 kW (kilowatts) per meter. If we hazard a guess of about 2000 km for the coastline of Britain, this corresponds to an energy of about 6×10^4 mW (megawatts), just for this rather low frequency wave (which of course could not be expected to be the same all around the island, so the actual energy flow is probably much higher than this).

KELVIN-HELMHOLTZ BILLOW CLOUDS

Most billow-type clouds and clear-air turbulence are believed to be the result of small-scale wave-like undulations that develop spontaneously when the vertical shear of a horizontal wind exceeds some critical value. The shear is a measure of how much the horizontal wind speed vector $U = (U(z), 0, 0)$ varies as a function of altitude (z), and this is usually defined by the derivative $U'(z)$, or its magnitude $|U'(z)|$. The process responsible for the development of these waves is called *shear* (or *Kelvin-Helmholtz*) *instability*. While growing, the waves must do work in exchanging denser fluid from below with lighter fluid (they smoke?) from above. Generally, the more stable the stratification, the more work that must be done by the waves against the force of gravity. If wave amplification is to occur, the waves must extract kinetic energy from the undisturbed shear flow at a rate *faster* than they lose energy by doing work against gravity. This means that the critical shear

$U'(z)$ required for wave amplification depends on the degree of density stratification; the greater the stratification, the larger the critical shear.

Both physical and mathematical arguments can be made to quantify this condition. For the former, suppose that two equal neighboring volumes of fluid, at heights z and $z + \delta z$, are interchanged. The work that must be done against gravity in order to effect this interchange is (per unit volume)

$$\delta W = -g\delta\rho\delta z,$$

where $\delta\rho$ is the (negative) density difference between the two heights (in a stably stratified fluid, where the density decreases upward). The kinetic energy available to do this work, again, per unit volume, is the difference in kinetic energy between the flow at the two heights and the mean flow (see the book by Chandrasekhar):

$$\frac{1}{2}\rho[U^2 + (U + \delta U)^2 - 2\{(U + U + \delta U)/2\}^2] = \frac{1}{4}\rho(\delta U)^2.$$

A *sufficient condition for stability* is therefore

$$\frac{1}{4}\rho(\delta U)^2 < -g\delta\rho\delta z.$$

When the appropriate limiting procedures are carried out (U and ρ, it is assumed, being differentiable functions of z), an equivalent form of this condition is

$$\left(\frac{dU}{dz}\right)^2 < -4\frac{g}{\rho}\frac{d\rho}{dz}.$$

In hydrodynamics, the *Richardson number J* is defined as

$$J = -\frac{g}{\rho}\frac{d\rho/dz}{(dU/dz)^2},$$

and as can be seen, this number measures the ratio of the stabilizing effect of density stratification to the destabilizing effect of shear in a fluid. Thus a sufficient condition for stability (and hence no billows formed by shear instability) is that

$$J \geq \frac{1}{4}$$

everywhere in the fluid medium. Alternatively, a *necessary condition for instability* (and hence billow formation) is that $J < 1/4$ at least *somewhere* in the fluid. (Note that this last condition does *not* mean that instability will occur if $J < 1/4$, merely that if instability does occur, then J will be $< 1/4$ at least somewhere in the domain.)

A mathematical demonstration of this "Richardson criterion" is rather involved, but the flavor of it is contained in the following much-abbreviated argument: "It can be shown that" *if instability occurs*, for $0 \leq z \leq d$, then

$$\int_0^d \rho\{|G'|^2 + k^2|G|^2\}\,dz + \int_0^d \rho(-g\rho'/\rho - [U']^2/4)\left|\frac{G}{W}\right|^2 dz = 0,$$

where G and W are quantities related to the wave motion and k is the wavenumber of the motion. This expression is equivalent to

$$\int_0^d \rho\{|G'|^2 + k^2|G|^2\}\,dz + \int_0^d \rho[U']^2\left(J - \frac{1}{4}\right)\left|\frac{G}{W}\right|^2 dz = 0,$$

which is clearly impossible if $J \geq 1/4$ everywhere, since for nontrivial motions all terms are positive on the left-hand side of the equation.

There are many other types of clouds to be seen. Shear-induced billows tend to be rather short-lived in general, but cloud patterns induced by *convective instability* are often seen for considerable periods of time. Again, the analysis presented here is merely meant to suggest the kind of length scales that may be appropriate for the atmosphere; much more complete (and complicated!) treatments are available in the scientific literature. In what follows we deal with an incompressible fluid; clearly the atmosphere is not, but the approach still has some value for our purposes. The mathematical model is essentially that of a viscous fluid (with coefficient of kinematic viscosity ν) uniformly heated from below (the sun-warmed ground). The fluid density is allowed to vary slightly with temperature (with α being the volume coefficient of thermal expansion). The ambient temperature is denoted by $T_0(z)$, where z is the altitude, that is, the distance from the lower heated surface. The constant κ is the thermal diffusivity of the fluid (it is proportional to the thermal conductivity), and the fluid is considered to be confined to the altitude range $[0, d]$ under what are called *free* boundary conditions, but with the horizontal x and y directions unbounded. Again this is not a completely realistic description of the atmosphere, but it does render the mathematics easier without totally destroying its relevance.

In this model, the ambient temperature decreases linearly from $z = 0$ to $z = d$ with a temperature gradient $-\Delta T/d$ (where $\Delta T > 0$). For small enough temperature differences, there is no fluid motion, since heat is transferred from the lower boundary to the top by thermal conduction alone. As the temperature difference between the boundaries increases, however, bulk motion of the fluid sets in. This phenomenon is termed *convection*. This is certainly what can happen when ground or concrete is warmed by the sun and a "bubble" of warm air begins to rise upward, but such heating is not uniform laterally in general, because of the varied terrain over which the sun shines.

As with the previous instability analyses, a solution in the form of a complex exponential function is sought; since this topic is based on the account by Acheson, his notation will be adopted so that the reader interested in consulting this text does not have to make tedious adjustments in notation. This is rather more complicated than previous analyses, yielding a sixth-order eigenvalue problem (a sixth-degree polynomial in the growth rate s). Some reasonable simplifications can be made to the boundary conditions, and a simpler but still useful system arises. We seek specific solutions for the vertical component of velocity $w(x, y, z, t)$ in the form

$$w = W(z) f(x, y) e^{st}.$$

Now the stability is determined by the sign of $\mathrm{Re}\,s$; the system is (linearly) stable or unstable depending on whether $\mathrm{Re}\,s$ is negative or positive, respectively. The functional form of $f(x, y)$ need not concern us here; it is sufficient to note that it is related to an as yet unknown constant a, which in turn is a measure of the horizontal length scale of the disturbance. The z-dependence of the solution is straightforward, being

$$W(z) = \sin\left(\frac{n\pi z}{d}\right), \quad n = 1, 2, 3, \ldots$$

(to within a multiplicative constant). Define the quantity

$$a_*^2 = a^2 + \frac{n^2 \pi^2}{d^2}.$$

This being stated, the governing algebraic dispersion relation is found to be

$$(s + \nu a_*^2)(s + \kappa a_*^2) a_*^2 - \frac{\alpha g \Delta T a^2}{d} = 0.$$

This quadratic equation in s has roots

$$s = -\frac{1}{2}(\nu + \kappa) a_*^2 \pm \left[\frac{1}{4}(\nu - \kappa)^2 a_*^4 + P\right]^{1/2},$$

where

$$P = \frac{\alpha g \Delta T a^2}{d a_*^2} > 0.$$

The quantity under the square root is obviously positive, so s is in both cases a real number; for one of these values of s to be positive (instability) it is necessary and sufficient that

$$P > \nu \kappa a_*^4$$

or

$$\frac{\alpha g \Delta T}{\nu \kappa d} > \frac{a_*^6}{a^2} = \frac{1}{a^2}\left(a^2 + \frac{n^2 \pi^2}{d^2}\right)^3.$$

Note that the right-hand side of this inequality depends on both a and n; it follows therefore that instability will occur as soon as the left-hand side exceeds the minimum of the right-hand side, this minimum being taken over both a and n. Obviously we choose $n = 1$, but what about a? Differentiation of this expression (with respect to a^2) is straightforward and yields that for a given value of n (here unity) the minimum occurs when

$$a = a_c = \frac{\pi}{\sqrt{2}d} \approx \frac{2.2}{d}.$$

(Check that this is indeed a minimum; major political battles have ensued because someone forgot to check the sign of the second derivative! See the article by Biddle.) The quantity a_c is essentially a critical horizontal wavenumber.

It is time to introduce yet another dimensionless quantity: the Rayleigh number $\mathcal{R}$, so named because in 1916 Lord Rayleigh published a paper on this problem of thermal convection. As he surmised, it captures the essentials of the problem. This number is defined as

$$\mathcal{R} = \frac{\alpha g \Delta T d^3}{\nu \kappa},$$

so that the criterion for instability may be elegantly expressed as

$$\mathcal{R} > \mathcal{R}_c = \frac{27 \pi^4}{4} \approx 658.$$

This analysis is for so-called stress-free boundaries (meaning in particular that the vertical velocity perturbation vanishes there). The corresponding calculation for *rigid* boundaries yields a critical wavenumber of $a_c \approx 3.1 d^{-1}$ and an instability criterion of $\mathcal{R} > \mathcal{R}_c = 1708$; the rigid boundary case is considerably more stable than the free boundary one.

What is this instability when it occurs? Convection! We shall have more to say about the shape of these convection cells below when we discuss some of the results of a more realistic nonlinear theory. For now, however, we may draw some limited conclusions from the definitions of a_c and $\mathcal{R}$. A typical horizontal length scale l associated with instability in the case of free boundaries is

$$l \sim \left(\frac{2\pi}{a_c}\right) \sim \left(\frac{\pi d}{1.1}\right) \sim (2.9d)$$

or about three times the vertical extent. In the case of rigid boundaries $l \sim (2d)$. In both cases the definition of the Rayleigh number enables us to conclude that both viscosity and thermal diffusivity play a stabilizing role against the onset of convection: the larger the value of v or κ, the greater the temperature difference ΔT must be before convection sets in. Also note that larger values of d tend to be a destabilizing influence (if other things remain unchanged).

A reminder is in order at this point. All the above analyses have been based on what we have referred to as *linear theory*. This means that we are in principle examining the stability of each system to *infinitesimal* disturbances, and clearly such things do not occur in reality—all measurable disturbances are non-infinitesimal, however small they may be! In view of this, it is both gratifying and surprising that predictions based on linear theory are usually consistent with what is actually observed and measured as far as the *onset* of instability is concerned. This difference can be put more precisely as follows, put here in the context of thermal convection, following Acheson. Linear theory answers (or attempts to) the question "Is there a critical value of ΔT above which infinitesimal disturbances do not remain infinitesimal as $t \rightarrow \infty$?" Nonlinear theory essentially tries to provide an answer to the question "Is there a critical value of ΔT *below* which the energy of *any* disturbance tends to zero as $t \rightarrow \infty$?" In the case of convection between two rigid plane boundaries this latter question has the answer yes, and the critical Rayleigh number is the same as for the linear question, 1708. This correspondence does not generally occur, so it should not be allowed to induce in us a false sense of security in matters nonlinear. Nevertheless, in this case good agreement has been found between this theoretical value of 1708 and experimental results.

Linear theory informs us that infinitesimal disturbances grow exponentially in time when (in the case of convection) $\mathscr{R} > \mathscr{R}_c$. In practice this does not happen (thank goodness for that!) because the nonlinear terms neglected in the linear equations soon become large enough to cry foul, invalidating the assumption of linearity, and the growth of the disturbance quickly saturates to some finite value for steady convection. A crucial quantity in nonlinear theory is the difference $\mathscr{R} - \mathscr{R}_c$, for this determines the manner in which the steady state is reached. Notice also that, apart from some general comments about length scales associated with convection (or shear-induced instability, for that matter), nothing specific (i.e., mathematical) has been stated about the *pattern* of instability (i.e., the "shape" of the solutions—the eigenfunctions for the problem).

This silence of linear theory is not conspiratorial; it is simply that the theory, like a child in the case of the missing cookies, has absolutely nothing to say. For example, recall the earlier handwaving about the nature of the function $f(x, y)$ (don't worry about it, I said). Well, now a lot more can be revealed: this function satisfies the equation

$$\frac{\partial^2 f}{\partial x^2} + \frac{\partial^2 f}{\partial y^2} + a^2 f = 0.$$

So *that's* where a^2 came from. There are many possible forms of solution to this partial differential equation (called a Helmholtz equation). In particular, both two-dimensional rolls and hexagonal cells are quite common, and are often seen as cloud formations under appropriate atmospheric conditions (even though the atmosphere clearly does not have a rigid upper lid).

This is not yet the end of the story. It has been found experimentally that if $\mathscr{R}$ is increased well beyond the critical value, the state of steady convection may itself become prone to instability, leading to a new pattern of convection. At still higher values of $\mathscr{R}$ this pattern may also exhibit instability, perhaps leading to a time-dependent set of fluid motions. The development of these patterns depends on several factors, including the temperature dependence of ν, the value of the so-called *Prandtl number* ν/κ, and the presence of side-wall boundaries. If the upper surface of the fluid is free, then surface tension may play a role (but not in the atmosphere; as will be seen below, it is more likely to be of importance in the kitchen). In 1900, for example, H. Bénard observed beautiful hexagonal convection cells in the laboratory, prompting Lord Rayleigh to develop the theory of convection in considerable detail; but the mechanism for producing those cells turned out to be a rather different form of convection, governed in part by the temperature dependence of surface tension. In Drazin and Reid we are encouraged to become kitchen experimentalists—their advice is to "Pour corn oil in a clean frying pan (i.e. skillet) so that there is a layer of oil about 2 mm deep. Heat the bottom of the pan gently and uniformly. To vizualize the instability, drop in a little powder (cocoa serves well). Sprinkling powder on the surface reveals the polygonal pattern of the steady cells. The movement of individual particles of powder may be seen, with rising near the centre of a cell and falling near the sides." Perhaps in order to avoid adverse comments on one's culinary or even gastronomical talents, it may be better to experiment alone, at least initially.

Here is another application of (nonlinear) convection theory that should be mentioned for completeness. A brief reference has been made to *plate tectonics* in chapter 2; cellular convection has also received some attention in chapter 6. What is the former and how is it related to the latter? This is not the place for a definitive account of either topic (and I am not the person to do it), but from a fluid dynamical perspective the early paper by Turcotte and Oxburgh is well worth reading. They develop a mathematical model to describe how "The hypothesis of large scale cellular convection is favoured as the driving mechanism of continental drift." Furthermore, they explain that

> Convective currents at present operating in the mantle are thought to have the form of elongated rolls, tens of thousands of kilometres long and several thousand kilometers wide. . . . The regions of converging, falling convection may be characterized by deep and intermediate focus earthquakes. Surface features may include island-arc areas, deep sea trenches, and other compressional phenomena. The regions of rising, diverging currents are associated with shallow seismicity. Surface features are the mid-oceanic ridges. . . . The ridges . . . display extensional features which suggest that the crust is moving away from them on either side. In addition the ridges are characterized by volcanic activity. . . . Dating of volcanic islands associated with mid-ocean ridges indicates that they have been convected away from their origin.

They further explain that because the convecting layer is thin (in relative terms) it is permissible to neglect any heat produced by radioactivity within the layer, and on the basis of their mathematical model they conclude that "the quantitative agreement between theory and measurement given in the paper strongly favours the hypothesis of mantle convection." The reader is asked to bear these ideas in mind when encountering the early "age of the earth" model toward the end of chapter 14; it is introduced primarily for historical interest, and while the mathematics is interesting, the conclusions are rendered obsolete by virtue of what is now known about plate tectonics.

With the exception of the previous paragraph, the basic instabilities considered here have relevance to atmospheric science via billow and convection clouds (one can often see large areas of roughly hexagonal cumulostratus clouds, both open and closed). But there is another important mechanism for driving wave motions in addition to shear and buoyancy that we have yet to consider: the *rotation* of the earth. This is not particularly significant over timescales short compared to one day, nor over length scales that are small compared to the radius of the earth; nevertheless it is responsible for large-scale waves (not to mention hurricanes, as we have seen in chapter 6) that affect the global weather and interact with the jet stream, namely, *Rossby waves*.

As with the convective stability problem, we shall pose and briefly discuss a well-defined laboratory situation and then draw some tentative conclusions for its applicability to the atmosphere. In the laboratory context, cylindrical geometry is imposed on the system. This does not present significant difficulties for application to rotating spheres (e.g., the planets and the sun) because local approximations may be made to accommodate the results from cylindrical geometry. Furthermore, the same types of analysis can be carried out for spherical shells (which for compressible fluids are often referred to as "atmospheres"), but such a level of sophistication is not necessary for our purposes. We shall again state results based on linear theory.

Consider a viscous fluid occupying the gap between two rigid, concentric, right circular cylinders of radius r_1 and r_2 $(> r_1)$ and angular velocities Ω_1 and Ω_2, respectively. In 1923 Taylor studied the stability of this flow to infinitesimal axisymmetric disturbances. If the cylinders rotate in the same sense (so that Ω_1 and Ω_2 are both positive, for example), then the onset of instability can be expressed in terms of yet another dimensionless number, the *Taylor number T*, defined as

$$T = \frac{2(\Omega_1 r_1^2 - \Omega_2 r_2^2)\overline{\Omega}d^3}{v^2 r_1}.$$

This form is based on the so-called thin gap approximation, whereby it is assumed that the gap d between the cylinders satisfies $r_2 - r_1 \ll r_1$, and $\overline{\Omega}$ is the mean of Ω_1 and Ω_2. There is a remarkable mathematical correspondence between the thermal convection problem and the present one when both are examined in terms of marginal stability by setting $s = 0$ in the governing dispersion relations: when allowances are made for the different dependent variables and boundaries, and so forth, they turn out to satisfy identical differential equations and boundary conditions with $\mathcal{R}$ replaced by T! Therefore the criterion for centrifugal instability in the narrow gap flow is $T > 1708$. The ensuing motion consists of counter-rotating Taylor vortices placed like smoke rings in a periodic manner with axes along the axis of symmetry of the cylinders. The cells are almost square in cross section, being of "height" $h \approx \pi d/3.1 \approx d$. This topic of instability discussed so briefly here is known as the *Taylor problem*.

SPIDERWEBS AND THE STABILITY OF CYLINDRICAL FILMS

Have you ever noticed beads of sticky substance regularly placed along the threads of a spiderweb? That substance is the "glue" that spiders use to keep the unfortunate insects that get caught in their webs from seeking pastures new.

This "pearling" phenomenon, as Philip Ball has noted in *The Self-Made Tapestry*, arises, not because the spider has painstakingly placed them there; rather it is a consequence of the instability of cylindrical columns of liquid to small departures from "cylindricity," or in other words, to undulations along its surface. The basic idea behind all this is that, given any small "waviness" on the surface, the surface tension of the liquid acts to accentuate this curvature, and the result is that each undulation is pulled into a rough blob isolated from its companions. They are strung out like pearls,

and the unwitting insects are caught, to be eaten at a later date. This instability was studied by Lord Rayleigh at the end of the nineteenth century,
and for that reason it is sometimes referred to as the *Rayleigh instability*.
It should not be confused with another phenomenon, the *Rayleigh-Taylor
instability*, which is associated with adjacent fluids of different densities in
a gravitational field. This was studied briefly in the chapter on clouds and
earlier in this one.

An interesting feature of many fluid-dynamical (and other) instabilities
is that, while they may occur (as here) for all wavelengths of undulation
(or at least a continuous subset), there is usually a particular wavelength λ
(or equivalently, a *wavenumber* $k = 2\pi/\lambda$) that is the *most* unstable. This
means that wave-like perturbations with this wavelength depart from the
cylindrical form (in this instance) most rapidly in time. It is this critical wavelength λ_c that determines the size and separation of the droplets along the
spider thread. It is essentially this same instability that is responsible for the
break-up into droplets of a thin nonturbulent stream of water issuing from
a tap (see figure 8.3a), although the presence of gravity does tend to accelerate the instability by "tearing" the droplets away from the stream. Melting
fuse wire is subject to the same effects (but check the wiring anyway).

Ball explains clearly how such pattern-forming processes may be initiated
by abruptly occurring instabilities: "Generally an instability sets in when
some critical parameter is surpassed. . . . Two common aspects of pattern-
forming instabilities are that they involve symmetry-breaking . . . and that
they have a characteristic *wavelength*, so that the features of the pattern
have a specific size." In the present case, the symmetry that is broken is the
cylindrical symmetry of the liquid about the axis of the thread.

To discuss this instability in mathematical terms, consider a cylinder of
liquid (or soap film) of radius *a* and ignore gravitational forces; the fact that
the figure is drawn vertically is of no significance. Let the axis of symmetry
be the *x*-axis and the radial direction be denoted by *r*. Now we allow the
column to be perturbed by a small periodic or undulatory disturbance of
wavelength λ (wavenumber *k*), so that its radius *r* is deformed as a function
of axial position *x* in the following manner:

$$r = a + b\cos\left(\frac{2\pi x}{\lambda}\right) = a + b\cos kx, \quad b \ll a.$$

The strong inequality $b \ll a$ means that the disturbance is really small; it
changes the ambient radius *a* by only a small relative amount (this essentially
reduces our problem to a linear one, as we shall see shortly). Referring to the
diagram of the now-corrugated cylinder (figure 8.3b), note that this system
is stable (the uniform cylindrical shape is *restored*) if the pressure at points
like *A* is greater than the pressure at points like *B*. This is because the pressure will tend to be equalized, and this will induce the opposite curvature at
these points, restoring the column toward its original nonperturbed shape.

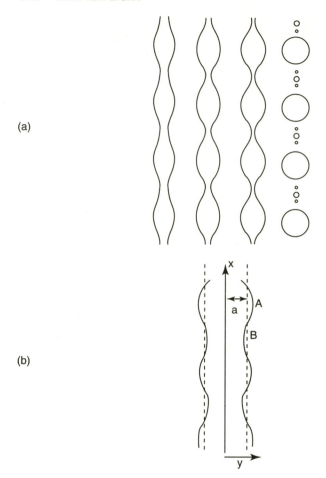

Figure 8.3. (a) The instability of a column of water (or soap film); it tends to "pearl" or break up into droplets. (b) The notation used in the derivation of the stability criterion. From *The Science of Soap Films and Soap Bubbles*, by C. Isenberg (1992), reprinted by permission of Dover Publications.

For any given point on the liquid surface, the maximum possible radial extension under this perturbation is $R_1 = a + b$; in the perpendicular direction (parallel to the axis of symmetry) the radius of curvature (as any calculus book will demonstrate) is

$$R_2 = \left[1 + \left(\frac{dr}{dx}\right)^2\right]^{3/2} \times \left(\frac{d^2r}{dx^2}\right)^{-1}.$$

For sufficiently small initial deformations we may assume that $|dr/dx| \ll 1$, so

$$R_2 \approx \left(\frac{d^2r}{dx^2}\right)^{-1}.$$

From the form of the perturbation it follows that

$$\frac{d^2r}{dx^2} = -k^2b\cos kx \le k^2b,$$

so at the point of greatest extension (e.g., A), the *maximum* pressure above the external (atmospheric) pressure is

$$\sigma\left(\frac{1}{R_1} + \frac{1}{R_2}\right) = \sigma\left(\frac{1}{a+b} + k^2b\right).$$

Similarly, at points like B, the *minimum* pressure difference above the external is

$$\sigma\left(\frac{1}{a-b} - k^2b\right).$$

Therefore the pressure difference between points like A and B is

$$\sigma\left(\frac{1}{a+b} - \frac{1}{a-b} + 2k^2b\right) = \sigma\left(2k^2b - \frac{2b}{a^2-b^2}\right) \approx 2b\sigma\left(k^2 - \frac{1}{a^2}\right),$$

since by hypothesis $b \ll a$. This pressure difference is positive when $ka > 1$, or $2\pi a = \lambda_c > \lambda$. From our earlier discussion this corresponds to a stable situation; any disturbance with $\lambda < \lambda_c$ tends to damp out. On the other hand, if $\lambda > \lambda_c$, the critical wavelength, then the cylindrical column becomes unstable, causing the sticky thread to fragment into well-defined drops.

Thus instability occurs if the wavelength is too large or the column is too narrow, according to the above criterion. It must be pointed out that this analysis is only sufficient to describe the onset of instability; a detailed study of the subsequent instability (for which we have only employed the waving of hands) requires a fully nonlinear analysis. It is interesting to note that this instability can be induced in a narrow stream of water flowing from a tap by placing a vibrating tuning fork close to the stream; if the frequency is low enough then the corresponding corrugations induced on the surface of the stream will break up into droplets.

CHAPTER NINE

Bores and Nonlinear Waves

> ...the waters piled up. The surging waters stood firm like a wall . . .
>
> —Exodus 15:8a

What is a bore? The answers will vary depending on whether one is at a cocktail party, the banks of the River Severn in England, or the Bay of Fundy in Nova Scotia (to name but two of many geographical locations). We will focus our attention on tidal bores, which, as David Lynch points out in an article of the same title, are remarkable hydrodynamic phenomena. A tidal bore is the incoming tide in the form of a wave (technically, it is called a *solitary* wave) moving upstream in a river that empties into the sea. Even after the bore passes a particular location, the river flows upstream for a while before the current reverts to its normal direction. Not every such river will support such bores; there are two conditions that are necessary for such phenomena to occur. The first is that the tides in the adjoining bay must be very high, with at least a 20 ft difference between high and low tides. (In the Bay of Fundy, for example, the difference can be as much as 55 ft. Seeing the tide come in is an impressive sight.) Second, the topography must be right: the river must be shallow with a gently sloping bottom and the estuary must be broad and funnel-shaped (the latter condition will be incorporated in a mathematical model below). This forces the incoming tide to build in height as it moves upstream until a "wavefront" is born and propagates in the same direction. This wave is one of several types of solitary wave (see the section on nonlinear waves later in this chapter); although water is a fundamental medium in which they are generated and propagated, they are known to occur in many other environments and contexts—the atmosphere (including that of Mars), optics, molecular biology, plasma physics, and many others outside the scope of this book. On rare occasions, a "non-bore" river may support a bore as a result of a tsunami propagating across the ocean, the latter occurring because of seismic or volcanic activity on the ocean floor.

The most commonly cited distinguishing feature of solitary waves is that they do not disperse; they can maintain their essential features (such as shape and speed) over large distances provided no significant changes in their environment (such as water depth) occur. The reason for this is that, being nonlinear, they have the capacity to "keep at bay" the effects of wave dispersion (and in some cases frictional dissipation); that is, there is a balance between the competing effects of growth and decay. They can be thought of as a nonlinear version of the linear surface gravity wave in shallow water, which as we have seen is nondispersive because every individual component of an initial disturbance (after a sufficiently long time) travels at the same speed (unlike the case of such waves traveling in deep water, where component waves of different wavelengths mutually separate). The interaction—or lack of it—between such waves determines what type of traveling waveform will occur. In practice, of course, solitary waves will not go on for ever. Energy is drained from them by friction on the bottom of the channel, and also by viscosity, which is akin to friction between the layers of fluid. Both effects ultimately generate heat with consequent loss of wave energy. Additionally, changes in the channel topography such as widening can reduce the speed of the bore to such a degree that it is eventually carried downstream by the stronger current.

What are some other characteristics of tidal bores? They are somewhat unpredictable insofar as they vary in time and space: from tide to tide and from river to river. They can be as high as 25 ft or as low as a few inches (the latter do not usually draw thousands of visitors each year, unlike some of the larger ones, the one that used to occur on the River Seine in France being one notable example). Typically they are higher near the banks than in mid-channel. Since they are harbingers of the turning tide, they are largest around the time of the new or full moon, when the tides are largest. As already noted, after the wave passes the river continues to flow upstream for a while, and so the wavefront represents a demarcation front between two parts of the river with opposite directions of flow. Since the water level behind the bore is higher than that ahead of it, the river flows upstream but downhill. The bore may be manifested as a breaking, foaming, turbulent crest like that on a shelving beach, or as a nonturbulent, gently rounded wave, known as an undular bore. The latter are frequently followed by smaller waves; Lynch observed as many as 55 such waves behind a bore on an Amazon tributary. Which type of bore occurs depends on the height of the tide, the wind speed and direction, the local depth of the river, and the shape of the bottom. Indeed, many bores are of mixed type, being undular in the middle region of the river and breaking toward the banks, and they may change as they propagate if the river topography changes. Bald eagles are known to prefer turbulent bores; they have been observed in Alaska to hunt fish by following bores moving inland from the Pacific Ocean— the fish are brought to the surface by the turbulence of the bore. The fish presumably prefer undular bores.

To appreciate the mechanism of bore formation, we must spend a little time discussing the tides. The predominant tide-producing object is the moon (the sun's tidal effect is less than half that of the moon, as will be shown later in this chapter). To simplify matters, we will ignore the solar tidal contribution, the complex interaction between land and sea (including the impeding effects of land mass), and also the much smaller land tides that are produced by the moon and sun. Tides are produced as a result of differential gravitational forces; on a planet totally covered by water, this would produce two tidal "bulges" with maxima lying on a line joining centers of mass of the earth and the moon, but on opposite sides of the earth. They travel around the earth with a period of about 28 days, which is the orbital period of the moon, so the rotation of the earth causes any particular location on the surface to "pass through" each tidal maximum (and minimum) about once each day.

Things are more complicated than this, of course, in particular because the moon revolves around the earth in the same direction as the earth rotates. This means that the tidal bulges have moved forward somewhat by the time a given point on the earth's surface has undergone a half rotation, so there is rather more than 12 hours between tides; it is closer to 12 hours 25 minutes. This is not the case everywhere, because of the above-mentioned interference of the land masses with these tidal oscillations; some places have only one tide per day, and some, such as Tahiti, have little or none. This is because Tahiti lies on a *node*—a stationary point about which the tidal standing wave oscillates. Obviously this is a very important place to visit and conduct further scientific research.

Now let us bring the sun back on board, so to speak. While less effective than the moon at raising tides, it is nonetheless significant. When the three bodies (sun, earth, and moon) are essentially collinear (ignoring differences in the orbital planes of the latter two bodies), the tide-producing forces, and hence the tides, are at their greatest. This occurs about twice a month at new and full moons; such a configuration is known as *syzygy*, a word that has been especially noted by Scrabble fanatics. These are "spring tides"; so-called "neap tides" occur when the earth-sun and earth-moon lines are at right angles (so that the moon is at first quarter or third quarter), and the sun and moon to some extent oppose the tidal forces of each other. This means that bores generally form at the times of spring tides.

There is another phenomenon that contributes to the formation of bores: *resonance*. Every container of water, be it lake, tidal basin, or bathtub, has a natural frequency of oscillation (put more evocatively, "sloshing"). In lakes, such as Lake Michigan, such oscillations are called *seiches*. Just as pushing a child on a swing at the end of each upswing can increase the amplitude of oscillation, so the amplitude of a bore can increase if the impulsive force of the ocean tide matches the fundamental period associated with the tidal basin. The wave of longest wavelength is called the fundamental mode, and for a closed "container" its period of oscillation is approximately twice the longest dimension of the container (or basin) divided by the wave speed, $2L/v$. If the basin is open at one end (corresponding to the mouth of a river), the period of oscillation is twice this amount, $4L/v$. If this period is close to 12 hours 25 minutes, then the basin will resonate with the incoming tide, thus reinforcing the high tide and correspondingly reducing the low tide.

This is not the only way that resonance can occur. As Lynch points out, an asymmetry between the tides at the end of the basin and near the river mouth can produce the same effect. About half the known bores, apparently, are associated with the phenomenon of tidal basin resonance, one example being that which occurs in the Bay of Fundy. The fundamental period there is about 11 hours (close enough to help!). Another strong resonance occurs at Cook Inlet in Alaska, so strong, in fact, that a 14 ft tide at the mouth of the river is amplified into at least a 30 ft tide at Anchorage.

We are nearly at the point where we may do some mathematics. To get there, let us follow the incoming tide in a tidal basin. It generally arrives in the form of waves with long wavelength. As they enter the narrowing estuary they are funneled, and as the depth decreases the wave height increases, thus facilitating the transition from tide to bore. Waves farther from the shore, being in deep water, move faster than those closer to the shore in shallow water, and, just as in wave refraction, the latter are overtaken by the former, thus steepening the leading side compared with the trailing side. Eventually, a bore may result if the difference in depth is great enough.

Up to this point we have considered only the speed of the waves themselves, and no mention has been made of the speed of the water "particles" (small volumes of water) themselves. In deep water, to a first approximation, the waves do not transport water itself, only energy. The trajectories of surface water particles are closed circular loops, as would be demonstrated by a cork bobbing up and down as a wave passes through its location. However, as the wave moves up the estuary into increasingly shallow water, the surface particle trajectories are flattened into ellipses, until eventually they move back and forth in an essentially linear fashion, and at this juncture water as well as energy begins to be carried toward the land. It is not surprising, therefore, that the relationship between the speeds of the wave (v) and the water particles (u) becomes paramount. When $u < v$ the fluid flow is said to be *subcritical*; when $u > v$ it is called *supercritical*. This leads to an alternative definition of a tidal bore: the transition between these two flow régimes is called a *hydraulic jump*, so *a tidal bore is just a hydraulic jump moving upstream*. The bore forms at the point where the particle speed is equal to the speed of the wave. A circular hydraulic jump can be seen in the kitchen sink when a column of water from the tap hits the base of the sink and spreads out with decreasing speed; the flow is supercritical because the shallow water wave speed is smaller than the flow speed. Waves are thus carried forward by the moving water. Farther out radially, the flow has slowed somewhat and becomes subcritical; here there is a circular region of elevated water arising because of the accumulation of standing waves there. The hydraulic jump occurs at the circular wall of water.

Returning to the bore on the river, it is clear that it must move faster than the downstream current, or it would not propagate upstream at all. Let us imagine that we are watching a bore move upstream. To us, it does so with a speed equal to the difference between that of the hydraulic jump and the river current. But the physics is unchanged if we convert to a coordinate system in which the bore appears stationary, that is, it moves with the coordinate system (this is an example of a *Galilean transformation*). Under these circumstances the formulation is a little easier and gets to the heart of the matter quickly, as we shall see.

Now to the mathematics (see figure 9.1); the book by Tricker provides many further details. Suppose that a tidal bore advances upstream with speed c (from left to right in the diagram) and the particle speed is u. If we superimpose an adverse flow (right to left) on the system, thus making the bore appear stationary, the water ahead of the bore is no longer still, but moves to the left with speed c, and the water behind the bore now moves to the left with speed $c - u$ (if $c > u$, but even if not, the mathematics will still be valid).

Now we examine the energetics of this new system. The loss of kinetic energy per unit mass for a surface fluid particle as it flows up the side of the bore must equal the work done against gravity in so doing, so that

$$\frac{1}{2}c^2 - \frac{1}{2}(c - u)^2 = gh \tag{1}$$

or upon simplification,

$$cu = gh + \frac{1}{2}u^2. \tag{2}$$

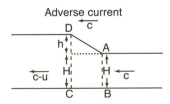

Figure 9.1. A tidal bore of height h in a uniform channel of depth H prior to the bore's passage. Redrawn from *Bores, Breakers, Waves and Wakes*, by R.A.R. Tricker (1964), M. & B. Elsevier: New York.

If H is the uniform depth of the water ahead of the bore and the bore itself is higher by a constant amount h, then by conservation of mass the following relationship exists for a stream of unit width

$$Hc = (H + h)(c - u),$$ (3)

which simplifies to

$$hc = u(H + h).$$ (4)

Let us first make an approximation: suppose that the flow is highly subcritical, such that $u \ll c$. Under these circumstances equation (2) becomes

$$cu \approx gh$$

which, together with equation (4) implies that

$$c \approx \sqrt{g(H + h)}.$$ (5)

Thus a river for which $H = 12$ ft and $h \ll H$ will support a bore of speed $c \approx 13.4$ mph; this is about right for the River Severn, where the freshwater current is about 2 mph, reducing the bore speed to about 11–12 mph for our intrepid observer. If we do not make the approximation $u \ll c$, then by substituting u from equation (4) into equation (2) we obtain, after a little algebra, the exact result

$$c = \sqrt{\frac{2g(H + h)^2}{2H + h}}.$$ (6)

A useful approximation between this exact result and that in (5) results if we expand the radicand to first order in h/H to obtain

$$c \approx \sqrt{g\left(H + \frac{3h}{2}\right)}.$$ (7)

Note that is readily deduced from the approximations (5) and (7) that both $c(h)$ (for fixed H) and $c(H)$ (for fixed h) are monotone increasing functions of their arguments. Thus the speed of the bore increases with both depth of the stream and height of the bore. What does equation (6) imply about this?

Now let us investigate what happens when there is a change in the river depth caused by a narrowing of the channel (see figure 9.2). Suppose without loss of generality that a uniform stream of width B and height H narrows to a uniform stream of width $B - b$ (where $b > 0$) and depth $H - h$ (where this last quantity may be negative). Again, mass conservation gives the relation

$$BHu = (H - h)(B - b)v$$

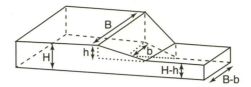

Figure 9.2. The geometry corresponding to a bore given a change in river depth produced by a narrowing channel. Redrawn from *Bores, Breakers, Waves and Wakes*, by R.A.R. Tricker (1964), M. & B. Elsevier: New York.

or

$$\frac{v}{u} = \frac{BH}{(H-h)(B-b)},$$

(8)

where v is the particle speed in the narrower channel. Since the difference in particle kinetic energies in the two parts of the stream is proportional to

$$v^2 - u^2 = u^2\left(\frac{v}{u}+1\right)\left(\frac{v}{u}-1\right) = 2gh,$$

(9)

it follows from equations (8) and (9) that, neglecting the products hb, b^2, and h^2 as squares of small quantities,

$$ghHB \approx u^2(Bh + bH)$$

or

$$\frac{h}{b} \approx \frac{u^2}{gB(1 - u^2/gH)}.$$

(10)

This is an important result, for it shows that when the particle speed u (in the wider part of the channel) is less than the shallow water wave speed $\sqrt{gH}$ (subcritical flow), then $h > 0$ and the surface is lowered in the narrower section (it is clearly undefined when $u = \sqrt{gH}$). Conversely, for supercritical flow ($u > \sqrt{gH}$), $h < 0$ and the surface is elevated there. So we are back to bores!

We can relate all this to an interesting dimensionless quantity—a number, therefore, but not just any old number. Consider the quantity

$$F = \sqrt{\frac{u^2}{gH}}.$$

Subcritical flow corresponds to $F < 1$ and supercritical flow to $F > 1$. This is the square root of a ratio of speeds and is called the *Froude number*, after a nineteenth-century civil engineer, William Froude. The Froude number is proportional to the ratio of the kinetic energy of the moving water to the gravitational energy in the waves, as is easily shown. Equivalently, it is

Range of F	Description of bore
$1.0 < F < 1.7$	Called an undular jump; the advancing tide moves gently upstream. It breaks only where F is higher, that is, perhaps, where the water is shallower (near the river bank). When $F > 1.7$, all bores break.
$1.7 < F < 2.5$	Weak jump; small waves slide down the leading edge; water behind is smooth.
$2.5 < F < 4.5$	Oscillating jump; large amplitude surface waves are generated.
$4.5 < F < 9.0$	Steady jump; leading edge is turbulent; surface of trailing edge can be smooth.
$F > 9.0$	Strong jump; its energy is rapidly dissipated; bore moves upstream in an irregular manner; high-speed jets may extend ahead of the main jump.

equal to the ratio of the speed of the waves just behind the bore (because at the hydraulic jump, the wave and particle speeds are equal) to the speed of the waves in the undisturbed river ahead of the bore. When $h \ll H$ it leads to the simple relationship that F is the square root of the ratio of depth behind the bore to the depth in front of it. This follows by virtue of the fact that the speeds are proportional to the square root of the channel depth. The Froude number is very useful, isn't it? This can be developed further, though only the results will be stated here. Table 9.1 summarizes the characteristics of bores for various ranges of Froude number, based on information in the article by Lynch.

NONLINEAR WAVES

A remarkable description of a solitary wave (though not so-called at the time) was provided by J. Scott-Russell in his "Report on Waves" to the 1844 meeting of the British Association for the Advancement of Science. It concerned a chance observation he had made while riding near the banks of the Edinburgh-Glasgow canal in 1834 and subsequent laboratory experiments he had carried out in the intervening years. According to his report:

> I was observing the motion of a boat which was rapidly drawn along a narrow channel by a pair of horses, when the boat suddenly stopped—not so the mass of water in the channel which it had put in motion; it accumulated round the prow of the vessel in a state of violent agitation, then suddenly leaving it behind, rolled forward with great velocity assuming the form of a large solitary elevation, a rounded, smooth and

well-defined heap of water, which continued its course along the channel apparently without change of form or diminution of speed. I followed it on horseback, and overtook it still rolling at a rate of some eight to nine miles an hour, preserving its original figure some thirty feet long and a foot to a foot and a half in height. Its height gradually diminished, and after a chase of one or two miles I lost it in the windings of the channel. Such, in the month of August 1834, was my first chance interview with that singular and beautiful phenomenon.

This phenomenon generated interest from experimental and theoretical scientists over the years since his report, and in 1895 Korteweg and de Vries published a detailed mathematical description of this type of solitary wave. Such a wave is now known to arise from a precise balance between the steepening effects of finite amplitude and the "flattening" effects of dispersion (for another description of this wave by Scott-Russell see the account in the book by Acheson).

Another effect that can counteract the steepening effects of finite-amplitude waves, is dissipation of energy due to fluid viscosity. This is essentially the mechanism maintaining the *tidal bore*, which, as already noted, is a type of *hydraulic jump*, analogous to a shock wave in a gas. It is this type of nonlinear wave that we shall investigate first. We shall take as our prototypical governing equation that known as *Burger's equation*, namely

$$\frac{\partial u}{\partial t} + u\frac{\partial u}{\partial x} - \mu\frac{\partial^2 u}{\partial x^2} = 0. \tag{11}$$

Some explanation of these terms is necessary. The dependent variable $u(x, t)$ represents the particle velocity amplitude and is considered to vary only in one space dimension and time; this will be a reasonable description of such a wave propagating along a uniform stream in the x-direction. The second term is a nonlinear "convection" term and represents the effects of wave steepening; the final term on the left represents the effects of diffusive dissipation forces, where $\mu > 0$ is for our purposes a generic coefficient of diffusion. It is these last two terms that essentially "compete" in the waveform and maintain its shape over long distances. After finding one appropriate solution to this equation, we shall consider its linear counterpart and see the limitations of linear theory in describing such phenomena.

As is customary in such studies, we shall look for *steady* solutions to (11) of the form

$$u(x, t) = U(\xi), \text{ where } \xi = x - ct. \tag{12}$$

This represents a wave of shape U moving to the right with constant speed $c > 0$ (a simple change in sign for c corresponds to a wave moving to the left). On substituting this form into equation (11) the following ordinary differential equation for $U(\xi)$ results:

$$-c\frac{dU}{d\xi} + U\frac{dU}{d\xi} - \mu\frac{d^2U}{d\xi^2} = 0. \tag{13}$$

This can be integrated directly to yield the nonlinear first-order differential equation

$$-cU + \frac{1}{2}U^2 - \mu\frac{dU}{d\xi} = K,$$

where K is a constant of integration. This can be rearranged into the form

$$\frac{dU}{d\xi} = \frac{1}{2\mu}(U - U_+)(U - U_-), \tag{14}$$

where

$$U_+ = c - \sqrt{c^2 + 2K} \quad \text{and} \quad U_- = c + \sqrt{c^2 + 2K} \tag{15}$$

are the roots of the quadratic equation

$$U^2 - 2cU - 2K = 0.$$

The subscripts on U do not relate the signs of the radical, they refer to the signs of ξ as explained below.

Clearly, at this stage the value of K is unknown, but to ensure that the roots $U_\pm$ are real, we shall impose the restriction that $c^2 + 2K > 0$. This of course implies that $U_- > U_+$. It is now only a matter of integrating equation (14); in these days of symbolic calculus software some of the demand for knowing techniques of integration has decreased, but in the true spirit of technophobia we shall proceed with the exercise nonetheless. Clearly

$$\int \frac{dU}{(U-U_+)(U-U_-)} = \frac{1}{B}\int\left(\frac{1}{(U-U_-)} - \frac{1}{(U-U_+)}\right)dU = \frac{\xi}{2\mu} + \text{constant},$$

where $B = U_- - U_+ = 2\sqrt{c^2 + 2K}$. Therefore,

$$\ln\left|\frac{U - U_-}{U - U_+}\right| = \frac{B\xi}{2\mu} + \ln D,$$

D being a positive constant.

Based on the physical observation that an idealized smooth bore (which we are attempting to describe here in mathematical terms) has a higher elevation behind than ahead, we shall for simplicity invoke the condition that $U'(\xi) < 0$. Then it follows from equation (14) that

$$U_+ < U < U_-.$$

Therefore we may write

$$\frac{U_- - U}{U - U_+} = De^{B\xi/2\mu},$$

which satisfies the conditions

$$\lim_{\xi \to \infty} U = U_+ \quad \text{and} \quad \lim_{\xi \to -\infty} U = U_-.$$

If we impose the natural symmetry condition that $U(0) = (U_+ + U_-)/2$, then $D = 1$, and after a little rearrangement (okay, quite a lot) we may write the solution U in the form

$$U(\xi) = c - \sqrt{c^2 + 2K} \tanh\left(\frac{\sqrt{c^2 + 2K}}{2\mu}\xi\right)$$

or in terms of the original independent variables and limiting states $U_\pm$

$$u(x,t) = \frac{1}{2}\left([U_+ + U_-] - [U_- - U_+]\tanh\left[\frac{U_- - U_+}{4\mu}\left\{x - \frac{U_- + U_+}{2}t\right\}\right]\right).$$
(16)

This represents, in particular, an idealized form for a tidal bore propagating upstream, in which two asymptotically uniform states are smoothly joined through a continuum of varying states (see figure 9.3).

In reality, of course, this smooth form is rarely achieved because there are other factors (e.g., topography of the channel) that have been neglected in the formulation of the problem in terms of Burger's equation. Nevertheless, it does provide some insight into the nature of nonlinear effects on simple types of wave. It is interesting to contrast this analysis with that which pertains to a linear version of this equation, namely

$$\frac{\partial u}{\partial t} + c\frac{\partial u}{\partial x} - \mu\frac{\partial^2 u}{\partial x^2} = 0,$$

Figure 9.3. A typical "hyperbolic tangent" wave shape for a solution of Burger's equation (1). For an advancing tidal bore, the profile as drawn moves from right to left.

where c is any real constant. Since this is linear, we can gain some information by examining the dispersion relation based on substituting a complex plane wave solution of the form

$$u(x,t) = ae^{i(kx-\omega t)}.$$

The governing dispersion relation is

$$\omega = ck - i\mu k^2, \omega \in \mathscr{C},$$

from which

$$\operatorname{Re}\omega = ck \quad \text{and} \quad \operatorname{Im}\omega = -\mu k^2 < 0.$$

Hence, according to linear theory, a particular Fourier component of the general solution may be written as

$$u(x,t) = ae^{-(t/t_0)}e^{i(kx-ckt)},$$

where $t_0 = (\mu k^2)^{-1}$, which of course attenuates exponentially with decay time t_0. Note that this time decreases as k increases (i.e., as the wavelength decreases). This means that only long waves tend to survive for large times. In addition, for fixed k it follows that the attenuation is greater if the coefficient μ is large than if it is small. It is also clear that this linearized version of Burger's equation cannot admit a solution joining the two states $U_{\pm}$. There has to be a nonlinear description, but one feature is retained in linear theory; since

$$\operatorname{Re}\left(\frac{\omega}{k}\right) = \operatorname{Re}\left(\frac{d\omega}{dk}\right) = c,$$

the wave and group speeds are equal and the linear wave is dispersionless, as is the nonlinear form we have derived.

We return briefly, to consideration of the prototypical solitary wave observed by Scott-Russell. The governing nonlinear partial differential equation is of third order in the spatial variable, so we will content ourselves here with merely stating the equation, one class of solution, and a brief excursion into the predictions of the corresponding linear theory. The equation of interest here is the Korteweg-de Vries (KdV) equation

$$\frac{\partial u}{\partial t} + u\frac{\partial u}{\partial x} + K\frac{\partial^3 u}{\partial x^3} = 0, \tag{17}$$

where $K > 0$. By proceeding in a manner similar to that used for Burger's equation (but involving properties of a *cubic* equation in u) it can be shown that there are basically three classes of solutions to this equation. One such solution type is called a *Cnoidal* wave (because of its association with the Jacobian elliptic function Cn) and in one sense it is the intermediate class. When the amplitude u is very small, the waves are effectively sinuosoidal, and, not surprisingly, the linear theory applies. As the amplitude increases, the crests become steeper and the troughs flatter than in sinusoidal waves (probably typical of the waves that form in association with some types of undular bores). As the amplitude increases still further, the wavelength also increases until eventually a solitary wave results, and this is the wave that concerns us here. It is a symmetric peaked traveling wave of the form (see figure 9.4)

$$u(x,t) = U_\infty + a \operatorname{sech}^2\left[\sqrt{\left(\frac{a}{12K}\right)}\left\{x - \left(U_\infty + \frac{a}{3}\right)t\right\}\right], \qquad (18)$$

where U_∞ is the limiting uniform state defined by

$$\lim_{|\xi|\to\infty} U(\xi) = U_\infty,$$

where, as before, $\xi = x - ct$ (try verifying that this is indeed a solution of the KdV equation by direct substitution). The quantity a is the amplitude of the wave; notice that the speed of the wave (defined as the multiplier of t above) is a linearly increasing function of a; relative to the uniform state at infinity it is directly proportional to a. The upshot is this: the bigger the wave, the faster it moves. We may somewhat arbitrarily define the width of the waveform to be twice the value of ξ when $u = a/2$; temporarily setting $U_\infty = 0$ for convenience we note that since

$$u(x,t) = a \operatorname{sech}^2 A\xi,$$

where

$$A = \sqrt{\left(\frac{a}{12K}\right)},$$

then

$$\frac{a}{2} = a \operatorname{sech}^2 A\xi_h,$$

where ξ_h is the half-width at half-height. From this it follows that

$$\xi_h = A^{-1} \operatorname{arccosh}\sqrt{2} \propto \left(\frac{K}{a}\right)^{1/2}.$$

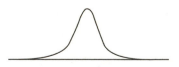

Figure 9.4. A typical "hyperbolic secant squared" wave shape for a solution of the Korteveg-de Vries equation (17).

Thus the width is inversely proportional to the square root of the wave amplitude; the higher the wave, the narrower it is. Since K is the coefficient of the dispersive term in the KdV equation, this result also indicates that the greater the dispersion, the flatter the wave.

Finally, what information, if any, does linear theory provide for such waves? Again, we utilize a linear form of equation (17), namely

$$\frac{\partial u}{\partial t} + c\frac{\partial u}{\partial x} + K\frac{\partial^3 u}{\partial x^3} = 0,$$

for which the dispersion relation is, following, as they say, standard procedure,

$$\omega = ck - Kk^3,$$

so that ω is a real function of k. The wave speed is

$$\frac{\omega}{k} = c - Kk^2$$

and the group speed is

$$\frac{d\omega}{dk} = c - 3Kk^2.$$

Since these are not equal, the system, like its nonlinear counterpart, is dispersive (but remember in that case that this is counteracted by the nonlinearity tending to steepen the wave).

In closing this section, I give the account of a large solitary wave, described in the book by Paterson. He states

> An earthquake in Alaska in 1958 loosened an unstable rock face above the head of Lituya Bay (which is a fjord in the coastal mountains), and an estimated 9×10^7 tonnes of rock fell into the water. Most of the rock motion was vertical, and most of the energy went into an enormous splash, which reached some 550 m up the opposite mountain. The displacement of water gave a huge solitary wave down the bay, which is roughly a channel of uniform depth. Eye-witnesses estimated the wave height to be about 15–30 m, and the speed to be around 45 m s^{-1}. The wave cleared the shore of trees to a height of over 30 m at a large distance from the original rock-fall.

This, it must be said, was clearly not a linear wave.

TIDES: DIFFERENTIAL GRAVITATIONAL FORCES

We have referred to tides and pointed out that they are caused, not by gravitational forces per se (though they would not exist without them), but by the *variation* of these forces with distance. It is readily shown that tidal forces "fall off" with distance considerably faster than these more familiar forces, varying inversely with the *cube* of the distance (as opposed to the inverse square of the distance for gravitational forces). Newton's law of universal gravitation states that any two objects of mass M and m say experience a mutual gravitational attraction of force F, where

$$F = G\frac{Mm}{R^2},$$

where G is the so-called gravitational constant ($G = 6.67259 \times 10^{-11}$ m^3 kg^{-1}s^{-2}) and R is the distance between their centers of mass. Deriving the result we seek algebraically at first, consider the difference in such gravitational forces if the distance R changes to $\widetilde{R} = R + \delta R$. Then the consequent change in the mutual gravitational force is

$$\delta F = F(\widetilde{R}) - F(R) = -GMm\left[\frac{(2R + \delta R)\delta R}{R^2\widetilde{R}^2}\right] \approx -2GMm\frac{\delta R}{R^3} \propto R^{-3},$$

the approximation being valid provided $R \gg \delta R$, as is usually the case. Note that if $\delta R > 0$ (the objects become farther apart) then $\delta F < 0$ (the force is reduced), as would be expected. Correspondingly obvious results follow if $\delta R < 0$. This means that for small relative changes in distance R, the change in the gravitational force varies inversely with the cube of R.

Not surprisingly, we can do this a little more quickly and elegantly via differential calculus by noting that if $T(R)$ is the tidal force defined as the negative derivative of $F(R)$, then

$$T = -\frac{dF}{dR} = \frac{2GMm}{R^3}.$$

Using the differential notation $dF = F'(R)\,dR$, where $dR = \delta R$, we may express the relative change in forces such that

$$\frac{dF}{F} = -2\frac{dR}{R},$$

whereas

$$\frac{dT}{T} = -3\frac{dR}{R}.$$

Thus, for a given percentage change in R, there is a corresponding *double* change in F and a *triple* change in T. Noting the sign changes, a 1 percent increase in the earth-moon distance gives rise to a 2 percent decrease in the resulting gravitational force and a 3 percent decrease in the tidal force, all at any given distance along the line joining the centers of mass (things get a little more complicated if we do not consider point masses!—see Tricker's book on waves). Let's put some more numbers to work.

The distance of the moon from the earth varies between $R_{\min} = $ 356,334 km (221,463 mi) and $R_{\max} = 406{,}610$ km (252,710 mi), so if $\delta R = R_{\max} - R_{\min}$, then

$$\frac{\delta R}{R_{\min}} = \frac{R_{\max}}{R_{\min}} - 1 \approx 0.14 \text{ or } 14 \text{ percent,}$$

so that at $R_{\max}$ the gravitational force and the tidal force are reduced by approximately 28 percent and 42 percent of their respective values at $R_{\min}$. This is quite significant!

Now let us bring the sun into the situation, noting some important relative physical characteristics between the sun, earth, and moon. The mass of the sun, $M_\odot$, in terms of the mass of the earth, M_e, is

$$M_\odot \approx 3.33 \times 10^5 M_e.$$

Since $M_e \approx 81 M_m$ (the mass of the moon being M_m),

$$\frac{M_\odot}{M_m} \approx 2.7 \times 10^7.$$

Similarly, for the earth-sun and earth-moon radii ratio,

$$\frac{R_\odot}{R_m} \approx \frac{9.3 \times 10^7}{2.37 \times 10^5} \approx 3.9 \times 10^2.$$

Therefore the tidal forces produced on earth by the sun are

$$\left(\frac{M_\odot}{M_m}\right)\left(\frac{R_\odot}{R_m}\right)^{-3} \approx 0.46$$

times as large as those produced by the moon—less than half! It is interesting to note that the corresponding ratio of *gravitational* forces on the earth is

$$\left(\frac{M_\odot}{M_m}\right)\left(\frac{R_\odot}{R_m}\right)^{-2} \approx 178,$$

that is, much greater for the sun. Let us now look at a simple model concerning the effects of tidal friction.

THE POWER OF "TIDE": THE SLOWING POWER OF TIDAL FRICTION

What is the power dissipation of the earth's rotation due to ocean tidal friction? This question was neatly addressed in a short note by Shea in 1987 and his calculations are adapted to the present context. According to one estimate from astronomers, this loss of rotational energy slows the "day" by about 16 seconds every million years. Following Shea, we make the assumption that this change is due primarily to power dissipated by the tides. Let's see where all this leads.

The rotational kinetic energy K of the earth is given by

$$K = \frac{1}{2}I\omega^2,$$

where I is the moment of inertia of the earth about its axis and ω is the angular speed of rotation of the earth. If the earth were a homogeneous sphere of radius R, then I would be equal to $\frac{2}{5}MR^2$, where the mass $M = \frac{4}{3}\pi R^3 \rho$, ρ being the uniform density, as is readily proved in applications of typical "Calculus I" classes. However, the earth has a low-density crust and a high-density core, so in fact $I = 0.3444MR^2$. Clearly $\omega = 2\pi/T$, where T is the length of one day in seconds. Thus

$$K = 0.6888M\frac{R^2\pi^2}{T^2}.$$

Since $M = 5.98 \times 10^{24}$ kg and the (average) value of $R = 6.38 \times 10^6$ m, it follows that $K = 1.655 \times 10^{39}\ T^{-2}$ J s^2 (energy is in units of joules, J; 1 joule = 1 newton-meter = 1 kg m^{2}s^{-2}). Letting $\Omega = 1.655 \times 10^{39}$ T^{-2} J s^2, it follows from the use of differentials that if

$$K = \Omega T^{-2}$$

then

$$\delta K = -2\Omega T^{-3}\delta T.$$

Clearly $\delta T = -16$ s, $T = 8.64 \times 10^4$ s, so

$$\delta K \approx 8.2 \times 10^{25} \text{ J}.$$

To estimate the energy dissipation rate (or power P), divide this by the million years ($\approx 3.15 \times 10^{13}$ s) to get

$$P = 2.6 \times 10^{12} \text{ W}$$

(power is in units of watts, W; 1 watt = 1 joule per second). For purposes of comparison, note that large commercial generating plants produce energy at a rate of 1×10^9 W.

A final point concerns the relationship between what we have been discussing and the rather surprising fact that we can infer the ratio of the densities of the sun and moon from a knowledge of their respective tidal contributions (and vice versa).

TIDES, ECLIPSES, AND THE SUN/MOON DENSITY RATIO

Generalizing the work by Hodges (1987) a little, we will show that with reasonable assumptions,

$$\frac{\text{amplitude of terrestrial tides due to the moon}}{\text{amplitude of terrestrial tides due to the sun}}$$

$$\approx \frac{\text{mean density of the moon}}{\text{mean density of the sun}}.$$

As noted above, tides are due to *differential* gravitational forces between mutually attracting objects, so that they are proportional to M/x^3, where M is the mass of the "other" object (sun or moon) and x is the earth-sun/moon distance (or more accurately, the distance between the centers of mass of the two; the letter R is now reserved for the radius of an object). To see this, note that the magnitude of the mutual gravitational force between two objects of mass M and m is

$$F = G\frac{Mm}{x^2},$$

where G is the constant of universal gravitation (6.67×10^{-11} N m^2 kg^{-2}). Then the differential gravitational force is

$$\frac{dF}{dx} = -2G\frac{Mm}{x^3} \propto \frac{M}{x^3} \propto \frac{R^3}{x^3}\bar{\rho},$$

where for a uniform spherical body of radius R and density $\bar{\rho}$, $M = \frac{4}{3}\pi R^3 \bar{\rho}$. Neither the sun nor the moon is uniform in composition or perfectly spherical; we shall not concern ourselves with the minor deviations from sphericity however. Suppose that the density distribution for the sun or moon, $\rho(r)$, is continuous in $[0.R]$. Then we define the mean density $\bar{\rho}$ by

$$\bar{\rho} = \left(\frac{4}{3}\pi R^3\right)^{-1}\int_0^R 4\pi r^2 \rho(r)\, dr$$

$$= \frac{3}{R^3}\int_0^R r^2 \rho(r)\, dr.$$

It is this that is to be used in the above formula for the relative tidal amplitudes. Essentially, a sphere the size of the sun (moon) with uniform density $\bar{\rho}$ will have exactly the same mass as the sun (moon), and that is how we define the mean density. It might be interesting to pursue the above expression for $\bar{\rho}$ for certain profiles $\rho(r)$ using the data from the paper by Hodges (1991) (homework for the reader!).

Okay, where were we? Ah yes; now we use the fact that total solar eclipses occur to imply that the angular sizes of the sun and moon are approximately equal (as we saw in chapter 4, both objects subtend an angle of about $\frac{1}{2}^{\circ}$ at the earth's surface; that is equivalent to a baby aspirin held at arm's length by an adult). This means in turn that R/x and $(R/x)^3$ are approximately the same for the sun and the moon; and this in turn means that the ratio of the respective tidal forces (and hence tidal amplitudes due to them) is simply the ratio of the mean densities. Our result is established! The ratio of mean densities (moon:sun) is about 3.3:1.4 or 2.4; the tidal forces due to the moon are more than twice those due to the sun. Of course, it also works in reverse: if we can measure the tidal amplitudes we can infer the relative mean densities. A more exact calculation based on the actual *mean* values of R/x for the moon (0.00452) and the sun (0.00465) yields the value 2.2 instead of the above 2.4. Note that the accurate values of R/x explain why there are more annular than total solar eclipses (do you see why?).

The Fibonacci Sequence and the Golden Ratio (τ)

> ... and let them measure the pattern.
> —Ezekiel 43:10 (KJV)

> See how the lilies of the field grow. They do not labor or spin. Yet I tell you that not even Solomon in all his splendor was dressed like one of these.
> —Matthew 6:28, 29

Although originally considered by Fibonacci of Pisa in 1202 in connection with (idealized) rabbit population growth, the infinite set of numbers 1, 1, 2, 3, 5, 8, 13, 21, 34, 55, 89, 144, ... has a wonderful variety of properties and applications (but not as many as some have claimed—the article by Markowsky, discussed below, deals with various misconceptions about the golden ratio τ in connection with many human "constructions"). Before discussing those properties, I should clarify what I mean by the word "idealized" above. The rabbits were supposed to reproduce according to strict rules concerning their numbers and frequency of producing offspring, and they were assumed to be immortal! While these assumptions left something to be desired, they did result in a fascinating and almost ubiquitous sequence of numbers (at least in nature), to which we now return.

With the exception of the first two terms in the above sequence, each term is the sum of the two immediately preceding terms, that is, if f_n represents the nth term in the sequence ($n = 1, 2, 3, \ldots$), then for $n \geq 3$

$$f_n = f_{n-1} + f_{n-2}.$$

As will be seen below, it can be shown that

$$\lim_{n \to \infty} \frac{f_n}{f_{n-1}} = \frac{1 + \sqrt{5}}{2} = 1.61803398\ldots$$

This number is called the golden ratio (or golden number, golden mean, divine proportion) and we denote it by the Greek letter τ. There are various geometric representations of this fascinating number, and some of them are depicted in figure 10.1a, b. The mathematical details of the results stated here will be provided in a later section. Consider first the line segment AB

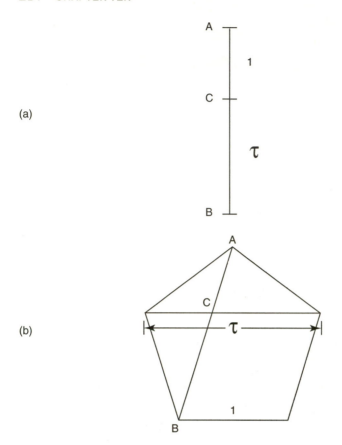

(a)

(b)

Figure 10.1. (a) The golden number τ is defined by the ratio $\tau : 1 = (\tau + 1) : \tau$; (b) in a regular pentagon with unit sides, each diagonal has length τ. Also, diagonals intersect in such a way that $BC : CA = \tau$. From "The Mathematical Daisy," by R. Dixon (1981), *New Scientist*, 17 December: 792–795. Reprinted by permission of *New Scientist*.

divided at C so that the ratio of the larger part (CB) to the smaller part (AC) is the same as the ratio of the whole (AB) to the larger segment (CB). If $AC = 1$ unit, and $CB = \tau$ units in length, then it is easily found that $\tau = (1 + \sqrt{5})/2$. The Greeks referred to this procedure as dividing a line in "extreme and mean ratio." This is also sometimes referred to as the golden section; it arose historically in connection with the properties of the regular pentagon, as will be established shortly, after reviewing some other general features of τ.

In particular, if we draw a rectangle whose adjacent sides are in the ratio $\tau : 1$, this "golden rectangle" can be decomposed into a nested set of ever-decreasing golden rectangles. By marking off a square of side 1 from the original figure, one is left with a smaller golden rectangle, and the process can be continued recursively. Corresponding points of each square spiral inward along a logarithmic or equiangular spiral (or approximately so: qualitatively similar spirals can be generated by using other rectangular forms, of course). This process naturally can be reversed, and since each step generates a larger (or smaller) geometrically similar version of the golden rectangle, we have a rectangular equivalent of the chambered Nautilus shell exhibiting uniform or *gnomonic* growth (figure 10.2). Peregrine falcons are believed to fly in such spiral paths when pursuing prey (see Tucker 2000b, c).

In a regular pentagon, any diagonal is τ times as long as a side, and furthermore two intersecting diagonals divide each other in the golden section $\tau : 1$, as will be proven later. In a circle, the golden angle is subtended by an arc that is the equivalent of "1" in the golden section: thus if α is the golden angle (in radians) then

$$\frac{2\pi - \alpha}{\alpha} = \frac{2\pi}{2\pi - \alpha},$$

yielding a quadratic equation in α with the smallest root being $\alpha = (3 - \sqrt{5})\pi$ radians, or $\approx 137.5°$ (see figure 10.3).

Another feature of the number τ, seemingly unrelated to the above, is its representation as a so-called *continued fraction*. We shall have more to

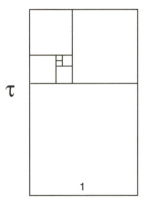

Figure 10.2. A "gnomon" consisting of successively smaller squares and golden rectangles. The largest golden rectangle is divided into a square, leaving another golden rectangle, and the process continues indefinitely (in principle). From "The Mathematical Daisy," by R. Dixon (1981), *New Scientist*, 17 December: 792–795. Reprinted by permission of *New Scientist*.

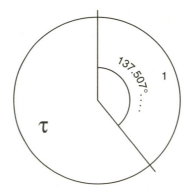

Figure 10.3. The golden angle, derived in the same manner for a circle as for a line segment. From "The Mathematical Daisy," by R. Dixon (1981), *New Scientist*, 17 December: 792–795. Reprinted by permission of *New Scientist*.

say about this in a later section, but it will suffice for now to note that

$$\tau = 1 + \cfrac{1}{1 + \cfrac{1}{1 + \cfrac{1}{1 + \cfrac{1}{1 + \cfrac{1}{1 + \cdots}}}}}.$$

PHYLLOTAXIS

Phyllotaxis is the distribution or arrangement of leaves on a stem and the mechanisms that govern it. The term is used by botanists and mathematicians to describe the repetitive arrangement of more than just leaves; petals, seeds, florets, and branches (sometimes) also qualify. These arrangements are closely related to the well-known and previously mentioned Fibonacci numbers 1, 1, 2, 3, 5, 8, 13, 21, 34, 55, 89, . . . and the related golden number or ratio $\tau = (1 + \sqrt{5})/2 \approx 1.618034$ (sometimes, the reciprocal of τ, $\tau^{-1} \approx 0.618034$ is referred to as the golden ratio). Numerical and geometric patterns based on these numbers abound in nature and have been studied for hundreds of years, and for that reason alone the basic features of phyllotaxis can be found in many elementary texts.

Examine some flower petals as examples of this. Lilies have 3 petals, buttercups 5, some delphiniums 8, marigolds 13, asters 21, and daisies 34, 55, or even 89. And there *are* exceptions: the geometer H.S.M. Coxeter has written in his book *Introduction to Geometry* that "phyllotaxis is really not a universal law but only a fascinatingly prevalent tendency." Plants in general face predicaments that we humans can identify with: how to occupy space, receive sunlight, and interact with the environment in an optimal fashion. As a branch grows upward it generates leaves at regular angular intervals that branch out from the stem. Obviously, if these angular intervals are *exact* rational multiples of 360°, then the leaves will grow directly above one another in a set of rays (as viewed from above), and would inhibit those below from sunlight and moisture to some extent (think about leaves sprouting every 180°, 1/2 revolution, or 90°, 1/4 revolution).

In practice, plants seem to choose rational approximations to the "most irrational" number in order to optimize leaf arrangement! We will examine that strange statement a little later on, but for now we note that, depending on the plant, leaves are generated after *approximately* (a very significant word here) 2/5 of a revolution or circle (for oak, cherry, apple, holly, plum), 1/2 (elm, some grasses, lime, linden), 1/3 (beech, hazel), 3/8 (poplar, rose, pear, willow), and 5/13 (almond). Other approximations are 3/5, 8/13, 5/13, ... ; these are called phyllotactic ratios, and, as you will have noticed, their numerators and denominators are Fibonacci numbers (though not necessarily consecutive ones). An example of a 3/8 phyllotactic ratio (along with many other examples) can be found in the book by Garland; 8 stems are generated in 3 complete turns, not counting the original stem.

Probably the most striking illustration of phyllotaxis is to be found in the arrangement of seeds on a large sunflower head. They are distributed in two families of spirals, perhaps 34 winding clockwise and 55 counterclockwise, or (55, 89) in the same order, or even (89, 144) in especially large specimens. Similar patterns occur in daisies also, though of course they are more compact than in sunflowers. Brousseau carried out a detailed examination of such spirals, particularly in pinecones. Again, there are in general two sets of parallel bract spirals, a steep one from lower right to upper left, and a shallower one from lower left to upper right, perhaps 8 of the former and 5 of the latter, or (3, 5), or (8, 13). According to Brousseau, at least 95 percent of the spiral numbers in pinecones are from the Fibonacci sequence. Such spirals also arise in the petals of artichokes and the "scales" of pineapples. In the latter case, three sets of spirals usually occur because the hexagonal-shaped scales have three pairs of opposite sides. The three sets of parallel spirals are usually Fibonacci numbers (e.g., 8, 13, 21) and are shallow, medium, and steep, respectively.

The question arises—how do these spirals come to be in the first place? In order to answer this question I shall draw on the very clear explanation given by Ian Stewart in his book *Life's Other Secret*. A plant grows upward by generating new cells at the tip of the sunlight-seeking shoot. The potential for spiral formation is already laid down at this stage, as the new cells take up their positions among their neighbors. At the center of the tip is a small circular region of tissue called the *apex*, and around this apex form small lumps called *primordia*. As the apex grows away from the primordia, they "do their own thing" by developing into petals, leaves, seeds, florets, branches, or whatever the case may be. However, the spirals we see—the *parastichies*—are "merely" by-products of the sequence of events in the plant's growth pattern, a striking optical illusion. The order in which the primordia appear is of fundamental importance here. They trace out a tightly wound spiral—the generative spiral. The farther a primordium is from the apex, the earlier it was formed.

Seen from above or from the center of the apex, the angle between successive primordia is always about the same ($\approx 137.5°$; see figure 10.5). This is called the divergence angle, and is intimately related to the golden ratio τ, as we have already seen and develop below in some detail. For now, however, if we take the ratios of consecutive numbers in the Fibonacci sequence, multiply by 360°, and then subtract that answer from 360° (because we use the internal angle, which is less than 180°) we get a set of approximations to $137.50776\ldots°$, a more accurate expression of the golden angle (which as we have seen is the mathematical equivalent of an idealized divergence angle). Thus, starting with the pair $(3, 5)$

$$360°(1 - 3/5) = 360° \times 2/5 = 144°,$$

$$360°(1 - 5/8) = 360° \times 3/8 = 135°,$$

$$360°(1 - 8/13) = 360° \times 5/13 \approx 138.5°,$$

$$360°(1 - 13/21) = 360° \times 8/21 \approx 137.1°,$$

$$360°(1 - 21/34) = 360° \times 13/34 \approx 137.6°,$$

$$360°(1 - 34/55) = 360° \times 21/55 \approx 137.5°.$$

Thus each successive ratio of Fibonacci numbers provides a better approximation to the golden angle. It turns out that if the divergence angle is *less* than 137.5°, gaps appear in the seed head, and only one family of spirals is seen. Gaps appear again if the divergence angle is *more* than 137.5°, but this time only the other family of spirals is noticeable (see figure 10.4). The golden angle thus hits the sunflower on the head, so to speak, for it is the *only* angle at which seeds pack together without gaps, and at this angle both spirals occur simultaneously. As Stewart points out, this most efficient packing makes for a solid and robust seed head.

Let us reintroduce the vertical dimension and remind ourselves that every *leaf* also needs its place in the sun, and this is best achieved in a direction not already occupied by a previous one. Thus, as one by one the remaining largest gaps get filled by a socially responsible leaf, things become crowded

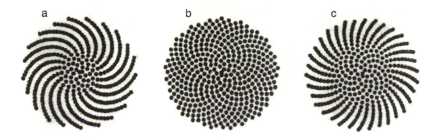

Figure 10.4. Spiral patterns based on the "nearly golden angles" (a) 137.3°;
(c) 137.6°. The golden angle 137.5° is used in (b). From *The Algorithmic Beauty of Plants*, by P. Prusinkiewicz and A. Lindenmayer (1990), reprinted by permission of Springer-Verlag.

but in an even and gradual manner, a manner that is only possible with the golden angle as the divergence angle (figure 10.5). As early as 1914, in the first edition of his book, Theodore Cook wrote "the fact that plants express their leaf arrangement in terms of Fibonacci numbers, so frequently that it passes for the normal case, is the proof that they are aiming at the utilisation of the Fibonacci angle which will give minimum superposition and maximum exposure to their assimilating members." While we might disagree with the imputation of motive to the plants, the basic ideas, if not the complete mechanism underlying them, are still quite appropriate today.

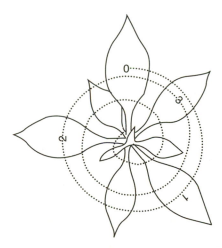

Figure 10.5. The arrangement of leaves on a stem according to the golden angle principle, viewed from above. From "The Mathematical Daisy," by R. Dixon (1981), *New Scientist*, 17 December: 792–795. Reprinted by permission of *New Scientist*.

We have already noted that any rational multiple of 360° (or 2π radians) will not produce an efficient seed or leaf arrangement, merely rays with gaps between them. For example, if p and q are positive integers, $(360p/q)°$ gives q radial lines. So the multiple of 360° must be *irrational*—not an exact fraction. But which irrational multiple is best (there are, after all, an infinity of them to choose from)? The quick answer is one that never settles down to a *rational approximation* for very long, but let us again follow Stewart in his discussion on this. There is a "most irrational" number, and it turns out to be (surprise, surprise) the golden number. The golden number (or its reciprocal, as we have defined it) is the limit of the sequence of ratios of consecutive Fibonacci numbers,

$$2/3, 3/5, 5/8, 8/13, 13/21, 21/34, 34/55, \ldots,$$

etc., each new ratio getting closer and closer to $0.6180339\ldots = \tau^{-1}$. These are rational approximations to τ^{-1}. A measure of how irrational a number is can be determined from the errors between the approximations and the number itself. How fast does this difference—the error—shrink to zero? It can be proved that for τ (or its reciprocal) the errors shrink more slowly than for *any* other irrational number. It is the most badly "approximable-by-rationals" number there is! Its "badness" is exceeded only by the awkwardness of the preceding sentence. So the golden number is very special, surprising, and strange.

Another, less noticeable feature of sunflower or daisy parastichies is worth pointing out. As one moves across the disk (the "capitulum"), the number of parastichies changes from one pair of Fibonacci numbers to another in a rather gradual fashion. During growth of the plant (according to Alan Turing) various such parastichy numbers come into prominence at different stages. An increasing number of florets is needed to pack around the growing circumference as the growth rate of the florets slows down. As Dixon points out, if the floret growth rate is exponential, the parastichy numbers stay the same, "but the smooth way in which the contacting systems switch from one set to another without the pattern departing too far from a regular packing is perhaps the crucial and unique property of Fibonacci phyllotaxis. The comparable phenomenon of being non-golden angles is always less regular." (But, as noted below, a "$\sqrt{2}$ flower" comes close to this.)

A phenomenological model describing the distribution of florets in a sunflower head was provided by Vogel. If θ is the angular position of the nth floret, measured from some reference direction in a polar coordinate system originating at the center of the capitulum, then, to a good approximation,

$$\theta \approx n \times 137.5°$$

and

$$r \approx c\sqrt{n},$$

r being the distance between the center of the capitulum and the center of the nth floret. The constant of proportionality c is called the scaling parameter (in this model the florets themselves do not grow in time). The nth floret is counted outward from the center, so it represents, in discrete terms, the "reverse age" of the floret. The square-root dependence of r on n is readily established in this model by noting that, if the florets are all the same size and are densely packed, the total number that can fit inside a circle of radius r is proportional to its area,

$$n \sim r^2 \quad \text{or} \quad r \sim \sqrt{n}.$$

This model is a useful one for describing the gross features of floret arrangement, but much more detailed mathematical models are required to adequately explain the divergence angle of 137.5° (see the bibliography). Naylor's paper "Golden, $\sqrt{2}$ and π Flowers: A Spiral Story," on the fact that spiral arms in such "golden flowers" seem to fall into certain families, one set with 8 spiral arms, one with 13, another with 21, and so on. By numbering seeds consecutively as they "move" outward (the most recent being number 1) Naylor shows that a Fibonacci-numbered seed will fall close to an arbitrarily chosen 0° line, since from the above expression for θ ($\theta \approx n \times 137.5°$) θ will be close to an integer multiple of 360°. Thus if k is a given Fibonacci number, there is a spiral composed of seeds that are multiples of k, because each such multiple $(k, 2k, 3k, \ldots, pk, \ldots)$ is rotated by the same amount from the previous one. This will also occur for seeds that *differ* by such multiples, so that in general members of the "k family" are seeds with numbers $pk + q$, where p and q are non-negative integers, q being fixed for that particular spiral arm. A natural question to ask (and Naylor does) is whether any other irrational number (such as π or $\sqrt{2}$) would work just as well as τ in reproducing these common spirals in nature. We have essentially addressed this issue above (see below also), and we leave it to the interested reader to identify the connections between the two approaches. Naylor explains why the first two dominant spirals in a "π-flower" have 7 and 113 arms, respectively, and why the "$\sqrt{2}$-flower" is a much closer rival to the golden one.

THE GOLDEN RATIO AND THE REGULAR PENTAGON

What have they to do with each other? Let's get straight into some geometry and find out, shall we? The following argument is adapted from that by Higgins. Recall that the sum of the interior angles of any n-gon is $(n-2)\pi$

radians, or $(n-2)180°$. For a *regular* polygon, therefore, each such angle is $(1-2/n)\pi$ radians. In particular, for $n = 5$ each interior angle (e.g., $\angle ABC$) is 108°. Referring to figure 10.6, note that, since $\triangle ABC$ is isosceles,

$$\angle BAC = 36° = \angle EAD = \angle CAD.$$

It follows that

$$\angle BAK = 72° = \angle BKA$$

(Triangles ABC and ABE being congruent), and hence that the line segment BK is of unit length (since ABK is isosceles). From the notation of the diagram we see that the length of the diagonal BE can be expressed as

$$BE = BK + KE = d = a + 1. \tag{1}$$

The triangles ABE and AKE are geometrically similar because their corresponding angles $(108°, 36°, 36°)$ are equal, so by taking appropriate ratios of corresponding sides we find that

$$\frac{d}{1} = \frac{1}{a} \quad \text{or} \quad ad = 1. \tag{2}$$

Combining these two relations between a and d, it follows in particular that

$$d^2 - d - 1 = 0, \tag{3}$$

the positive root of which is

$$d = \frac{1 + \sqrt{5}}{2} = 1.61803398\ldots \equiv \tau,$$

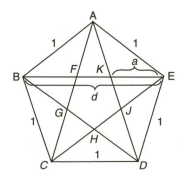

Figure 10.6. The figure used in the proof of the relationships between the properties of a regular pentagon and the golden number.

so now we have revealed to the world that d is τ. Note of course that

$$a = d - 1 = \tau - 1 = \frac{\sqrt{5} - 1}{2} = 0.61803398\ldots$$

Returning to the geometry of the pentagon, equation (2) is equivalent to the statement that

$$\frac{d}{BK} = \frac{BK}{a},$$

that is, the diagonals of a regular pentagon divide each other in the golden ratio, so the comments made in connection with figure 10.1b have been proved. Also we have seen that, in a regular pentagon with unit sides, the length of a diagonal is the golden ratio, often denoted by ϕ as well as τ. This number is irrational, as can be seen in several ways, two of which we mention here. Both theorems (and the corollary that follows) use the *reductio ad absurdum* method of proof, which assumes initially that the theorem as stated is *not* true, and then establishes a contradiction, demonstrating that the original premise was false.

Theorem 1. τ is an irrational number.

Proof. τ is a solution of equation (3) above. Suppose that it is a rational number, i.e., $\tau = p/q$ in lowest terms, where p and $q < p$ are positive integers. Then substituting this in equation (3) and rearranging the result as

$$p(p - q) = q^2 \tag{4}$$

it follows that p divides q^2, but since p and q possess no common factor greater than unity, it must be the case that $p = 1$. But equation (4) may also be written in the form

$$p^2 = q(p + q), \tag{5}$$

which means that q divides p^2; and for the same reason as before, $q = 1$. This means that $\tau = 1$, but this is obviously not a solution of equation (3), so we have established a contradiction. Therefore τ is irrational.

This is certainly a succinct little proof (see Ron Knott's extensive website on Fibonacci numbers: *http://www.mcs.surrey.ac.uk/Personal/R.Knott/Fibonacci/fib.html*), but there is another (stated as a corollary below), based on the *fundamental theorem of arithmetic,* which we state without proof. This theorem states that a composite natural number can be written uniquely as a product of prime numbers in order of increasing size. We start with a general result which will be used below:

Theorem 2. The square root of any prime number is irrational.

Proof. Let p be any prime number, and suppose that its square root is rational, that is, $\sqrt{p} = a/b$ or

$$pb^2 = a^2, \tag{6}$$

where a and b are natural numbers with no common factors greater than unity. Suppose that a is the product of d prime numbers

$$a = a_1 \times a_2 \times a_3 \times \cdots \times a_d$$

and that b is the product of e prime numbers

$$b = b_1 \times b_2 \times b_3 \times \cdots \times b_e$$

Then it follows that a^2 is the product of $2d$ primes, and b^2 the product of $2e$ primes. But if equation (6) is correct, the left side of that equation is the product of an odd number of primes while the right side is the product of an even number of primes. This contradicts the above-stated fundamental theorem of arithmetic, so we have our sought-for contradiction, thus establishing the theorem, and in particular, that $\sqrt{5}$ is irrational.

Then we have the following result:

Corollary 1. τ is an irrational number.

Proof. As in theorem 1, let $\tau = p/q$ in lowest terms, where p and $q < p$ are positive integers. Then, from the definition of τ,

$$\frac{p}{q} = \frac{1 + \sqrt{5}}{2}$$

or

$$\sqrt{5} = \frac{2p - q}{q},$$

which is a contradiction, thus establishing the result for a second time.

There are two more particular results I would like to share since we are no doubt by now well into a theorem-proving mode. As above, we let the Fibonacci sequence be represented as

$$f_1, f_2, f_3, \ldots, f_n, \ldots$$

Then we state

Theorem 3. The number f_n is the nearest integer to the nth term a_n of the geometric progression whose first term is $\tau/\sqrt{5}$ and whose ratio is τ.

Before proving this, let us ask what it means (which is always a good idea). Essentially, it means that we may approximate a_n by f_n (and more important, vice versa) and that this approximation gets better with increasing n (from the corollary). We need to find an explicit expression for f_n. This can be established from the relationship satisfied by any three consecutive terms $\{f_n, f_{n+1}, f_{n+2}\}$ in the Fibonacci sequence (for $n \geq 1$), namely

$$f_{n+2} = f_{n+1} + f_n. \tag{7}$$

This is a linear second-order *difference equation* with solution characterized by linear combinations of the form λ^n for some numbers λ. Substituting $f_n = \lambda^n$ into equation (7) gives by now familiar equation $\lambda^2 - \lambda - 1 = 0$ with roots

$$\lambda_1 = \frac{1 - \sqrt{5}}{2} \equiv \beta \quad \text{and} \quad \lambda_2 = \frac{1 + \sqrt{5}}{2} \equiv \tau.$$

Note that

$$-1 < \lambda_1 < 0 \quad \text{and} \quad \lambda_2 > 1,$$

so λ_2 is the dominant eigenvalue. Note further that

$$\lambda_1 = 1 - \lambda_2 = -\tau^{-1}.$$

The general solution to equation (7) is

$$f_n = A_1 \beta^n + A_2 \tau^n, \tag{8}$$

where A_1 and A_2 are arbitrary constants. They can be fixed for the Fibonacci sequence

$$(0), 1, 1, 2, 3, 5, 8, 13, 21, 34, 55, 89, \ldots$$

by choosing any two terms as "initial conditions."

Note that a term $f_0 = 0$ has been placed at the beginning of the sequence; while it is certainly not necessary to do this, it still permits the difference equation to be satisfied (now for $n \geq 0$), and it has the advantage of simplifying the ensuing algebra. Thus since $f_0 = 0$ and $f_1 = 1$ the following two equations must be consequences of equation (8):

$$A_1 + A_2 = 0 \quad \text{and} \quad \beta A_1 + \tau A_2 = 1. \tag{9}$$

Solving these gives the coefficients as

$$A_1 = -\frac{1}{\sqrt{5}} = -A_2,$$

and so

$$f_n = \frac{1}{\sqrt{5}} (\tau^n - \beta^n). \tag{10}$$

This is called Binet's formula. For large values of n this is cumbersome to calculate by hand (not that we need to do this any more), so a simple approximation to f_n would be nice to have. This is readily available because now we are in a position to prove theorem 3.

Proof.

$$|f_n - a_n| = \frac{1}{\sqrt{5}} |(\tau^n - \beta^n) - \tau^n| = \frac{|\beta^n|}{\sqrt{5}} = \frac{|\beta|^n}{\sqrt{5}} < 1.$$

Indeed, since

$$\frac{|\beta|^n}{\sqrt{5}} < \frac{1}{2} \quad \text{for all } n \geq 1,$$

it follows that

$$|f_n - a_n| < \frac{1}{2}$$

always, and so the theorem is proved.

In other words, the magnitude of the difference between any term in the Fibonacci sequence and the corresponding monomial in τ, namely a_n, is less than 1/2, and so a_n will approximate f_n to the nearest integer (and vice versa). Furthermore, this approximation gets better with n as corollary 2 states (the proof being obvious):

Corollary 2.

$$\lim_{n \to \infty} |f_n - a_n| = 0.$$

How good is this approximation? Pick your favorite integer: 14 you say? Well

$$a_{14} = \frac{\tau^{14}}{\sqrt{5}} \approx 377.0006$$

compares rather well with exact value of $f_{14} = 377$. What about a real test, though? The sequence $\{a_n\}$ is approximately given by

$$0.73, 1.17, 1.89, 3.07, 4.96, \ldots$$

whereas the sequence $\{f_n\}$ is

$$1, 1, 2, 3, 5, \ldots$$

(having now ignored the "zeroth" term introduced earlier for convenience). Things begin to look rather good at the fourth term, but it's actually better than that, because we only need to round to the nearest integer, as stated in the theorem.

In order to come full circle on some of the earlier comments made about the mathematical nature of the golden ratio τ, it is necessary to spend a little time on the subject (mentioned earlier) of *continued fractions*. We shall concentrate on a special class called simple continued fractions, which are of the form

$$a_1 + \cfrac{1}{a_2 + \cfrac{1}{a_3 + \cfrac{1}{a_4 + \cdots}}}$$

in which a_1 is usually an integer (or may be zero) and all the other a_i are positive integers. Such continued fractions may be finite $(1 \le i \le N)$ or infinite. The former class consists of terminating continued fractions and is commonly written in terms of the partial quotients a_i as $[a_1, a_2, a_3, \ldots, a_N]$. The process is well illustrated with reference to a rational approximation for π, but first we consider the case of a rational number, no doubt your favorite and mine: 67/29. The procedure is similar to that of the Euclidean algorithm. We write

$$\frac{67}{29} = 2 + \frac{9}{29} = 2 + \frac{1}{29/9} = 2 + \frac{1}{3 + 2/9} = 2 + \frac{1}{3 + \frac{1}{9/2}} = 2 + \frac{1}{3 + \frac{1}{4 + 1/2}},$$

at which point the process terminates because the reciprocal of the last entry (1/2) is 2. Hence

$$\frac{67}{29} = 2 + \cfrac{1}{3 + \cfrac{1}{4 + \frac{1}{2}}} = [2, 3, 4, 2].$$

(Since in the last fraction we may write

$$\frac{1}{a_4} = \frac{1}{2} = \frac{1}{1 + \frac{1}{1}},$$

it is clear that we may also write

$$\frac{67}{29} = [2, 3, 4, 1, 1],$$

but this "slight" lack of uniqueness need not concern us here.)

If we carry out the same procedure for the approximation $\pi \approx \pi_{app} = 3.14159265$ (accurate to eight decimal places, the accuracy that follows), then

$$\pi_{app} = 3 + 0.14159265 = 3 + \frac{1}{1/0.14159265} = 3 + \frac{1}{7 + 0.06251349}$$

$$= 3 + \frac{1}{7 + \frac{1}{1/0.06251349}} = 3 + \frac{1}{7 + \frac{1}{15.99654731}} \approx 3 + \frac{1}{7 + \frac{1}{16}} = \frac{355}{113}.$$

Thus in the process of finding the partial quotients of the continued fraction for π_{app}, a rational approximation for π of 355/113 (≈ 3.1415929) has been constructed, which is only three ten-millionths different from the actual value of π to this accuracy. The partial quotients are $\pi_{app} = [3, 7, 16]$ for this approximated approximation! Compare this with the first seven quotients of the *infinite* continued fraction for π, namely $\pi = [3, 7, 15, 1, 292, 1, 1, \ldots]$.

Associated with the partial quotients of a continued fraction are the corresponding *convergents*. Referring to the general form $[a_1, a_2, a_3, \ldots, a_N]$ for a finite simple continued fraction, the convergents are obtained in succession by terminating the expansion procedure after the first, second, third, $\ldots$, steps. The first five convergents for π are

$$\frac{3}{1}, \frac{22}{7}, \frac{333}{106}, \frac{355}{113}, \frac{103993}{33102}.$$

Now let us return to the quadratic equation satisfied by the golden ratio τ, $\tau^2 = \tau + 1$. Since this may be written as

$$\tau = 1 + \frac{1}{\tau},$$

the continued fraction for τ can be readily deduced to obtain

$$\tau = 1 + \frac{1}{1 + \frac{1}{1 + \frac{1}{1 + \cdots}}} = [1, 1, 1, \ldots],$$

so all the partial quotients are one! This is the simplest of all infinite simple continued fractions. The convergents to τ very are interesting: they are

$$\frac{1}{1}, \frac{2}{1}, \frac{3}{2}, \frac{5}{3}, \frac{8}{5}, \frac{13}{8}, \frac{21}{13}, \ldots,$$

that is, both numerators and denominators are formed from consecutive integers in the famous Fibonacci sequence $1, 1, 2, 3, 5, 8, 13, 21, 34, 55, 89, 144, \ldots$

The infinite continued fraction for $\sqrt{2}$ can be shown to be $\sqrt{2} = [1, 2, 2, 2, \ldots] \equiv [1, \bar{2}]$, and the corresponding convergents are

$$\frac{1}{1}, \frac{3}{2}, \frac{7}{5}, \frac{17}{12}, \ldots$$

It can be shown that $e - 1 = [1, 1, 2, 1, 1, 4, 1, 1, 6, 1, 1, 8, \ldots]$; this and many other fascinating properties of irrational numbers and their relationships to continued fractions can be found in the book by Olds. However, for the beauty of simplicity the golden number just cannot be beaten! And as a precursor to the next chapter, consider the following interesting feature of the Fibonacci sequence (we'll meet yet another in chapter 12).

Many people are interested in their genealogy or family tree. The male bee has quite a remarkable one, as pointed out by Jean and Johnson. The eggs laid by a queen bee produce female bees if fertilized, and drones if unfertilized. The only ancestor of the drone is therefore the queen, who of course has a male and a female immediate ancestor. This is all the information we need to build up a family tree for the male drone, M. His ancestor is the queen, F, and hers are M, F. Continuing this backward we find 2 female and 1 male ancestor, preceded (eventually) by 8 females and 5 males at the seventh "level"; indeed, the first seven levels contain $1, 1, 2, 3, 5, 8, 13$ bees. At each level there is a Fibonacci number of males, a Fibonacci number of females, and a Fibonacci number of bees. So if you prefer not to use the difference equation (7) to generate Fibonacci numbers, go and count bees, as we shall do, in a manner of speaking, in the very next chapter.

However, before we do so, there is one loose end that requires tying up (isn't that impossible with only one end?), namely, the list of common misconceptions about τ that is discussed by Markowsky. In one sense there is a fine line between mathematics and numerology, and from time to time overzealous individuals cross it! Markowsky reminds us (and especially golden ratio enthusiasts) that "measurements of real objects can only be approximations." While such a reminder should be unnecessary, he goes on to note, "it is unfortunate that many writers on mathematical subjects treat measurements of real objects as if they were exact numbers." In what he calls the *Pyramidology fallacy* (following Martin Gardner), it is easy to find all sorts of apparent relationships between numbers (such as τ), dates, lengths, and so on if one substitutes wishful thinking for a systematic search pattern and consistency of approximation.

The following misconceptions are discussed in considerable detail in Markowsky's article: (i) the name "Golden Ratio" was used in antiquity; (ii) the Great Pyramid was designed to conform to τ (denoted by Φ in the article); (iii) the Greeks used τ in the Parthenon; (iv) many painters, including Leonardo da Vinci, used τ; (v) the United Nations building embodies the ratio τ; (vi) a golden rectangle is the most aesthetically pleasing rectangle; (vii) the human body exhibits τ, and finally, a claim that was new to me: (viii) Virgil's *Aeneid* exhibits τ. It is all fascinating stuff, and should be required reading for all fans of the golden number!

CHAPTER ELEVEN

Bees, Honeycombs, Bubbles, and Mud Cracks

> . . . sweeter also than honey and the honeycomb.
> (KJV)
>
> —Psalm 19:10b

A polygon, as the old joke goes, is not a dead parrot (but then neither do geometers move in the best circles). There is something fascinating about the symmetries of a regular polygon, and when polygons are approximated in nature, such configurations capture the attention of even the most casual observer. When outside, look up, and on occasion you may see hexagonal convection-cell clouds; look around, and you may see mud cracks or salt flats exhibiting the same type of pattern; the Giant's Causeway on the Antrim coast in Northern Ireland is composed of some 40,000 basalt column formations, most of them with hexagonal cross sections (though some have four, five, seven, or eight sides). Carefully examine a wasps' nest or a beehive, and again, approximately regular hexagonal cells (in cross section) abound.

Indeed, without stealing too much thunder from the study of the honeycomb below, it is readily appreciated that circular cells produce wasted space, whereas hexagonal cells are more efficient. In fact, if six regular hexagons are placed in a "ring" with each pair sharing one common side, a free hexagon appears in the center; and as pointed out by William Roberts, this sharing of sides by the original six effectively yields yet another, because the perimeters of five hexagons are used to "cover" the area of seven such hexagons. Consequently, the 7-hexagon honeycomb array, and by implication the whole honeycomb, maximizes the region enclosed while minimizing the perimeter (and hence the lateral surface area), and therefore the wax needed for construction of the array. This will be verified below, and in passing we will obtain some crude bounds on the value of π, including the lower bound of 3 cited in the Bible (though not as a lower bound!).

What is the area A and perimeter P of a regular n-sided polygon (or in brief, a regular n-gon)? Clearly such an n-gon always can be inscribed in a circle with appropriate radius. From the diagram (figure 11.1a) it is clear that n interior angles of 2θ each must total 2π radians, so $\theta = \pi/n$. Furthermore, if the radius of the exscribed circle is r, then the total area of

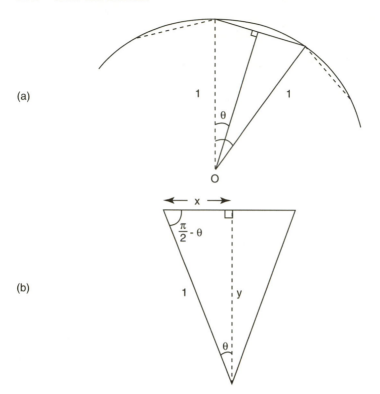

(a)

(b)

Figure 11.1. Geometric details for a regular n-gon inscribed in a circle of unit radius.

n interior triangular "cells" is from figure 11.1b

$$A = n \times \frac{1}{2} \times 2x \times y = nr^2 \sin \frac{\pi}{n} \cos \frac{\pi}{n}$$

and the perimeter P of the n-gon is

$$P = 2nr \sin \frac{\pi}{n}.$$

Now consider this n-gon to be the cross section of an open right prism of height h. The volume and surface area of this prism (excluding the polygonal bases) are, respectively, $V = Ah$ and $S = Ph$. Of interest to us here is the *volume-to-surface-area ratio* R (or the reciprocal of this, for that matter; we have already encountered that in an earlier chapter). This is independent of h; in fact

$$R = \frac{V}{S} = \frac{A}{P} = \frac{r}{2} \cos \frac{\pi}{n} = R(n).$$

Obviously n must be an integer greater than two.

n	$\cos\left(\frac{\pi}{n}\right)$	$R(n)$
3	0.5	0.25r
4	0.707	0.354r
6	0.866	0.433r
10	0.951	0.476r

The table lists four such values of n. It would appear from this table and basic geometrical intuition that $R(n) \to 0.5r$ as $n \to \infty$, which of course means that the n-gon becomes less and less distinguishable (to the eye) from a circle. In this limiting case, of course,

$$R = \frac{\pi r^2}{2\pi r} = 0.5r.$$

Although of course n is an integer, a graph of $R(n)/r$ is shown in figure 11.2 for n extended to continuous values, showing the asymptotic approach to 0.5. Obviously the quantity R is a maximum for a circular cross section; this can be interpreted as a maximum enclosed area for a given perimeter, or, in a dualistic sense, as the minimum perimeter for a given enclosed area. The problem is that if we wish to tessellate the plane with such shapes, a circle just will not do! In three dimensions the corresponding problem is that circular cylinders are not space-filling, which is not helpful if we wish to construct a honeycomb. What then are the criteria for regular polygons (regular prisms) to tessellate the plane (or fill two-dimensional space)? The answers readily spring to mind—only equilateral triangles, squares, and

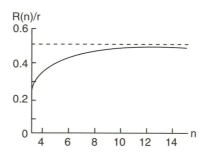

Figure 11.2. The volume-to-lateral surface area ratio for regular polygonal prisms (or equivalently the area-to-perimeter ratio for regular n-gons). Both ratios are normalized by the radius r of the escribed circle. Obviously, n is a positive integer, but the graph is drawn treating n as continuously varying; the horizontal asymptote defines the limiting value for a cylinder (or a circle).

regular hexagons can do this. While this may seen obvious, let's check it out a little further.

By examining the points of intersection of the vertices of adjacent regular polygons using the notation above, it is clear that an integer multiple N of the interior vertex angle ϕ for each polygon must equal 2π radians (so there is no overlap or "underlap"). Thus

$$\phi = 2\left(\frac{\pi}{2} - \theta\right) = \frac{\pi}{n}(n-2) = \frac{2\pi}{N},$$

so that

$$N = \frac{2n}{n-2} = 2 + \frac{4}{n-2}.$$

This result requires that the quantity $4/(n-2)$ should be a positive integer exceeding two. The only values, of n for which this can occur are $n = 3$ (equilateral triangles), $n = 4$ (squares), and $n = 6$ (regular hexagons), so these are the only regular n-gons that can tessellate the plane. From the earlier volume/surface area discussion, we can note that the closest such polygon to a circle (which maximizes the ratio $R(n)$) consistent with these spatial constraints is the regular hexagon.

Now no one is suggesting that bees have to take a course in optimization theory in order to make a honeycomb, because the same hexagonal pattern is evident in a collection of identical coins placed in contact on a flat surface (try it and see!). When the bees scurry around and wiggle together in the soft honeycomb, their bodies (which have approximately circular cross sections) automatically shape the same type of pattern in the wax. Pappus of Alexandria considered that bees are endowed with "a certain geometrical forethought" and (as quoted by D'Arcy Thompson) "There being, then, three figures which of themselves can fill up the space round a point, viz. the triangle, the square and the hexagon, and the bees have wisely selected for their structure that which contains most angles, suspecting that it could hold more honey than either of the other two." Later writers have suggested that this "economical" hypothesis was unjustified, and that the hexagonal cell was "no more than the necessary result of equal pressures, each bee striving to make its own little circle as large as possible."

All this leads naturally to a consideration of the area A and perimeter P of a regular n-gon escribing the circle of radius r mentioned earlier. Arguments similar to those above yield the following expressions for A and P:

$$A = nr^2 \tan \frac{\pi}{n}$$

and

$$P = 2nr \tan \frac{\pi}{n}.$$

Using the fact that an inner hexagon, circle, and outer hexagon have increasing P- and A-values, the following crude bound for π are readily obtained using the respective expressions for both P and A:

$$n \sin \frac{\pi}{n} < \pi < n \tan \frac{\pi}{n}$$

and

$$n \sin \frac{\pi}{n} \cos \frac{\pi}{n} < \pi < n \tan \frac{\pi}{n}.$$

For the particular case of $n = 6$ these reduce to what shall be termed "crude bee bounds," with no disrespect to bees, namely,

$$3 < \pi < 2\sqrt{3} \approx 3.464$$

and

$$\frac{3\sqrt{3}}{2} \approx 2.598 < \pi < 2\sqrt{3}.$$

Obviously from these bounds the closest integer approximation we may use is that $\pi \approx 3$, which may also be inferred from 1 Kings 7:23 and 2 Chronicles 4:2 in the Bible. Archimedes (b. 287 B.C.) did a much better job than either we or the bees have done; he started with the bounding regular hexagons and doubled the number of sides four times over. His resulting 96-sided polygons enabled him to determine that

$$3^{10/71} < \pi < 3^{1/7}, \text{ or } 3.140845\ldots < \pi < 3.142857\ldots.$$

And this was done without using trigonometry! A highly readable and somewhat pithy account of this and many other investigations of π can be found in the book by Beckmann.

Bear in mind that each cell of the honeycomb is a three-dimensional entity; in fact it is approximately a regular hexagonal prism with one open end and one trihedral apex: three plane surfaces meet at a point. It is related to the rhombic dodecahedron. A fascinating and detailed account of the structure and formation of the "bee's cell" can be found in the classic work by D'Arcy Thompson; we will content ourselves here with a much condensed account (see Batschelet for other details). It was believed by many that bees form their three-dimensional cells in such a way as to minimize the surface area for a given volume; this would use the least amount of wax in cell construction. But this is not the case: their design is not the most efficient possible (see the article by Hales and further comments below). Nevertheless, we can go some way to further illustrate this principle for "cells" in the plane. As noted in the introduction to this chapter, a judicious arrangement of the perimeters for five hexagons may be used to obtain an area equivalent to seven. By comparing the resulting total perimeter and area with those of the seven inscribed circles (P_c, A_c) we can find the "efficiency" of the hexagonal structure (P_h, A_h) relative to the circles. For the perimeter this ratio is

$$\frac{P_h}{P_c} = \frac{5(4\sqrt{3}r)}{7(2\pi r)} = \frac{10\sqrt{3}}{7\pi} \approx 0.788,$$

so that the 7-hexagonal structure has about 21 percent *less* perimeter than the seven inscribed circles. The corresponding area ratio is

$$\frac{A_h}{A_c} = \frac{7(2\sqrt{3}r^2)}{7(\pi r^2)} = \frac{2\sqrt{3}}{\pi} \approx 1.103,$$

so that the 7-hexagonal system has about 10 percent *more* area than the seven inscribed circular areas. This much more area for a lot less perimeter is rather impressive (see the article by Roberts).

There is another loose end to tie up. Recall the *A:P* ratio for the inscribed regular polygon, defined earlier as

$$R(n) = \frac{r}{2} \cos \frac{\pi}{n} = R_i, \text{ say.}$$

For the escribed hexagon it can be seen directly from the definitions of A and P that the corresponding ratio is *independent* of the number of sides n:

$$R_e = \frac{r}{2}.$$

This is only attained in the limiting case $n \to \infty$ for the inscribed polygon. Since we have noted that wiggling bees with approximately circular cross sections generate the escribed hexagons, it appears that this process is optimal in practical terms.

For the real, three-dimensional honeycomb cell, it has been noted above that despite earlier claims to the contrary, that cell is not the most efficient design possible (in an isoperimetric sense). This was shown by Tóth, but the most economical form has not yet been determined. However, we know that the bees do a good job in two dimensions. The first general proof of the so-called *Honeycomb Conjecture* was supplied by Hales; it states that *any partition of the plane into regions of equal area has a perimeter at least that of the regular hexagonal honeycomb tiling.* In his article, Hales states that "the honeycomb problem is preparation for the real challenge, the Kelvin problem. What is the most efficient partition of space into equal volumes?" Again, the word *efficient* as used here refers to the minimization of the surface area of the boundary. The Kelvin problem has implications for the structure of *foam,* some aspects of which are discussed later in this chapter.

Despite the fact that the honeycomb is not optimally efficient in the above sense, it is nevertheless amenable to optimization! By this I mean that given a regular hexagonal prism, open at one end and sealed at the other by three rhombuses, what is the apex angle θ that minimizes the total surface area of the cell? According to D'Arcy Thompson, the angle θ is remarkably consistent within each cell (seldom differing from the amount calculated below by more than 2°; see figure 11.4). Based on the cell geometry, it will be shown below that the surface area S of the cell is given by

$$S = 6sh - \frac{3}{2}s^2 \cot\theta + \frac{3\sqrt{3}s^2}{2}\csc\theta$$

$$= 6sh + \frac{3s^2}{2}(\sqrt{3}\csc\theta - \cot\theta), \tag{1}$$

where s, the length of the sides of the hexagon, and h, the height, are constants. To aid the bees in their analysis, we will carry out the following steps to minimize, with respect to the apex angle θ, the surface area (and hence, presumably, the wax used in construction of the cell, though again bear in mind that because sides are shared with neighboring cells it is not an isolated prism). Here is what we will do:

(i) Find $dS/d\theta$.
(ii) Hence find the angle θ corresponding to a *minimum* of S.
(iii) Find the minimum value of S in terms of s and h.

Clearly, we may write

$$S = 6sh + \frac{3s^2}{2}f(\theta),$$

where

$$f(\theta) = \sqrt{3}\csc\theta - \cot\theta, \tag{2}$$

so that if a minimum of S occurs, the derivative

$$f'(\theta) = \csc\theta(\csc\theta - \sqrt{3}\cot\theta) = 0.$$

This occurs of course when

$$\theta = \theta_c = \arccos\frac{1}{\sqrt{3}} \approx 54.7°.$$

Figure 11.3 shows the quantity $f(\theta)$. The mathematical verification that this extremum is indeed a minimum follows from either the use of the first derivative test (the easiest way) or a further differentiation (with a soupçon of algebraic reduction) to show that

$$f''(\theta) = \csc\theta(2\sqrt{3}\csc^2\theta + \sqrt{3} - 2\csc\theta\cot\theta),$$

and so $f''(\theta_c) = 3\sqrt{3} > 0$ as required. How does this compare with the angle (or small range of angles in actuality) in bee cells? According to Batschelet,

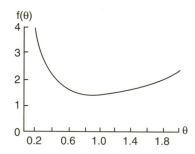

Figure 11.3. The dependence of the term $f(\theta)$ defined by equation (2), on the trihedral angle θ for a honeycomb cell.

this angle is difficult to measure (which is rather easy to believe) but the average of many measurements "does not differ significantly from the theoretical value."

How does the formula for the surface area of the cell arise? The following is based on the accounts by D'Arcy Thompson and, much later, by Batschelet. Consider a right prism with regular hexagonal bases and height h, with the vertices of the top base denoted by $ABCDEF$ and the open base denoted by $abcdef$, with side s as shown in the diagram (as in figure 11.4). Now let the alternate corners B, D, and F be cut off by planes through the lines AC, CE, and EA, which meet at the point V on the prism axis VN. These planes intersect the vertical edges Bb, Dd, and Ff at the points X, Y, and Z. The pieces that have been cut off are (by symmetry) regular

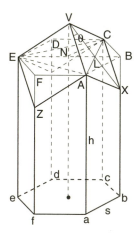

Figure 11.4. The geometry of the honeycomb cell. From *Introduction to Mathematics for Life Scientists*, by E. Batschelet (1975), reprinted by permission of Springer-Verlag.

tetrahedra: $ABCX$, $CDEY$, and $EFAZ$. These are then placed on top of the remaining solid in such a way that the vertices X, Y, and Z coincide with the point V. This would make a good party game, but is perhaps easier to appreciate if we regard the line segments AC, CE and EA as "hinges" from which these tetrahedra can be "folded over." The faces $AXCV$, $CYEV$, and $EZAV$ are rhombuses (again, by symmetry), and now the bees have a refurbished cell with the same volume as the original one.

The faces are formed from wax, and for a given volume it is certainly an economy to minimize the amount used, which translates into minimizing the surface area of the cell, though as noted above we must beware of attributing to the bees any particular mathematical ability. Given that the dimensions of the cell are approximately constant from cell to cell (neglecting slight variations due to bee size and cell manufacture) the variable that is best used as independent is the angle $\theta = \angle NVX$.

Now what? Some mathematics! If L is the point of intersection of CA and VX, then L bisects the segment NB and so $NL = s/2$. Hence CL, which is the height of the equilateral triangle BCN, is equal to $\sqrt{3}s/2$. In the triangle NLV it follows that $VL = (s \csc \theta)/2$. The segments CL and VL are the legs of the right triangle VLC, and four such triangles comprise the rhombus $AXCV$, which therefore has area $\sqrt{3}s^2/2 \sin \theta$. There are three such areas to include as well as the six lateral faces (e.g., $abXA$) of the prism. They are all congruent trapezoids, and since $BX = VN$ it follows from the geometry of triangle VNL that $BX = (s \cot \theta)/2$. This allows us to compute the area of the face $abXA$ as

$$\frac{s}{2}(aA + bX) = \frac{s}{2}(2h - BX) = sh - \frac{s^2}{4}\cot \theta.$$

Hence the total area

$$S = 6sh - \frac{3}{2}s^2 \cot \theta + \frac{3\sqrt{3}s^2}{2} \csc \theta,$$

which is precisely what was stated at the beginning of this section. There are more interesting angles and properties of the bee cell than we have related here; we leave this to the interested reader to pursue.

Let us now examine another part of a worker bee's job description: *collecting nectar*. This is an interesting exercise in elementary calculus; I discovered this application in Fred Adler's book. After the bee finds a flower she sucks it up at a slower and slower rate as the nectar is depleted. Again using anthropomorphic language, let us suggest that she "wants" to optimize the combination of travel time plus feeding time, and ask the question: *When should she give up and move on to the next flower?* This depends of course on how far away the next flower is, but we can make some general observations nonetheless. If she stays too long in one place she will visit fewer flowers, but if she stays only briefly she will spend most of her time flying. Let us further simplify matters by assuming that the flowers are uniformly spaced and hence that her travel time τ between flowers is constant (alternatively, we could treat τ as an *average* travel time between differently spaced flowers). We also assume that the bee needs to bring back as much nectar as possible before she "clocks out" for the day. In order to do this she should maximize the rate of nectar collection per visit (including travel time between flowers). Let $F(t)$ be the amount of nectar collected in time t at a given flower (F is assumed for our purposes to be a differentiable function). In view of the gradually depleting supply, we envisage that this function will be concave down with $F(0) = 0$. We also define $R(t)$ as the rate at which nectar is collected, in the following manner (R is also assumed to be differentiable):

$$R(t) = \frac{\text{food/visit}}{\text{time/visit}} = \frac{F(t)}{t + \tau},$$

where "visit" includes the time to reach the flower (from the previous flower) and the time spent at the flower. It is reasonable to suppose that R should be maximized, which means that we require $R'(t_*) = 0$ and $R''(t_*) < 0$ for some time $t_* > 0$. The first condition implies that

$$F'(t_*) = R(t_*). \tag{3}$$

The second condition follows for many reasonable forms of F (e.g., a limited growth function like $F = 1 - \exp(-\lambda t)$, where $\lambda > 0$). Indeed, since

$$R''(t_*) = \frac{F''(t_*)}{t_* + \tau},$$

it follows from the second condition that a local maximum of R occurs at $t = t_*$ provided $F(t_*)$ is concave down. But what does the first condition mean to us and imply for the bee? Since R as defined is an average rate of nectar collection over the visit to one flower, it means that when the instantaneous rate of nectar collection equals this average rate, she should move on to pastures new! This type of result is sometimes referred to as the *marginal value theorem*: it has the pithy interpretation for the bee—"leave when you can do better elsewhere!"

These ideas are easily appreciated when illustrated graphically (see figure 11.5); note that when τ is smaller, so too in general is the optimal feeding time t_*. For a given value of τ, there is a unique value of t_* provided that $F(t)$ is concave down.

In conclusion, we note some interesting facts regarding nectar collection (more can be found at the website *http://www.pbs.org/wgbh/nova/bees/buzz.html*). It seems that in order to make one pound of honey, workers in a hive fly 55,000 miles and tap two million flowers; furthermore, in a

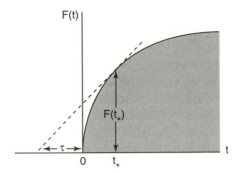

Figure 11.5. The graphical interpretation of equation (3) for a maximum of $R(t)$, occurring at $t = t_*$.

single collecting trip, a worker will visit between 50 and 100 flowers. She will return to the hive carrying over half her weight in pollen and nectar. Theoretically, according to the Nova website, the energy in one ounce of honey would provide our bee with enough energy to fly around the world! However, at an average speed of only 15 mph this would take about 70 days without pit stops!

SOAP BUBBLES, FOAM, AND MINIMAL SURFACES

Foam is an agglomeration of a large number of different bubbles. Generally speaking, each such bubble is a polyhedral cell with several different faces, curved as a result of excess pressure across it. By *Plateau's rules* (see below), the foam will contain three film surfaces intersecting along lines at 120°, and/or four lines of soap film meeting at a point with adjacent lines intersecting at 109°28′16″. Joseph Plateau (1801–1883) was a Belgian physicist who carried out extensive experimental work on the subject of soap films, all the more remarkable when one realizes that he was almost blind, having observed the sun for more than 25 seconds with the naked eye during an experiment carried out in 1843 (there appear to be some discrepancies regarding these historical details in the literature; I have taken my information from Hildebrandt and Tromba).

Plateau's Rules: (sometimes rules 2 and 3 are combined in the literature)

1. Three smooth surfaces of soap film intersect along a line.
2. The angle between any two tangent planes to the intersecting surfaces, at any point along the line of intersection of three surfaces, is 120°.

3. Four of the lines, each formed by the intersection of three surfaces, meet at a point, and the angle between any pair of adjacent lines is $109°28'16''(\approx 109.47°)$.

From where does this strange angle come? It is in fact $\arccos(-1/3)$; while this may not help directly, it does indicate that the origins of the angle may be fairly simple. This may be illustrated by considering a cube with the central interior point P connected to the vertices by line segments (figure 11.6). Only four vertices need to be considered here; they define the vertices of a regular tetrahedron. Because this configuration is identical to the interior structure of a soap film formed within a regular tetrahedral frame, we need only calculate the angle α between the lines joining the central point to two diagonally opposite vertices on one face of the cube. Let the vertices be identified by their components in Cartesian 3-space as in the diagram. The tetrahedron vertices are $(0,0,0), (2,2,0), (2,0,2)$ and $(0,2,2)$. Also identified are the vertices $(2,0,0)$ and $(2,2,2)$. The vector a joins the origin to the point P; clearly P has components $(1,1,1)$. Then the vectors $\mathbf{a} = \langle 1,1,1 \rangle$, $\mathbf{c} = \langle 2,2,0 \rangle$ and $\mathbf{a} + \mathbf{b} = \mathbf{c}$, so $\mathbf{b} = \langle 1,1,-1 \rangle$. By the definition of the vector inner (or dot) product,

$$\mathbf{a} \cdot \mathbf{b} = 1 = |\mathbf{a}||\mathbf{b}| \cos\theta = 3\cos\theta$$

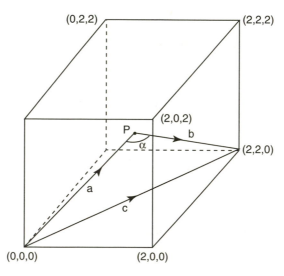

Figure 11.6. The basic configuration used in determining the "tetrahedral angle" α for soap films. According to Plateau's rules (see text), three such film surfaces intersect in a smooth line at an angle of $120°$, and only four such lines can intersect at a single point. At this point, the angle α between any two adjacent lines is $109° 28' 16''$.

but since $\alpha = \pi - \theta$, $\cos\alpha = -\cos\theta = -1/3$, so

$$\alpha = \arccos\left(-\frac{1}{3}\right) = 109°28'16''.$$

As an aside: a nostalgic bonus (for me) from the above inner product is the famous *Cauchy-Schwarz inequality* for vectors **a** and **b** in an inner product space, one form of which is

$$(\mathbf{a} \cdot \mathbf{b})^2 \leq (\mathbf{a} \cdot \mathbf{a})(\mathbf{b} \cdot \mathbf{b}).$$

Why nostalgia? On one occasion while I was employed at what was then called the New University of Ulster in Coleraine, Northern Ireland, I received an amusing postcard from the students in one of my classes. The card merely stated that "By the Cauchy-Schwarz inequality, (the attentiveness of the class)^2 ≤ (the quality of the jokes)^2 × (the brightness of the ties)^2." Since the class had obviously been paying considerable attention, the reader will correctly infer from having read this far that the ties I wore in those days must have been *extremely* bright!

Like so much else in nature, the patterns, configurations, and properties of foam are manifested in an average sense; foam does not consist of a neat array of cubes (say) laid out in a rectangular grid. But what might an "average" foam cell look like? To answer this question, consider an approximation for the *average number of faces, vertices, and edges* associated with a typical polyhedral cell in the foam (figure 11.7) by supposing it to be a regular polyhedron with planar faces (not too bad an approximation, in fact, because the curvatures of the faces are frequently small). If each face of the regular polyhedron has n edges, and the interior vertex angles of each face are θ, then $n(180 - \theta) = 360$, so $n\theta = 180(n - 2)$ and

$$n = \frac{360}{180 - \theta}.$$

For $\theta \approx 109.47°$ this gives the average number of edges as $n \approx 5.1$.

Now we introduce some further notation: let ϕ be the sum of face angles surrounding any vertex, v the number of vertices, and f the number of faces of the regular polyhedron that represents our "average" soap bubble. Thus for a cube, the faces being squares, these properties are $n = 4, \theta = 90°, \phi = 270°, v = 8$, and $f = 6$. For a dodecahedron, $n = 5, \theta = 120°, \phi = 360°, v = 20$, and $f = 12$. If we focus our attention on any particular vertex, we see that by summing over all the internal angles of the faces the following relationship must be true:

$$v\phi = fn\theta = 180f(n - 2).$$

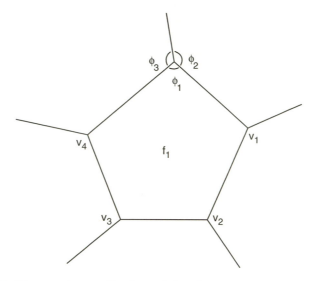

Figure 11.7. The notation used in determining the average properties of foam. The figure is drawn here for an obliquely viewed dodecahedron, for which $n = 5$, $f = 12$, $v = 20$, and $\theta = 120°$.

The total number of edges e of the polyhedron is half the product of the number of faces and the number of edges per face (each face shares a common edge), so

$$2e = fn$$

and hence

$$v\phi = 360(e - f).$$

Now for the beautiful part. Euler's theorem (there are many such theorems, but this one is preeminent in solid geometry, and has been mentioned already in chapter 3) states that

$$v - e + f = 2,$$

so that

$$v\phi = 360(v - 2),$$

from which we find that

$$v = \frac{720}{360 - \phi}.$$

It has been noted above that, for soap films, one of Plateau's rules states that adjacent lines meeting at a point have an angle $\theta = 109.47°$ between them.

The soap-film pattern within a regular tetrahedral framework illustrates this nicely: there are six triangular sections, the angle between each being 120°, and the angle between lines of intersection being θ. Since three faces constitute a vertex in this configuration (and all others formed by soap films) it follows that $\phi = 3\theta$, so the average number of vertices is

$$v = \frac{240}{120 - \theta} \approx 22.8.$$

Using Euler's theorem again along with $e = fn/2$, we find that $v + f - fn/2 = 2$, or

$$f = \frac{2(v - 2)}{n - 2}.$$

Substituting the values $v \approx 22.8$ and $n \approx 5.1$ in this expression, we find that the average number of faces is $f \approx 13.4$. Thus the "average" soap bubble in foam is closest in form to the dodecahedron if we compare it with the Platonic solids.

Let us consider now a *single* bubble. If it is not large enough to become mis-shapen by breezes or burst by sadistic adults, it will be spherical in shape as it floats along. Why spherical? The basic reason is that when the surface of a soap bubble is closed, it encloses a certain volume of air, and the forces of surface tension will seek to minimize the work done in enclosing this air. Thus the surface minimum area will be "sought" and it can be shown using the calculus of variations that this surface must necessarily be spherical. A "dual" result is that, for a given area of closed surface, the maximum volume is contained when the surface is that of a sphere. Just thinking of how much mathematics and physics is involved should prompt us all to offer to wash the dishes on a regular basis. These features involve two-dimensional surfaces enclosing three-dimensional volumes, so it should not come as a surprise to learn that the corresponding results hold true in lower dimensional configurations, namely, areas enclosed by closed curves.

THE ISOPERIMETRIC PROPERTY OF THE CIRCLE AND THE "SAME AREA" THEOREM

The word "isoperimetric" means "constant perimeter" and here concerns the largest area that can be enclosed by a closed curve of constant length. We shall state the result as a theorem.

> **Theorem 1 (the isoperimetric theorem).** Among all planar figures of equal perimeter, the circle (and only the circle) has *maximum area*.

We shall demonstrate a proof of this result below using some theory from the calculus of variations. There is also an equivalent "dual" statement known as the "same area" theorem.

Theorem 2 (the same area theorem). Among all planar figures of equal area, the circle (and only the circle) has *minimum perimeter*.

This theorem is a consequence of theorem 1, as is readily shown. Consider first a closed curve $\mathscr{C}$ of perimeter L enclosing a domain of area A. Let r be the radius of a circle of perimeter L: then $L = 2\pi r$. This encloses a disk of area πr^2. By the isoperimetric property of the circle, the area enclosed by $\mathscr{C} \not> \pi r^2$, and it equals πr^2 if and only if $\mathscr{C}$ is a circle. On the other hand,

$$\pi r^2 = \frac{(2\pi r)^2}{4\pi} = \frac{L^2}{4\pi}.$$

Then we obtain the *isoperimetric inequality*

$$A \le \frac{L^2}{4\pi},$$

with the equality holding only for a circle. Now we are ready for the proof of theorem 2 assuming the isoperimetric theorem (theorem 1). Consider an arbitrary planar figure with perimeter L and area A. Let $\mathscr{D}$ be a circular disk with the same area A and perimeter l: then $l^2 = 4\pi A$. If $L < l$, the isoperimetric inequality would be violated, so $L \ge l$.

Now follows an interesting historical application of these properties.

Queen Dido's Isoperimetric Problem

The story of Princess Dido is related in detail in the book by Hildebrandt and Tromba; it is briefly summarized here. A Phoenician princess from the city of Tyre about 900 B.C. she fled by ship from her ruthless brother King Pygmalion to Africa at the place that later became Carthage. In trying to buy land from a local ruler, the strange bargain was struck that she could have only that land which could be enclosed by the skin of an ox. While she may not have taken a class in the calculus of variations (it would not be formulated until 2500 years later) she undoubtedly had great intuition, for according to the account rendered in Virgil's *Aeneid* (the national epic of Rome), she used a straight part of the Mediterranean coastline as one boundary, and used thin strips cut from the oxhide to enclose the maximum possible area in a semicircular region. If she had not used the coastline, then according to the isoperimetric theorem, a circular region would have enclosed the maximum possible area. We shall do some arithmetic for both situations.

Growing up in rural England as the son of a farmworker, I encountered cows many times, and also the occasional unfriendly bull. "Oxen" as such were not part of my experience, so I'll estimate the surface area of a really big Guernsey or Jersey cow! Ignoring the head, legs, and tail, let us treat the cow's torso as a rectangular box about 4 ft high, 2 ft wide, and 5 ft long; the surface area of the box is approximately 80 sq ft. If the strips cut from the hide were as thin as 0.1 inch wide (as Hildebrandt and Tromba suggest), then the combined length L of such strips is given in feet by

$$L \approx \frac{80 \times 12}{0.1} \approx 10^4 \text{ ft} = \pi r$$

for a semicircular contour of radius r. The corresponding area of the region is

$$A_s = \frac{\pi}{2}r^2 = \frac{\pi}{2}\left(\frac{L}{\pi}\right)^2 = \frac{L^2}{2\pi} \approx 1.6 \times 10^7 \text{ sq ft} \approx 350 \text{ acres} \approx 0.6 \text{ sq mi.}$$

If the boundary were circular instead of semicircular, the corresponding area would be half this amount, or about one-third of a square mile.

Consider a generalization of this problem. Suppose that a curve of length L that is attached to a line segment of length l, forming a closed curve. What should be the shape of L in order for the area enclosed to be a maximum? Consider two situations, the first in which L is a circular arc (enclosing an area A_c), and the second in which it is not (and enclosing an area A)! Imagine also that in both cases a circular arc below the line segment l is added to give an additional segment of area a. By the isoperimetric theorem,

$$A + a \le A_c + a \qquad \text{so that } A \le A_c,$$

so that Princess Dido knew what she was doing!

There is a corresponding isoperimetric inequality in three dimensions, by the way, which is of course appropriate for our isolated soap bubble. For a solid of volume V and surface area S, this inequality is

$$V^2 \le \frac{S^3}{36\pi}$$

with the equality holding only for a sphere, since

$$\left(\frac{4}{3}\pi r^3\right)^2 = \frac{(4\pi r^2)^3}{36\pi}.$$

Note also that, of all enclosed volumes with a given surface area, the sphere has the largest volume. Returning to two dimensions, we study an interesting and commonly observed phenomenon that imposes a mechanical component to the above theory: cracks in the surface of dry (or drying) mud.

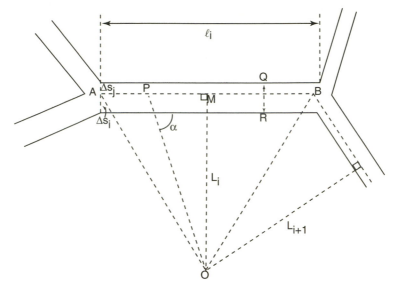

Figure 11.8. The notation and configuration used in the "mud crack" problem (see text for details). From "A Theory of Surface Cracks in Mud and Lava and Resulting Geometrical Relations" by L. I. Hewes (1948), *American Journal of Science*, 246: 138–149, reprinted by permission of the *American Journal of Science*.

MUD CRACKS AND RELATED GEOMETRIC THEOREMS

What follows, based on the work of Hewes (1948), applies to a variety of situations, from drying mud to cooling basalt, and even patterns of cracking in some enamels and chinaware. The basic physical assumptions are that the cooling material (whatever it is) is level, homogeneous, chemically inert, and with a plane exposed (or contact) surface. The surface tension is assumed to be uniform throughout, but it increases progressively as the mud (say) dries, until it exceeds the "rupture strength" of the material. The surface cracks that form are most frequently polygonal in shape, with sides that are approximately rectilinear. Some interesting mathematical analyses of these crack patterns can be carried out, one of which is based on a principle of "least work." It is by no means obvious that this approach is appropriate, however, and so like many other mathematical models, it does not necessarily "explain" the observations to the satisfaction of everyone.

Let the interior tension T per unit length at any instant be constant in every direction at every point, and consider the rupture to occur for unit depth. Under this tension the rectilinear crack has opened uniformly along its length by an amount RQ, the width of the crack (see figure 11.8; O is the point of zero displacement, in some sense the center of the polygonal crack). This width is a direct result of shrinkage of material perpendicular to the axis of the crack, which, while varying from crack to crack, is assumed to be constant for each individual crack. Consider now a crack of length l_i arising from a shrinkage (or displacement) Δs_i on one side of the axis AB (there is a corresponding displacement on the other side, of course). From Hooke's law for elastic bodies, the tension T per unit length and the displacement are related by

$$T = \frac{E \Delta s_i}{L_i},$$

where E is the modulus of elasticity (or better, the average stress-strain ratio during contraction, since the material is not in general truly elastic), and L_i is the original length OM of a material "fiber" under tension perpendicular to the crack. However, in view of the assumption stated earlier regarding the isotropy of the tension, it follows that at the instant of rupture the tension along any line OP is also equal to T. Thus the displacement in this direction (on one side of the axis) is $\Delta s_i / \sin \alpha$. Since $\sin \alpha = L_i/OP = E\Delta s_i/T(OP)$ it follows that

$$\frac{\Delta s_i}{\sin \alpha} = \frac{(OP)T}{E} = \frac{L_i}{\sin \alpha} \times \frac{T}{E}$$

or, as above,

$$T = \frac{E \Delta s_i}{L_i}.$$

This establishes that the tension T satisfies this relationship at any point along the rectilinear crack. Hence the amount of work done W_i by the tension acting across the entire length l_i of the crack is based on the standard result (prove it!) that the energy per unit length of a stretched fiber is $\frac{1}{2} \times T \times \Delta s_i$. For the whole crack this means that

$$W_i = \frac{1}{2} T l_i \Delta s_i = \frac{T^2 l_i L_i}{2E}.$$

This can be expressed conveniently in terms of the area A_i of the triangle AOB as

$$W_i = \frac{T^2 A_i}{E}.$$

If for the adjoining side (BC, say) the corresponding triangle has area A_{i+1} then a similar relationship holds for W_{i+1}, and so on. The *total* internal work done by the shrinking material adjacent to and within an n-sided closed convex polygon "cell" is thus given by

$$W = \frac{T^2}{E} \sum_{i=1}^{n} A_i = \frac{T^2 A}{E},$$

where A is the area of the polygonal cell. *The work done per unit area* is then just T^2/E. Invoking the "principle of least work" (well known to students) this internal work done must a *minimum* per unit area, or putting this another way, the area A must be the largest possible area that can be enclosed by the n sides l_i of a given plane polygon. When does this occur? It can be proved that this occurs *when the polygon is inscribed in a circle*. We will approach this proof somewhat obliquely by first establishing a useful result for triangles.

Consider any triangle ABC with one side of fixed length AC and with fixed perimeter P. It is clear that there are many possible combinations of side lengths that can give the same perimeter, so a natural question to ask is—*Which triangle has the largest area?* The answer is an isosceles triangle with base AC. The reason is that the locus of the vertex B is an ellipse with foci A and C with the sum of lengths $AB + BC =$ constant. By symmetry the triangle with largest area must be the one in which $AB = BC$.

Now consider an n-gon with perimeter P, the area of which exceeds that of any other n-gon with the same perimeter, and let a pair of adjacent sides be denoted AB and BC (thus B is a vertex of the n-gon). By virtue of the previous discussion of triangles, this particular piece of the polygon will have maximum area when $AB = BC$. This forces the adjacent sides BC and CD to be equal, and so on, which means that we have proved the following result, which we state in the form of a theorem:

Theorem 3. Of all possible n-sided polygons with perimeter P, the one with the greatest area is the one with equal sides.

This theorem has the obvious corollary: If the polygons are *inscribed in a circle*, the polygon is regular.

Now we are in a position to state a result with some application to the mud-crack problem:

Theorem 4. Of all n-gons with side lengths $a_1, a_2, a_3, \ldots, a_n$ arranged in that order, the one inscribed in a circle has the greatest area.

To prove this I follow the approach made by David Gay. Consider such a polygon P inscribed in a circle C; then "cut off" the n segments and add them to the corresponding sides of any other polygon P' not inscribed in a circle (figure 11.9(a)). The union of the circular arcs now forms a continuous (but not in general smooth) closed curve C', equal in length to the perimeter of the circle C (being composed of the same circular arcs). Now the area of circle C equals the total area of the n segments plus the area of the n-gon P (figure 11.9(b)). By the isoperimetric theorem, the area of circle C exceeds that surrounded by the curve C', which in turn equals the total area of the

(a)

(b)

(c)

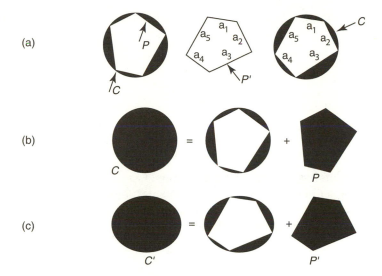

Figure 11.9. Polygons P, P' and closed curves C (a circle), C' used in establishing the proof of theorem 4. From *Geometry by Discovery* (1998), Wiley: New York, with permission of David Gay.

same n segments plus the area of the n-gon P' (figure 11.9c). Therefore we have

$$\text{area of } P > \text{ area of } P',$$

which is the desired result.

A question worth asking at this stage is that, given a set of n side lengths in a given order, does there exist a polygon composed of those sides which is inscribable in a circle? Let us play with this a little (get a pencil and paper—oh, and an eraser, since we're not philosophers).

Consider any point Q in the interior of an irregular convex n-gon, where Q is free to move, and construct all n triangles $\Delta_1, \Delta_2, \Delta_3, \ldots, \Delta_n$, with sides $a_1, a_2, a_3, \ldots, a_n$ as bases and with vertex Q. Let the sides joining the edge a_i of the n-gon be of length r_i and r_{i+1}, respectively (moving clockwise, say, around the perimeter from side a_1). Triangle Δ_1 has maximum area when $r_1 = r_2$, triangle Δ_2 when $r_2 = r_3$, and so on. This forces Q to move to the center O of a circle of radius $r = r_1 = r_2 = \cdots = r_n$ in order that the polygon have maximum area. This is merely a demonstration of the existence of such a circle; the constructive problem of identifying the location of the center is now posed (but not answered in general terms) by considering an irregular n-gon inscribed in this circle of radius r, with each side a_i subtending an angle $2\theta_i$ at the center O. If P is the perimeter of the n-gon, it follows that (i)

$$P = \sum_{i=1}^{n} a_i = 2r \sum_{i=1}^{n} \sin \theta_i$$

and (ii)

$$\sum_{i=1}^{n} \theta_i = \pi \quad \text{or} \quad \sum_{i=1}^{n} \arcsin \left(\frac{a_i}{2r} \right) = \pi.$$

Is there a unique solution set $\theta_i(r)$ to these equations? Certainly for a regular polygon. Dropping subscripts, note that for a regular n-gon (which gives the maximum area for a given perimeter if the sides are not prescribed) these equations reduce nicely to the well-known results

$$\theta = \frac{\pi}{n} \text{ and } r = \frac{a}{2} \csc \frac{\pi}{n}.$$

So armed with these results (and others like them; see the bibliography) we can go out into the dried-mud world and examine the cracks very carefully, in river beds, salt flats, or even (I have noticed) the surface of old roads paved with tarmac (see plate 17). . . .

APPENDIX: THE ISOPERIMETRIC PROPERTY OF THE CIRCLE

We wish to establish the result that the area enclosed by a closed curve of given length $2L$ is a maximum when the curve is a circle. Following Isenberg, we examine the somewhat less general case for which the curve is symmetric about a line, the x-axis, say. Furthermore, let the curve be chosen to pass through the origin without loss of generality. Thus we are interested in a segment of the x-axis defined by $0 \leq x \leq d$, say. The area enclosed by the closed curve is (in terms of arc length s)

$$A = 2 \int_0^d y(x) dx = 2 \int_0^L y \sqrt{1 - \left(\frac{dy}{ds} \right)^2} \, ds = \int_0^L F(s, y, y_s) ds,$$

where $y_s \equiv dy/ds$. Since F does not depend explicitly on s, a simpler form of the standard Euler-Lagrange equation for F may be used, namely,

$$F - y_s \frac{\partial F}{\partial y_s} = c$$

(c being a constant to be determined). This equation for F then simplifies to

$$cy_s = \pm \sqrt{c^2 - y^2}$$

and on taking the positive root it follows, using obvious dummy variables, that

$$\int_0^y \frac{c}{\sqrt{c^2 - \xi^2}}\, d\xi = \int_0^s d\sigma,$$

so

$$c \arcsin\left(\frac{y}{c}\right) = s$$

or

$$y = c \sin\left(\frac{s}{c}\right).$$

Applying the condition $y(L) = 0$ gives $c = L/\pi$ (taking the smallest value of L that satisfies this condition), and so

$$y = \frac{L}{\pi} \sin\left(\frac{\pi s}{L}\right).$$

We need to find $x(s)$, however; but since in differential form

$$dx = \sqrt{(ds)^2 - (dy)^2} = \sin\left(\frac{\pi s}{L}\right) ds,$$

it follows upon integration that

$$x = \frac{L}{\pi}\left(1 - \cos\left(\frac{\pi s}{L}\right)\right)$$

(since $x(0) = 0$). Thus the equation of the curve is

$$\left(x - \frac{L}{\pi}\right)^2 + y^2 = \left(\frac{L}{\pi}\right)^2,$$

which is of course the equation of a circle of radius L/π centered at $(L/\pi, 0)$, with area $A = L^2/\pi$.

CHAPTER TWELVE

River Meanders, Branching Patterns, and Trees

> . . . I saw a great number of trees on each side of the river.
>
> —Ezekiel 47:7b

As a caveat to some of this chapter (river meanders), and to a lesser extent part of the previous one (mud cracks), it should be reemphasized that, while the mechanisms and principles inherent in these mathematical models may represent part of the "truth" behind the observed patterns, they certainly do not qualify as "the whole truth and nothing but the truth." Indeed, as discussed at some length in chapter 1, this is the case to a greater or lesser degree for all mathematical models. In any field of scientific endeavor, pragmatically speaking, no model can fully encapsulate all the data gathered thus far from "out there" and predict all new phenomena (though it can be argued that General Relativity comes pretty close!). Therefore it should come as no surprise to learn that such "principles of least action" or "least work" as presented here are not universally accepted as to why they should work (or as some might argue, whether in fact they do work). Having made this point, we will now immerse ourselves in the exciting possibilities afforded by these models as potential descriptions of natural phenomena.

RIVER MEANDERS

The following observations, obtained from a study of more than fifty rivers by Luna Leopold and coworkers, are of interest in connection with the apparent regular "sinuosity" of rivers the world over. From their study the following statements can be made:

(i) No river, regardless of size, runs straight for more than 10 times its width.
(ii) The radius of a bend is nearly always 2–3 times the width of the river at that point.
(iii) The wavelength (distance between analogous points of analogous bends) is 7–10 times the (average) width.

The conclusion is that, despite considerable, even dramatic, variations and size in bed conditions, rivers are strikingly similar in their characteristics. Furthermore, it transpires that meanders are not "accidents" of nature; according to one theory, they define the form in which a river does the least work in turning (as in proceeding from a point *A* uphill to a point *B* downhill), which in turn defines the most probable form a river can take.

The name *meander* is from a winding stream in Turkey that was in ancient Greek times known as the Maiandros (today it is known as the Menderes). It might seem at first sight that local irregularities both within the river and at its boundaries are sufficient to divert the river from a straight course. Irregularities such as boulders, fallen trees, or harder rock through which the river flows may indeed be sufficient to cause meanders, but they are not *necessary* conditions, being unable to account for the consistent geometry of meanders, for example, both in nature and in the laboratory. Meanders are observed in ocean currents (such as the Gulf Stream) and in water channels on the surface of a glacier.

In rivers, the processes of continuous erosion, transportation, and deposition construct the ever-changing shape of rivers and their meanders. Imagine a slight bend in a river. As water flows around it the "centrifugal force" (a fictitious force, but one corresponding to a real phenomenon; it is the consequence of Newton's third law and the existence of the centripetal force) will force scouring of the outer, concave bank. The water will then move downward at this bank (it has nowhere else to go) and then across the river bed toward the other bank. As it does, friction with the bottom slows down the flow, allowing particles of silt and rock to be deposited there, though somewhat downstream due to the forward velocity of the river superimposed on this "crosswise" or "spanwise" flow. The actual fluid "particles" will execute helical paths as a result, and the scoured bank will be steeper than the gentler slopes on the other side of the river. Over time the meander will become more and more pronounced until to continue would result in the river moving uphill—a no-no for rivers! Thus the turning process begins again as gravity continues to pull the river water to a level of lower potential energy. Such meanders will also "move" downstream over time. The best conditions under which meanders arise is when the river traverses a gentle slope comprised of fine-grained, easily erodable material that has sufficient cohesiveness to provide firm banks (according to Leopold and Langbein).

As Leopold and coworkers found out, a given series of meanders tends to have a constant ratio between the "wavelength" of the curve (in its commonly understood sense) and its radius of curvature. If this ratio is fairly constant, the meanders appear, not surprisingly, to be rather regular also. Typical values range from about 5:1 for sine-type curves to 3:1 for more tightly looped meanders, the average over a sample of 50 typical meanders on different rivers and streams being about 4.7:1. The "tightness" of a bend (its sinuosity) is defined to be the ratio of the length of a channel in a given curve to its wavelength. This is rather variable, somewhere (for most meandering rivers) in the range 1.3:1 to 4:1.

Some work has been carried out to find the average (meaning most probable in this context) path taken by a random walk of fixed length between two given points A and B. The solution to this problem can be expressed exactly in terms of elliptic integrals (an analogous elastic-mechanical mathematical "simile" appears below), but to a good approximation it is reasonable to state that the most probable geometry for a river or stream is one in which the angular direction of the channel with respect to the mean down-valley direction is a sinusoidal function of the distance measured along the channel (see equation (1) and figure 12.1a below).

This family of curves is called "sine-generated" by Leopold, and does mimic the shape of real rivers. When considered in a group with (a) parabolic curves, (b) sine curves, and (c) circular (or arcs thereof) curves, the sine-generated curves were found to have the smallest total variation (the sum of the squares of the changes in direction were minimized) of the changes of direction along the curve. However, we are not yet done; such a curve is a curve of minimum total work. In the context of elasticity, this curve minimizes the total work in bending a steel strip between two fixed points. The strip will seek to minimize the bending by avoiding any regions of concentration, thus making the bending as uniform as possible. Since the work done in bending each element of length is proportional to the square of the angular deflection, we see that the sine-generated curve arises in this context also as the most natural configuration.

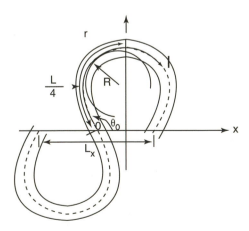

Figure 12.1. (a) The "river meander" configuration and notation used in deriving equation (1) Redrawn from "River Meanders" by L. B. Leopold and M. G. Wolman (1960), *Bulletin of The Geological Society of America*, 71: 769–794.

From related models in the literature the following curve is considered a good approximation to the shape of the center line for many meandering rivers:

$$\theta(l) = \theta_0 \cos\left(\frac{2\pi l}{L}\right), \tag{1}$$

where l is the length of the river from some initial point, L is a characteristic length associated with that particular river, and θ is the angle the central line makes with the down-river direction at each particular value of l. It has been shown that this equation arises from a type of random walk argument (see chapter 14 for a brief explanation of random walks) that can be summarized as follows. If we fix two points A and B in a plane, representing the "beginning" and "end" of a river, and have a point P moving between them, starting at A, then we ask that it should be able to arrive at B after a fixed number N of equidistant steps, where at each point P can change direction by any angle α (where α is a normally distributed variable). If the possible paths are smoothed by taking the step size smaller in a limiting sense, then among all such curves Von Schelling has proved that the likeliest to occur is the one that has the minimum overall (curvature)^2, that is, one for which

$$\int_a^b \frac{dl}{R^2}$$

is a minimum. The reason for the square of the curvature is that the curvature itself $(1/R)$ and its derivative can vary in sign along l, and because of cancellations, therefore, the integrals of these quantities may yield little or nothing of significance to us. This is best avoided by using non-negative quantities as above.

Now we are in a position to derive equation (1). (I have been unable to identify the source of the following argument, which broadly follows that of Langbein and Leopold [1960].) It has been stated by Leopold and Langbein that "meanders are not mere accidents of nature, but the form in which a river does the least work in turning." Based on this, consider the mean squared curvature defined by

$$\frac{1}{R_m^2} = \frac{1}{L} \int_0^L \frac{dl}{R^2}, \tag{2}$$

where L is the length of the double loop (figure 12.1a) and $l = 0$ corresponds to choosing the origin at the central point of inflection. In light of the above quotation, we require that the mean squared derivative of the curvature is a minimum, because this is one measure of the work done in turning (another would be to use the absolute value of the curvature and its derivative, but I am not aware that this has been done). Thus we seek curves such that

$$\frac{1}{L}\int_0^L \left[\frac{d}{dl}\left(\frac{1}{R}\right)\right]^2 dl \qquad (3)$$

is a minimum. A common procedure in mathematical modeling is to introduce dimensionless variables. It is not necessary to do so, but it has the advantage not only of tidying up the equations, but, more important, of bringing to the fore dimensionless combinations of terms that are significant in interpreting the underlying physics or biology of the system (e.g., the Richardson number or the Rayleigh number, both discussed in chapter 8). In the case here this does not happen, but we will proceed nonetheless by defining $\xi = l/L$, $\eta = L/R$, and $\eta_m = L/R_m$. Equations (2) and (3) may now be written as

$$\eta_m^2 = \int_0^1 \eta^2 \, d\xi \text{ and } \int_0^1 (\eta')^2 \, d\xi \qquad (4)$$

is a minimum. This is another type of isoperimetric problem, and rather than summarizing the basics of the variational calculus, we shall wave our hands a little and just state the relevant formulas (just as we did in chapter 5 in discussing the governing equations for refraction of light). In terms of what is called a Lagrange multiplier ($\lambda > 0$), we wish to minimize the integral

$$\int_0^1 F(\eta, \eta') \, d\xi = \int_0^1 [(\eta')^2 - \lambda \eta^2] \, d\xi, \qquad (5)$$

where $\eta = \eta(\xi)$ and $F(\eta, \eta')$ must satisfy the Euler-Lagrange equation

$$\frac{\partial F}{\partial \eta} - \frac{d}{d\xi}\left(\frac{\partial F}{\partial \eta'}\right) = 0. \qquad (6)$$

Upon substituting the integrand F into equation (6) there results a familiar ordinary differential equation in the form of the simple harmonic motion equation (but with the independent variable being length, not time)

$$\frac{d^2\eta}{d\xi^2} + \lambda\eta = 0.$$

The general solution of this equation is

$$\eta = C_1 \cos(\sqrt{\lambda}\xi) + C_2 \sin(\sqrt{\lambda}\xi) \equiv -\frac{d\theta}{d\xi}, \qquad (7)$$

where C_1 and C_2 are constants (as is λ), and the negative sign is introduced because if we start $l = 0$ at an inflection point (see the figure) then θ is a decreasing function of length. By direct integration of equation (7) it follows that

$$\theta = -\frac{C_1}{\sqrt{\lambda}} \sin(\sqrt{\lambda}\xi) + \frac{C_2}{\sqrt{\lambda}} \cos(\sqrt{\lambda}\xi). \tag{8}$$

But what about the constants; how do we determine them? This presents no problem—we just invoke some sensible boundary conditions and apply them to this equation. There must be three conditions to determine the three constants: (i) $\eta(0) = 0$; (ii) $\theta = 0$ when $\xi = 1/4$; and (iii) $\theta = \theta_0$ when $\xi = 0$. This gives $C_1 = 0$, $\sqrt{\lambda} = 2\pi$, and $C_2 = 2\pi\theta_0$. Thus

$$\theta = \theta_0 \cos\left(\frac{2\pi l}{L}\right),$$

which is equation (1) as required.

We can take this a little further. From the figure, the x-length of the S-shaped meander is denoted by L_x, and the *sinuosity* is defined as $\sigma = L/L_x$. From equation (1) we obtain

$$\frac{l}{L} = \xi = \frac{1}{2\pi} \arccos \frac{\theta}{\theta_0},$$

from which we have that

$$d\xi = -\frac{1}{2\pi (\theta_0^2 - \theta^2)^{1/2}} d\theta, \tag{9}$$

but since $dx = dl \cos\theta$ it follows that

$$d\left(\frac{x}{L_x}\right) = \sigma \cos\theta \, d\xi. \tag{10}$$

Now we are in a position to eliminate the differential $d\xi$ from (9) and (10) and integrate to obtain

$$\int_0^{1/4} d\left(\frac{x}{L_x}\right) = \frac{1}{4} = -\frac{\sigma}{2\pi} \int_{\theta_0}^0 \frac{\cos\theta}{(\theta_0^2 - \theta^2)^{1/2}} d\theta,$$

whence

$$\frac{1}{\sigma} = \frac{2}{\pi} \int_0^1 \frac{\cos(\theta_0 t)}{(1 - t^2)^{1/2}} dt, \tag{11}$$

where $t = \theta/\theta_0$. This is an interesting integral, related to a Bessel function, and we could follow many rabbit trails backward to establish this, but we won't. It is shown in many standard texts (e.g., Arfken) that $J_0(x)$, the Bessel function of the first kind of order zero, can be expressed as

$$J_0(x) = \frac{1}{\pi} \int_0^\pi \cos(x \sin\alpha) \, d\alpha = \frac{2}{\pi} \int_0^{\pi/2} \cos(x \sin\alpha) \, d\alpha$$

because $\sin \alpha$ is symmetric about $\alpha = \pi/2$ in the interval $[0, \pi]$. If we let $\sin \alpha = t$ in this integral, it reduces to

$$J_0(x) = \frac{2}{\pi} \int_0^1 \frac{\cos(xt)}{(1-t^2)^{1/2}} \, dt. \tag{12}$$

Therefore it follows from (11) that the sinuosity

$$\sigma = [J_0(\theta_0)]^{-1}, \tag{13}$$

this result being valid up to the first zero of the Bessel function, that is, for $0 \leq \theta_0 < 2.4048$ (in radians; in degrees, $\theta_0 < 137.8°$). In fact, θ_0 is further limited to less than about $126°$ (corresponding to $\sigma \leq 8.5$) because at this point the meander loops touch each other (see figure 12.1b). What the result (13) means is quite fascinating: for a sinusoidal curve of type (1), the sinuosity is determined entirely by the initial angle θ_0. In actuality, $\sigma \lesssim 5$ for the majority of rivers, and the majority of these have $1 \lesssim \sigma \lesssim 2$.

From equation (7),

$$\eta = 2\pi \theta_0 \sin(2\pi \xi),$$

so that the curvature

$$\frac{1}{R} = \frac{\eta}{L} = \frac{2\pi \theta_0}{L} \sin\left(\frac{2\pi l}{L}\right). \tag{14}$$

This has a maximum value (and therefore smallest radius of curvature, $R_{\min}$) when $\xi = l/L = 1/4$, that is, at the top or bottom of the loop. Thus

$$\frac{1}{R_{\min}} = \frac{2\pi \theta_0}{L} = \frac{2\pi \theta_0}{\sigma L_x},$$

so that

$$\frac{L_x}{2\pi R_{\min}} = \theta_0 J_0(\theta_0). \tag{15}$$

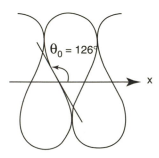

$\theta_0 = 126°$

x

Figure 12.1. (b) The limiting case of extreme meanders in the Bessel function model.

The graph of this function increases from zero to a maximum at $\theta_0 \approx 70°$, decreasing thereafter to zero as the angle θ_0 increases toward $\approx 138°$. As with the sinuosity, the maximum curvature (and minimum radius) is determined solely by the initial angle θ_0.

Another important feature of the geometry of river meanders is that it determines the rate at which the banks will be eroded. That rate is in general determined by the degree to which the river channel is bent. From what we have noted already about the properties of a sine-generated curve, rivers meandering in this manner tend to minimize the erosion at each point, and hence the total erosion. Several other interesting features of river meanders were noted by Leopold and Langbein based on observations carried out in 1959. They found unexpected contrasts between the profiles of both the water surface and stream bed in the cases of "straight reaches" and "curved reaches" (or bends) of a stream. In the latter case the slope of the water surface profile was steeper than that of the straight reach, and it was more nearly a uniformly decreasing profile than was the stepped profile for the straight reach. In this latter case, the profile was steep over "riffles," that is, regions above humps in the stream bed where the water flows faster, and much flatter over the intervening pools of deep water. The slope of the water surface is an approximate indicator of the rate at which energy is lost in the form of friction along the length of the stream (though of course the fluid velocity also plays a role in energy loss estimates). To a good approximation, then, a uniform downstream water surface slope is indicative of uniform energy expenditure per unit distance along the channel.

The meandering form of a river, then, tends to eliminate concentrations of energy loss and at the same time tends to reduce the total energy loss rate to a minimum. We can do no better than quote directly from an article by Leopold and Langbein on this point:

> It is important to recognize that these principles and tendencies operate in both river systems as a whole and in a given section of a river, and meanders must be considered in both contexts. The stream's cross-sectional shape, alluded to in the context of scouring and deposition clearly influences the nature of the flow and the velocity of the fluid elements in any given region. Velocity is greatest, other things being equal, in channels that offer the least resistance to flow, and resistance to flow depends on the surface area in contact with each unit volume of water. When this is minimized, so too is the resistance to flow (i.e. the frictional losses). Both wide, shallow channels and narrow, deep channels have relatively large surface area/unit volume ratios, and retard the flow of water more than say a semicircular channel of the same perimeter. Obviously, as we have seen, the flow in a curved channel produces an asymmetric cross section, which is not optimal, but the general principle is still valid as this feature is common to all meanders, and straight reaches do tend to

aquire more symmetrical cross-sections compared to their neighboring bends.

Mention has already been made of the analogy between river meanders and stresses in elastic wires, and now we shall pursue this a little further for both mathematical interest and historical completeness. Based on the account by MacMillan, we consider wires with applied stresses at the ends only. Consider, as in figure 12.2, a uniform elastic rod (naturally straight) that has been bent, with its two ends connected by a tight string. The rod is in equilibrium under the action of the two external forces T and −T, which are acting on its ends along the straight string. Let x and y be the coordinates of any point B of the arc AE, with the line EA being coincident with the x-axis. By a well-known theorem in statics, the tension −T acting at A is equivalent to a force −T acting at B and a couple, the moment of which is $-y_B T$, where T is the magnitude of the constant tension throughout the arc and string. The bending moment at the point (x, y) is the value of the couple at that point, that is, $D = T \times y$. But this bending moment is just EI/ρ, where E is Young's modulus for the elastic material, I is the moment of inertia of the cross section of the rod with respect to a line through its center, and ρ is the radius of curvature at the point (x, y). We have used a different notation (ρ instead of R) to avoid confusion with the meander problem: this is a mechanical analogy. Thus

$$y = \frac{EI}{T\rho} = \frac{a^2}{4\rho},$$

where $a^2 = 4EI/T$. We know from the calculus of plane curves that the curvature here is

$$\frac{1}{\rho} = \frac{-d^2y/dx^2}{[1 + (dy/dx)^2]^{3/2}}$$

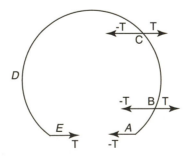

Figure 12.2. The analogy of the elastic wire under $ABCDE$ under tension T at the ends and uniformly throughout its length. From *Statics and the Dynamics of a Particle*, by W. D. MacMillan (1958), reprinted by permission of Dover Publications.

so the nonlinear ordinary differential equation satisfied by y is

$$a^2 \frac{d^2y}{dx^2} + 4\left[1 + \left(\frac{dy}{dx}\right)^2\right]^{3/2} y = 0.$$

We multiply through by $2dy/dx$ and rearrange the expression to get

$$\frac{d}{dx} y^2 = \frac{a^2}{2} \frac{d}{dx}(1 + y'^2)^{-1/2} \qquad \left(y' = \frac{dy}{dx}\right)$$

or

$$y^2 = \frac{a^2}{2}(1 + y'^2)^{-1/2} + C.$$

We will examine this solution for different classes of constant C. If θ is the smaller angle a tangent to the curve makes with the x-axis, so that

$$\frac{dx}{ds} = \cos\theta, \frac{dy}{ds} = \sin\theta \text{ and } \sqrt{1+y'^2} = |\sec\theta| = \sec\theta \text{ since } \theta \in \left(-\frac{\pi}{2}, \frac{\pi}{2}\right),$$

then the above equation can be rewritten as

$$y^2 = C + \frac{a^2}{2}\cos\theta = b^2 - a^2\sin^2\frac{\theta}{2},$$

where $b^2 = C + a^2/2$. The various possibilities to consider are $b^2 < a^2$, $b^2 > a^2$, and $b^2 = a^2$; the first of these three cases is of most significance for the river meander problem, and is the one on which we shall concentrate here. To this end, let $b^2 = k^2 a^2 (k^2 < 1)$ and let $-\sin\theta/2 = k\sin\phi$, so that $y = h\cos\phi$ (the positive root being chosen since $y > 0$). From figure 12.2 it can be shown that

$$\sin\theta = \frac{dy}{d\phi}\frac{d\phi}{ds} = -h\sin\phi\frac{d\phi}{ds},$$

from which it follows that

$$\frac{ds}{d\phi} = -h\frac{\sin\phi}{\sin\theta}.$$

Using this and the above definition of $\sin\phi$, the following result is obtained:

$$\frac{ds}{d\phi} = \frac{a}{2\sqrt{1 - k^2\sin^2\phi}},$$

and hence

$$s = \frac{a}{2}\int_0^\phi \frac{d\phi}{2\sqrt{1 - k^2\sin^2\phi}} \equiv \frac{1}{2}aF(k, \phi), \qquad (16)$$

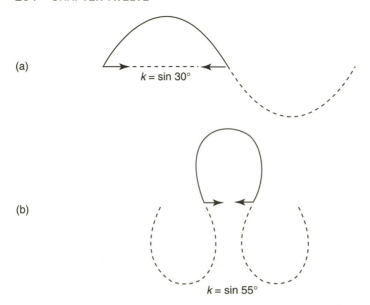

(a)

$k = \sin 30°$

(b)

$k = \sin 55°$

Figure 12.3. Basic meanders in the elastic wire according to the locus (18) for the following values of the parameter k: (a) $\sin 30° = 0.5$; (b) $\sin 55° \approx 0.82$. From *Statics and the Dynamics of a Particle*, by W. D. MacMillan (1958), reprinted by permission of Dover Publications.

where $F(k, \phi)$ is Legendre's elliptic integral of the first kind. There is more;

$$\frac{dx}{d\phi} = \frac{dx}{ds} \times \frac{ds}{d\phi} = \frac{a \cos \theta}{2\sqrt{1 - k^2 \sin^2 \phi}},$$

and therefore

$$\frac{dx}{d\phi} = \frac{a(1 - 2k^2 \sin^2 \phi)}{2\sqrt{1 - k^2 \sin^2 \phi}},$$

whence

$$x = a \int_0^\phi \sqrt{1 - k^2 \sin^2 \phi}\, d\phi - \frac{a}{2} \int_0^\phi \frac{d\phi}{2\sqrt{1 - k^2 \sin^2 \phi}}$$

$$\equiv aE(k, \phi) - \frac{1}{2} aF(k, \phi), \tag{17}$$

where $E(k, \phi)$ is Legendre's elliptic integral of the second kind.

Two examples are shown in figure 12.3a,b for different values of k. Since $y = h \cos \phi$, the parametric equations of the curve are represented by the ordered pair

$$\left(aE(k, \phi) - \frac{1}{2}aF(k, \phi), h \cos \phi \right). \tag{18}$$

We note that since y vanishes when $\phi = \pi/2$, and $y'' = 0$ when $y = 0$, the "meander" curve has a change of concavity and hence a point of inflection wherever it crosses the x-axis (as in figure 12.1 also). In the case for which $h^2 = a^2$ we have

$$y = a \cos \phi$$

and

$$y = a \int_0^\phi \left[\cos \phi - \frac{1}{2} \sec \phi \right] d\phi = a \sin \phi - \frac{a}{2} \ln(\sec \phi + \tan \phi)$$

(since $0 < \phi < \pi/2$). An alternate but equivalent representation for y is

$$y = a \sin \phi - \frac{a}{2} \ln \tan \left(\frac{\phi}{2} + \frac{\pi}{4} \right),$$

as is readily demonstrated. Finally, it follows that

$$s = \frac{a}{2} \int_0^\phi \sec \phi \, d\phi = \frac{a}{2} \ln \tan \left(\frac{\phi}{2} + \frac{\pi}{4} \right).$$

In this case the curve has a single loop with branches asymptotic to the x-axis from above. This follows because x approaches infinity as ϕ approaches $\pi/2$ and $y > 0$.

For the remaining case ($h^2 > a^2$), y never vanishes (even asymptotically), and there are no points of inflection on the curve. This case is less relevant to the problem of river meanders, but can be expressed parametrically in terms of Legendre elliptic integrals of the first and second kind as in the first case. It is interesting to play around with both the Bessel function model and this elastic analog and compare the resulting shapes, both being very reminscent of the meanders actually observed in rivers.

MORE ABOUT MEANDER MODELS

As noted at the beginning of this chapter, Langbein and Leopold discuss four different meander curves: a circular curve, in which two portions of circles are joined; a pure sine curve; a parabolic curve, in which two portions of parabolas are joined, and finally, a sine-generated curve, in which the change of direction of the meander bears a sinusoidal relation to the distance along the curve (as in equation (1)). All four curves have the same wavelength and sinuosity (the ratio of the curve length to distance along the x-axis). Using a discrete version of the mean square curvature integral (2), they show that the sums of the squares of the changes in direction differ greatly (as measured in degrees over 10 equally spaced lengths along the curves). In the order listed above they are as follows for their particular choice of curves: 4840 (circular curve), 5200 (sine curve), 5210 (parabolic curve), and 3940 (sine-generated curve).

Specifically, in the notation of this chapter, Langbein and Leopold represent an approximation to the sine-generated curve by the "sine" version of equation (1). They provide an approximate relation between the sinuosity σ and the initial angle θ_0 (in radians) as

$$\theta_0 = 2.2\sqrt{1 - \sigma^{-1}}.$$

In degrees, the numerical factor of 2.2 becomes about 126°, which is the same as the limiting value of θ_0 for the "gooseneck" meander discussed in connection with the Bessel function model. Inverting this expression, we have

$$\sigma = \left[1 - \left(\frac{\theta_0}{2.2} \right)^2 \right]^{-1}$$

(compare this with the Bessel function version, equation (13)). In the Langbein and Leopold model, the approximate expression for the *bend radius* R is found in terms of the wavelength L_x to be

$$R = \frac{L_x}{13} \frac{\sigma^{3/2}}{\sqrt{\sigma - 1}},$$

from which the ratio

$$\Omega(\sigma) = \frac{L_x}{R} = 13\sqrt{\sigma - 1}\,\sigma^{-3/2}.$$

The graph of $\Omega(\sigma)$ increases from a value of zero on the interval $[1, \infty)$ to a maximum value of approximately 5 at $\sigma = 1.5$ and decreases toward zero thereafter. It is readily (and somewhat tediously) shown that, after a sequence of variable changes equivalent to $\sqrt{\sigma - 1} = \tan\theta$,

$$\int \Omega(\sigma)\, d\sigma = 26\ln\left|\sqrt{\sigma} + \sqrt{\sigma - 1}\right| - 26\sqrt{1 - \sigma^{-1}} + C,$$

from which it follows that the average value of Ω on $[1.1, 2]$ is

$$\frac{10}{9} \int_{1.1}^{2} \Omega(\sigma)\, d\sigma \approx 4.76.$$

According to data used in their paper, for meanders in which sinuosities lie between 1.1 and 2.0, the average value of meander length to radius of curvature Ω is 4.7, showing good agreement between this model and actual measurements. Commenting on their sine-generated model, Langbein and Leopold write

> That actual meanders are often irregular is well known, but as observed above, those meanders that are regular in geometry conform to this equation. Deviations (or "noise"), it is surmised, are due to two principal causes: (1) shifts from unstable to a stable form caused by random actions and varying flow, and (2) non-homogeneities such as rock outcrops, differences in alluvium, or even trees.
>
> The irregularities are more to be observed in "free" (that is, "living") meanders than in incised meanders. During incision, irregularities apparently tend to be averaged out and only the regular form is preserved, and where these intrench homogeneous rocks, the result is often a beautiful series of meanders as in the San Juan River or the Colorado River.

BRANCHING SYSTEMS

Observers of nature have long wondered whether the principles and rules that appear relevant to one type of branching system, say, river drainage systems, would apply to others—trees, the blood circulatory system of animals, or their bronchial and alveolar networks. In particular, do the ordering systems (in a somewhat quantitative fashion) also apply to tree "networks"? There are several different systems that have been developed, but one of the most widely used is that due to *Strahler*. (Other schemes are those of Gravelius, Horton, Shreve, Weibel, and Horsfield. With the exception of the Horton system, discussed briefly below, we do not pursue these here.)

According to the Strahler system, the lowest-order units or links (e.g., outermost twigs in a tree or initial feeder streams in a drainage pattern) are assigned the numeral or rank 1. Whenever two of these first-order links meet to produce a larger link (branch or stream), that link is assigned the numeral 2. This link is in the direction of the trunk or main stream, depending on the context. Since in general the patterns of "confluence" (joining) are asymmetrical, when a link of order $n - 1$ meets a link of order n, the continuation is defined to be of order n. Two links of order n produce a link of order $n + 1$.

What is the purpose of such an ordering scheme? Well, it does enable consistent patterns of branching to become more easily recognizable in very complex branching systems (when one might not otherwise be able to recognize the wood, as it were, from the trees). Such a pattern, as pointed out by McMahon, is the *branching ratio*: the average ratio between the number of segments belonging to one order and the number of segments in the next higher order (a segment is a link, or set of adjoining links, of the same order). When the number of segments in a given order is plotted logarithmically against that order, there results (approximately) a descending straight line, indicating a roughly similar ratio between orders $n + 1$ and n. In a small poplar tree analyzed by the Strahler system, there were, from first to sixth order (the latter being the trunk), $960, 256, 68, 15, 2$, and 1 segments, respectively. The corresponding ratios are approximately $3.75, 3.76, 4.53, 7.5$, and 2.

Interestingly enough, when one plots the mean diameters of the segments in successive orders against order number, a "straight-ish" ascending line arises, and in several instances (for pulmonary arteries, the bronchial network, and several species of trees) the diameter ratio is approximately half the branching ratio. Such ordering systems demonstrate the (statistical) *self-similarity* of tree, arterial, and bronchial systems, and hence the fact that these systems are, down to a very small scale, *fractal-like* in their structure (fractals are mathematical objects that exhibit geometric or statistical self-similarity at all length scales; see the appendix at the end of the book).

There is an interesting connection between the Strahler and Horton schemes and the Fibonacci numbers. If a Fibonacci tree (a branching pattern according to the Fibonacci sequence) is treated as a drainage pattern, Sharp has shown that all the important branching characteristics of the basin are expressible as Fibonacci numbers. This is extremely useful because so much is known about this sequence, and the exact branching behavior can be found for very high orders. Furthermore, the Fibonacci model can be used to test any hypothesis formulated about natural patterns, and can be generalized to include a wide variety of drainage patterns. Some general features of such models are described below.

RIVER DRAINAGE PATTERNS

If the external tributaries (those that have an obvious beginning; see figure (12.4a,b) of a river or stream are of class 1, then when any two merge they form a branch of class 2. This idea can be generalized by defining the following rules concerning the merging of tributaries in a stream or river:

I   If two tributaries of the same class i merge, the resulting branch is of class $i + 1$.
II If two tributaries of different classes i and j merge, where $j > i$, the resulting branch is of class j.

If we denote by N_i the total number of tributaries of class i, and by m the class of the main stream, where m is the largest class that occurs based on the above rules, then we may establish some very interesting results

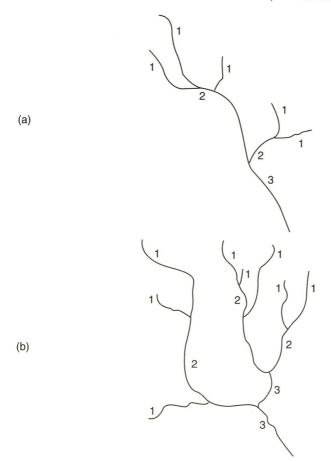

(a)

(b)

Figure 12.4. The branching patterns of two rivers in South Carolina according to the Strahler ordering scheme. Redrawn from "An Adventure into Applied Mathematics with Fibonacci Numbers," by R. V. Jean and M. Johnson (1976), *School Science and Mathematics* 89: 487–498.

about drainage patterns (see the paper by Sharp, for example). Following Jean and Johnson we will simplify the analysis a little and concentrate on a simple example. If we define the mean bifurcation ratio b for any river as the average number of tributaries emanating from a branch (e.g., if $N_1 = \alpha$ and $N_2 = \beta < \alpha$ then $b = \alpha/\beta$), then it has been demonstrated that

$$N_i = b^{m-i}.$$

Drainage networks for which $N_1 = F_k$ for some k, where F_k is the kth Fibonacci number, are called Fibonacci networks, and k is defined as the order of the network. In the figure shown, $k = 5$ ($F_k = 5$) and $k = 6$ ($F_k = 8$), respectively. In such cases, it is possible to prove that m is equal to the largest integer in $(k + 1)/2$ (denoted by $[(k + 1)/2]$) which is 3 in each of these cases, and that

$$N_i = F_{k-2i+2}. \tag{19}$$

Using these definitions and properties, we can calculate the total number of tributaries (of all orders) for Fibonacci networks,

$$N_t = \sum_{i=1}^{m} N_i = F_k + F_{k-2} + \cdots + F_{k-2m+2}, \tag{20}$$

upon using equation (19). If k is odd, $k = 2p + 1$, say, then this may be written as

$$N_t = F_{2p+1} + F_{2p-1} + \cdots + F_{2p-2m+3}. \tag{21}$$

Now let $2a + 1 = 2p - 2m + 3$, where a runs from 0 to $p - 1$ so that

$$N_t = F_{2p+1} + \sum_{a=0}^{p-1} F_{2a+1}. \tag{22}$$

By a well-known property of Fibonacci numbers,

$$\sum_{k=0}^{n-1} F_{2k+1} = F_{2n}, \tag{23}$$

so that (22) becomes

$$N_t = F_{2p+1} + F_{2p} = F_{2p+2} = F_{k+1}. \tag{24}$$

By means of a similar calculation when $k = 2p$ is even, writing $a = p - m + 1$ it follows that

$$N_t = \sum_{a=1}^{p} F_{2a} = F_{2p+1} - 1 = F_{k+1} - 1, \tag{25}$$

this time using the property that

$$\sum_{k=1}^{n} F_{2k} = F_{2n+1} - 1. \tag{26}$$

Applying these results to the streams in the figure, we readily see that for $k = 5$, $N_t = F_6 = 8$, and for $k = 6$, $N_t = F_7 - 1 = 12$. This is readily checked from the figure (noting that two contiguous branches of a given order count as one branch). The advantages of such mathematical schemes are most easily appreciated when there are many branches and classes present in a drainage network. For mathematical details the paper by Sharp should be consulted.

TREES

Thomas A. McMahon has said that "The branching forms of trees are deeply pleasing to us and would seem to serve the needs of the trees themselves. In their proportions there is a harmony that makes us wonder if we could discover a principle for their mechanical design."

There are other factors to be considered in the structure of trees, factors that unlike the branching schemes mentioned above are not topological in nature. The mechanical or elastic properties of trees determine whether or not a tree trunk will buckle under its own weight, and whether or not a branch will bend or break under a load. Thus stiffness and strength are properties determined by the elasticity of the structure. Consider some facts concerning trees: (i) In relation to its density, wood is stiffer and stronger in both bending and twisting, than concrete, cast iron, aluminum alloy, or steel. (ii) Trees are very frugal in their use of resources for growth. (iii) They "use" the "principle of minimum weight." This self-explanatory principle means that because every material used to stiffen or support a structure also adds to the total load the structure must bear, adding height (i.e., as in growth) must involve a trade-off between the strength of a structure and its weight (as noted in chapter 3, this is why King Kong and giant ants will not take over the world, for they cannot exist). This also explains, in part at least, why the earth's tallest, most massive, and longest living organisms are trees. Thus, the largest living animal, the blue whale, rarely exceeds 110 feet in length and 180 tons in weight, whereas giant sequoias (*Sequoiadendron giganteum!*) have grown to heights of 360 feet (110 meters) with girths of as much as 100 feet (30 meters). California's General Sherman sequoia is over 270 feet high and estimated to weigh at least 2500 tons. The oldest sequoias are more than 4000 years old (and if left undisturbed, trees will stand for many decades after their death).

The mechanical forces generated within a vertical stem as it bends under its own weight, or as it bends and twists in the wind, are most intense just beneath the surface of the stem. So to build a tree it is a good idea to place the strongest and stiffest available materials around or under the surfaces of vertical stems, because this is the best way for a tree to cope with the different types of mechanical forces that trees typically face during their lifetime. And that is just the way it's done! In fact trees, with their remarkable combination of strength and flexibility, are models of excellent design. The design of a tree trunk or limb employs both tension and compression—pull and push. This allows saplings to become vertical again after being buried by snow. When wind blows on a tree, the windward side is in tension and the leeward side is in compression. When a branch bends under a load of snow, the cells on the limb's upper side may shrink to allow for recovery.

Interestingly, in light of this, there is a related "new" science known as *biomimetics*. This subject is the (increasingly interdisciplinary) study of how human technology can be improved by copying designs from nature, by focusing on successful "patterns in nature." An example of this is found in the mathematical patterns (mentioned above) known as fractals: the branching, bifurcation, and repetition on increasingly smaller scales are known to occur in trees, streams, lungs, and capillaries. What works well in nature may work well elsewhere—leading to principles of efficient flow of fuel and optimal stress distribution in steel pins and plates used in orthopedic surgery, among other applications. Biomimeticists are aware that trees grow in ways that tend to minimize stress and distribute stress uniformly (shades of river meanders?). Thus every tree is shaped by the loads it must carry, and part of a tree's ability to adapt to its loads lies in the *cambium*, the growth layer that makes new cells just below the bark. The cambium can actually sense areas of stress, in a wound for example, and then produce more wood at that point, until the stress is reduced.

Mathematically, the stress is often related to the *curvature* of the wood fibers. Nearly 90 percent of wood cells are arranged in (guess what?) a honeycomb fashion. This pattern, as we have seen, occurs repeatedly in nature, and it results in lightweight strength. Few materials are as tough as cellulose (the main constituent of wood), whose strength derives from its ability to store energy like a spring. Indeed, Lorna Gibson (as quoted in Niklas) points out that "Wood's exceptional performance in bending reflects the fact that the trunk and branches of a tree are often loaded in bending: the trunk by the wind and the branches by their own weight. It arises from the honeycomb-like structure of the wood cells as well as the great stiffness and strength of the cellulose fibers." Claus Mattheck writes that "Trees are smart structures. They grow wood in a load-adapted way; they use the strongest wood where the highest internal stresses are. They also optimize their external shape."

THE GEOMETRIC PROPORTIONS OF TREES

The elastic properties of trees are at least as important as their strength properties as far as their ability to support weight is concerned. In winter one often sees birches and pines bent over under the weight of snow, their branches and even tops touching the ground. After the snow loads are lost, many will spring up again. Some that are less flexible do not: they have broken under their snowy burden.

It seems that the actual danger of breaking under added weight is not as great as the danger of *buckling*. As has been mentioned already in chapter 3, it can be shown that for cylindrical elastic columns, the square of the diameter is proportional to the cube of the height,

$$d^2 \propto h^3 \text{ or } d \propto h^{3/2},$$

provided certain physical properties remain constant. Indeed, in 1881 the mathematician G. Greenhill noticed the flagstaff in Kew Gardens (London), which was 221 feet high and 21 inches in diameter at the base. He asked himself how tall a cylindrical flagstaff of the same diameter could be before buckling under its own weight. Applying principles from solid mechanics, and some beautiful mathematics, he was able to show that the Kew flagstaff could not exceed about 300 feet in height. Furthermore, he was able to show that in general the diameter of tall cylindrical columns should increase as the 3/2 power of the height (the result stated above, under the same assumptions; see D'Arcy Thompson, p. 28). It turns out that similar principles and relationships apply to branches of trees, not just the trunk. The above thickness/height relationship has been verified experimentally.

We will examine the mathematics behind these statements on two fronts. The first will be on the basis of mechanical and dimensional arguments; the second will be more analytical, based on the kind of mathematics used by Greenhill. The latter approach will be found in the section below entitled "How high can trees grow?" If the ratio of the height to radius of a self-supporting vertical column is too large, it will become unstable to lateral deflections, as seems eminently reasonable to anyone who has experimented with flexible rods. The reason is that if there is a "bend" in the previously vertical beam, then there is a torque (or moment tending to cause rotation) due to the weight of the column above the point of maximum sideways deflection. In equilibrium, this is balanced by the respective compressive and tensile stresses on either side of the column. When this elastic torque (τ_e) is not sufficient to balance the so-called gravitational torque (τ_g), the column buckles. Thus for stability

$$\tau_e > \tau_g.$$

We follow the arguments of Lin (1982) to translate this inequality into one explicitly relating the height h of the vertical column to its diameter d. From figure 12.5 it follows that if W is the weight of the column, $\tau_g \sim Wa$ (a being the total lateral deflection and ignoring numerical constants) and the stresses σ give rise to forces $\sim \sigma d^2$ and moments $\tau_e \sim \sigma d^3$. From Hooke's law it follows that the maximum stress

$$\sigma_{max} \sim E\left(\frac{\Delta h_{max}}{h}\right),$$

where E is Young's modulus and Δh_{max} is the maximum amount that the sides can be stretched and compressed, respectively. Thus the above inequality becomes

$$Wa < E\left(\frac{\Delta h_{max}}{h}\right)d^3. \tag{27}$$

From the geometry of the figure, if R is the (large) radius of the approximately circular arc of the neutral line AB, then

$$h = 2\theta R \text{ and } h + \Delta h = 2\theta\left(R + \frac{d}{2}\right),$$

so that

$$\frac{\Delta h}{h} = \frac{\theta d}{h} \sim \frac{ad}{h^2}, \tag{28}$$

since $\sin\theta \sim \theta$ and hence $\theta \sim 2a/h$. From equations (27) and (28) therefore,

$$Wa < E\left(\frac{ad}{h^2}\right)d^3$$

or

$$h < \left(\frac{Ehd^4}{W}\right)^{1/3} = \left(\frac{E}{\rho}\right)^{1/3} d^{2/3}, \tag{29}$$

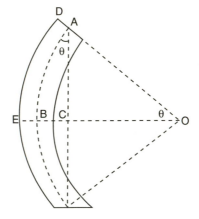

Figure 12.5. Schematic (and highly exaggerated) representation of the buckling of a beam, used in establishing the basic height/thickness ratio.

where the fact that $W \sim hd^2\rho$ has been used, ρ being the density of the column. Remember that this is essentially a dimensional argument, but it is valid to within multiplicative constants of order unity. This will be verified using somewhat more sophisticated mathematics a little later on.

SHAKING OF TREES

Theoretical studies predict that there is a simple relationship between the natural frequency of vibration (as in wind) of a whole tree (or individual branches) and its height: the frequency of vibration of the tree is inversely proportional to the square root of its height (or length). There is close experimental agreement also. The relationship is valid whether the tree is "naked" or "clothed"—though the leaves do increase the air resistance and approximately double the period of vibration. Now here's an interesting thing: silviculturists (people—but not lumberjacks—who work with trees and forests) report that trees raised in a greenhouse will grow more rapidly in girth if given as few as 5 or 10 modest "shakings" a day. Furthermore, trees grown outdoors should not be supported by guy wires for too long, or they will grow so long and slender that they will be unable to stand up by themselves when unsupported.

Consider first a branch extending from its base perpendicular to a vertical trunk. As it grows outward, it will sag more and more under its own weight. After a certain critical length is exceeded, the tip of the branch no longer extends radially any farther from the trunk, but actually moves closer as the length increases because of the greater sag. It turns out that the angle of the chord of a critical length beam is the same in the laboratory for beams of any diameter. This critical length corresponds to the greatest distance from a tree trunk that is available to a leaf, allowing the leaf to avoid being in the shade (for the most part) of leaves on higher branches for at least some of the time. Because of the constancy of the above-mentioned chord angle, despite the increased critical length for larger beams (or in principle, branches also), McMahon characterizes such beams as possessing elastic similarity. The diameter of such beams (assuming they are made of the same material) is proportional to their (critical) length raised to the 3/2 power: a result we have met before.

If elastically similar beams are rigidly supported at the lower end when placed vertically, and the free end is deflected sideways, the column will either buckle under its own weight (stay bent) or generally return to its upright position after some oscillations about that position. This will depend on the length of the beam, its modulus of elasticity, and its density. Short beams will vibrate back and forth with higher frequencies than longer beams of the same diameter. For a given diameter there is a beam of critical length for which the tendency to buckle is exactly counterbalanced by the tendency to return to a vertical equilibrium position. Such a beam is said to be neutrally stable; it does not oscillate, but smoothly returns to an upright position. This critical length is analogous to (but in general not the same as) the critical length for a sagging horizontal beam: the latter is a static configuration while the former is a dynamic one. Nonetheless, the diameters of neutrally stable beams are again proportional to their (length)$^{3/2}$.

We can characterize the relationship between the length l and diameter d of beams in a rather general way, namely,

$$d = Kl^{\beta},$$

where K and β are constants. In the elastic similarity model, $\beta = 3/2$. This relationship is rather closely adhered to in trees: McMahon found from an analysis of specimens of more than 600 species of trees growing in the United States not only that this was so, but that the corresponding mean height for each diameter was about 25 percent of the theoretical neutral stability height for a uniform column of the same diameter. There were of course variations for individual trees. What about alternative values of β in the equation above? For geometrically similar beams (no change in shape or proportion), $\beta = 1$, whereas values of $\beta > 3/2$ correspond to an even larger increase of diameter with length. For six trees representing four species (white pine, white oak, red oak, and cherry) McMahon found that $1.37 \leq \beta \leq 1.66$, with $\beta = 1.51$ and 1.50 for the red oak and cherry, respectively.

Similar considerations apply to the natural frequencies of vibration of trees. The frequencies of vibration of springs and masses depend on the spring stiffness and attached mass: increasing as the stiffness increases and decreasing as the mass increases. This is true in general for trees, although of course both the mass and stiffness are quantities distributed throughout the structure of the tree. For a beam clamped at one end, the natural frequency of vibration f is proportional to its (length)$^{-1/2}$, and this in fact applies to both whole trees and lateral branches. Thus small trees should vibrate faster than large trees (of the same species), and individual branches more so than whole trees. Theory predicts

$$f = K_1 l^{\beta - 2},$$

where K_1 is another constant. Thus for $\beta = 1$ (geometric similarity), $f \propto l^{-1}$; for $\beta = 3/2$ (elastic similarity), $f \propto l^{-1/2}$, and for $\beta = 2$ (static stress similarity), f is independent of length. From laboratory tests on branches, McMahon found an average value for $\beta - 2$ of about -0.59, thus confirming the elastic similarity model as the most realistic of the above three. Among trees of the same species, the main stem turned out to be more mechanically stiff than the branches, and the constant K_1 (and hence f) was between two and three times larger when the tree was leafless than when it was not. Obviously there is increased air resistance when the tree is in leaf, resulting in damping of vibration and hence lower frequency, though its frequency is still approximately proportional to $l^{-1/2}$.

In summary, then, it transpires that the principle underlying the structure and growth of most trees is elastic similarity (and its maintenance). Clearly as a tree grows in height its girth also increases, as we have seen, in proportion to (height)$^\beta$, where $\beta \approx 3/2$. The curvature of the tree induced by bending must have something to do with inducing girth, or trees would be buckled and bent everywhere in response to the normal environmental stresses acting on them. It appears that when the curvature induced by weight exceeds some threshold value the limb or trunk diameter is "encouraged" to increase. But how is this encouragement mediated? What mechanism is responsible for this?

In a beam bending under its own weight, the fibers of the wood are under either tensile or compressive longitudinal stress. The rate of change of stress perpendicular to the central axis of the beam or trunk is proportional to the local curvature of the beam. It is possible that radial segments (wood rays) in the cambium, which is responsible for new wood growth, are sensitive to stress gradients along them. If this is the case, then when these gradients exceed a threshold level, plant hormones known as auxins (which stimulate growth) may be released and cause radial growth of the tree, which would in turn maintain the weight-induced curvature of the tree within acceptable limits to avoid buckling or bending. Now, if the appendix at the end of this chapter has been consulted, we are in a position to discuss the heights of trees.

HOW HIGH CAN TREES GROW?

What follows will be a discussion based on purely mechanical considerations; it will be limited therefore by ignoring one (if not *the*) major aspect of tree growth, namely, that trees are limited by the height to which nutrient-rich water can be drawn by the tree (via transpiration). That itself is a very interesting problem, and one to which we shall devote a little time later in this chapter. We shall content ourselves here with an equally interesting problem in mechanical engineering: *the buckling of a uniform column* (our tree) *under its own weight*. Consider therefore such a column rigidly embedded in the ground (via the root system, in the case of a mature tree) with cross-sectional area A and weight per unit volume denoted by γ (see figure 13.7). The central axis point at the top of the tree serves as the origin of coordinates for the system, $x = 0$, and the ground level corresponds to

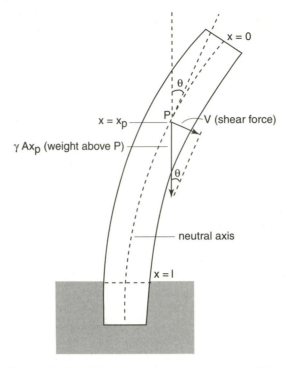

Figure 12.6. The detailed configuration used in deriving the differential equation (30) for the beam buckling problem. Redrawn from *Mathematical Methods in Engineering*, by T. von Karman and M. A. Biot (1940), McGraw-Hill: New York.

$x = l$. The fact that many trees taper in height (though many do not) is not significant here because the taper is usually relatively small (trees are usually not pointed at the top) and the branches of course compensate for the somewhat reduced weight of the trunk per unit height as we proceed upward. Indeed, it could be argued that the taper should consist of a widening with height to account for the canopy (branches and foliage). This will not concern us here, however; we shall essentially therefore derive an upper bound to the height of a generic tree based on the mechanical properties of wood (again, neglecting the vascular structure of the tree).

In the buckled position (see figure 12.6), the weight of the portion of the beam between the top and the cross section at P (a distance x_p along the axis) is in equilibrium with the resultant of the axial stresses and the shear force V acting at P. If the inclination of the column axis from the vertical at this point is θ, then the shear force is given by

$$V = \gamma A x_p \sin\theta \approx \gamma A x_p \theta$$

if θ is small, which will be the case at the point of buckling (in what follows the subscript on x will be dropped for simplicity of notation). The bending moment M at P is usually expressed in terms of the displacement from equilibrium η such that for small displacements

$$M = -EI\frac{d^2\eta}{dx^2}$$

(noting that $d\eta/dx < 0$ here). For small angles this can be expressed in terms of θ since $\theta \approx \sin\theta \approx d\eta/dx$, so

$$M = -EI\frac{d\theta}{dx}$$

at this level of approximation. The quantity EI is called the *flexural rigidity* of the column; it is the product of Young's modulus E for the material and the moment of inertia (sometimes called the second moment of area) I of the column cross section about the central axis. Both quantities are assumed constant here. The shear force $V = dM/dx$, so equating the two expressions for V we obtain the second-order ordinary differential equation

$$\frac{d^2\theta}{dx^2} + \frac{\gamma A}{EI}x\theta = 0. \tag{30}$$

The simple change of independent variable (plus a modicum of chain-rule) allows us to clean up this equation a little: if we define

$$x = \left(\frac{\gamma A}{EI}\right)^{-1/3}\xi,$$

then

$$\frac{d^2\theta}{d\xi^2} + \xi\theta = 0 \tag{31}$$

results, which is the canonical form of Airy's differential equation. And yes, this is the same equation that arises in connection with Airy's theory of the rainbow (and occurs in many other situations). The general solution to this equation may be written in terms of (surprise, surprise) Airy functions or, equivalently, Bessel functions of the first and second kinds of order one-third. We shall actually use a power series representation for these functions. Writing the general solution as

$$\theta = B\xi^{1/2}J_{1/3}\left(\frac{2}{3}\xi^{3/2}\right) + C\xi^{1/2}Y_{1/3}\left(\frac{2}{3}\xi^{3/2}\right)$$
$$= B\theta_1(\xi) + C\theta_2(\xi),$$

where B and C are arbitrary constants, the following series (convergent for all ξ) may be found:

$$\theta_1 = \xi(a_0 + a_1\xi^3 + a_2\xi^6 + \cdots) \text{ and}$$

$$\theta_2 = b_0 + b_1\xi^3 + b_2\xi^6 + \cdots$$

for constants a_i, b_i, $i = 0, 1, 2, 3, \ldots$. We will apply suitable boundary conditions to determine these constants for approximate solutions to the problem. Since there is no weight above $x = 0$ it follows that the bending moment is zero, which implies that at $\xi = 0$,

$$\frac{d\theta}{d\xi} = 0.$$

Since the column is rigidly "clamped" at $x = l$, which is equivalent to $\xi = (\gamma A/EI)^{1/3} l$, it follows that $\theta = 0$ at that point. Applying the first boundary condition to the general solution forces $a_0 = 0$, and it can be shown (using Airy's equation for the solution $\theta_1(\xi)$) from the recursion relation for the coefficients a_i, that all of them are proportional to a_0, for example,

$$a_1 = -\frac{a_0}{3 \times 4} = -\frac{a_0}{12};$$

$$a_2 = \frac{a_0}{3 \times 4 \times 6 \times 7} = \frac{a_0}{504};$$

$$a_3 = -\frac{a_0}{3 \times 4 \times 6 \times 7 \times 9 \times 10} = -\frac{a_0}{45360}, \text{ etc.,}$$

which means that all of them are zero, and so $\theta_1(\xi) \equiv 0$. Thus taking $C = 1$ without loss of generality, the solution to the problem has the form

$$\theta = \theta_2 = b_0 + b_1\xi^3 + b_2\xi^6 + \cdots.$$

Again, by substituting this into the differential equation, the recursion relation for the coefficients b_i gives the following equations in terms of b_0:

$$b_1 = -\frac{b_0}{2 \times 3} = -\frac{a_0}{6},$$

$$b_2 = \frac{b_0}{2 \times 3 \times 5 \times 6} = \frac{b_0}{180}, \text{ etc.}$$

Therefore we may write

$$\theta = \theta_2 = b_0\left(1 - \frac{\xi^3}{6} + \frac{\xi^6}{180} - \cdots\right) = 0$$

when $\xi = (\gamma A/EI)^{1/3} l$, by the second boundary condition.

We now examine two approximate solutions, the simplest of which arises if we neglect the third term above in comparison with the second, that is, if $\xi^3 \ll 30$. Strong inequalities like this one always lead us to ask "What is meant by *much less than?*" and the answer often depends on the context. If for the sake of argument here we say that this means less than one-fourth of the larger quantity, then we can write that $\xi \lesssim 2$. The solution then becomes $\xi \approx \sqrt[3]{6} \approx 1.8$. To the next order of approximation we find that

$$\theta = b_0 \left(1 - \frac{\xi^3}{6} + \frac{\xi^6}{180} - \frac{\xi^9}{12960} + \cdots \right) = 0.$$

Hence, if we neglect the fourth term in comparison with the fifth one, this is equivalent to requiring that $\xi^3 \ll 72$. Using the same criterion as above we need $\xi \lesssim 2.6$, and this yields the following (approximate) biquadratic equation in $z = \xi^3$:

$$z^2 - 30z + 180 = 0,$$

which has roots given by

$$z_1 = 3\sqrt{5}(\sqrt{5} + 1) \approx 21.7 \text{ and}$$
$$z_2 = 3\sqrt{5}(\sqrt{5} - 1) \approx 8.3.$$

We take the smaller root for obvious reasons and obtain $\xi \approx 2.0$, which is the smallest critical root above which buckling will occur. In terms of physical variables, this corresponds to a critical length of the column given by

$$l_c \approx 2 \left(\frac{EI}{\gamma A} \right)^{1/3}.$$

For a cylindrical column of circular cross section and radius r, the second moment of area I is $I = \pi r^4 / 4$, so that $I/A = r^2/4$. Hence

$$l_c \approx 2 \left(\frac{E}{\gamma} \right)^{1/3} \left(\frac{r^2}{4} \right)^{1/3} = \left(\frac{2E}{\gamma} \right)^{1/3} r^{2/3}, \tag{32}$$

that is, *the critical length is approximately proportional to the two-thirds power of the radius.* As noted earlier, this is known as the law of elastic similarity, a term already encountered in chapter 3. Equivalently, we may write this relationship for a given type of material as

$$r \propto l_c^{3/2}.$$

Obviously the quantities E and γ will vary with the material under consideration. We may note that E has dimensions of stress (which are the same as pressure, or force per unit area) because in the simplest terms E is defined as stress divided by strain, and the latter, being a ratio of lengths, is dimensionless. Similarly, γ has dimensions of force per unit volume, so E/γ has dimensions of length and may be interpreted as a characteristic length associated with each type of material. Denoting this length by L we may summarize the above fundamental result as

$$l_c \approx 2^{1/3} L^{1/3} r^{2/3} \propto L^{1/3} r^{2/3}.$$

As we have seen, E is Young's modulus and γ is the weight per unit volume for the material under consideration. The radius of the cylindrical column is r. In the table below (adapted from the book *Plant Biomechanics* by J. Niklas) the values of E and ρ (density) for various trees are stated, from which γ can be calculated: weight per unit volume = *mass* × *gravity* ÷ *volume* = *density* × *gravity*, so $\gamma = \rho g$, where $g = 9.81$ m/s^2. (A giganewton (GN) is 10^9 newtons.) The last column gives the critical height in feet. Typical (guessed!) r-values are provided in column one. Thus

Tree (and estimate of typical r)	E (GN/m²)	ρ (kg/m³)	γ (N/m³)	L (m)	l_c (m)	l_c (ft)
Ash ($r = 9$ in $= 0.23$ m)	11.0	526	5160	128.7	60.8	200
Black oak ($r = 18$ in $= 0.46$ m)	11.3	669	6562	120.0	90.2	296
Redwood ($r = 1$ m)	9.4	436	4277	130.0	164.0	538

Clearly, according to this model, trees are built with a large safety factor! The tallest known redwood tree, the California redwood, is 366 ft. The corresponding *records* for some other trees are Douglas fir (302 ft), giant sequoia (272 ft), ponderosa pine (223 ft), cedar (219 ft), Sitka spruce (216 ft), and beech (161 ft). Typical values for the black oak and white ash are 50–80 ft (diameter 1–21/2 ft) and 80 ft (diameter about 2 ft), respectively. The redwood is typically about 350 feet high.

Note that this upper limit l_c has been derived purely on the basis of mechanical considerations. As noted at the beginning of this section, a related but different constraint, totally ignored here, is that of the ability of the tree to move water through the trunk to the branches and leaves. This is done by transpiration, and clearly there is a limit to how high a tree can be for this mechanism to work efficiently. This will provide an upper limit *less* than l_c (I suspect), and hence a more realistic one. Engineers like to build in safety factors as well! Furthermore, it is interesting that on the basis of a purely engineering type of analysis we have been able to deduce the basic observations concerning the relationship of tree radius to tree height.

I close this section with a few comments about getting sap up a tree, based on Steven Vogel's lovely article on dimensionless numbers. A useful dimensionless quantity in this connection is the *Bond number*, which measures the relative size of the competing forces of weight (downward) and surface tension (upward), and is defined as

$$B = \frac{mg}{\gamma l}$$

for a body of mass m, where γ is the coefficient of surface tension and l is the perimeter of the wetted surface (typically, this is just the circumference of the conduit, pipe, or straw, depending on context). If $B < 1$ surface tension wins; if $B > 1$ gravity does (perhaps that's why 007 is able to scale heights so easily—his Bond number is smaller than one).

Now in a tree the columns of sap extend continuously from the roots to the leaves. The question arises as to whether the capillarity (surface tension) force alone is sufficient to accomplish this. The upward pressure $p = 2\gamma/r$, r being the radius of the vascular conduit, so now

$$B = \frac{\text{weight of column}}{\text{pressure} \times \text{area}} = \frac{\pi r^2 h \rho g}{(2\gamma/r)\pi r^2} = \frac{r h \rho g}{2\gamma},$$

so that for $B < 1$ it follows that

$$h < \frac{2\gamma}{r\rho g}.$$

Taking $\gamma \approx 70$ dynes/cm, $\rho = 1$ gm/cm^3, $g \approx 980$ cm/s^2 and a typical radius as $r \approx 1/20$ mm $= 1/200$ cm, we find that h can be at most about 30 cm, so clearly, it isn't enough (except for bonsai trees!).

What is the resolution of this problem? Vogel points out that the columns of sap in trees essentially hang from the tops of trees by virtue of the tremendous internal cohesion of the water drawn up by the process of transpiration, which is evaporative water loss from the leaf canopy. In this situation water can exit the leaves but air cannot enter, which means that at the leaf/air interface the appropriate radius is that for the pores in the cell walls within the leaves. This is typically 10^{-4} mm, or about 500 times less than the conduit radii, which means that the height limit increases by this factor, that is, $h < 150$ m, this upper limit being greater (by nearly a factor of two) than the height of any tree known. This is a *strict* upper limit of course, since no account was taken of pressure losses due to flow in the conduits and in extracting water from the soil. There is therefore a large margin left over for the tree heights we observe!

THE INTERCEPTION OF LIGHT BY LEAVES

How much shade does a layer of leaves provide for the layer below? This is an important question, because ideally, parts of all leaves should be in

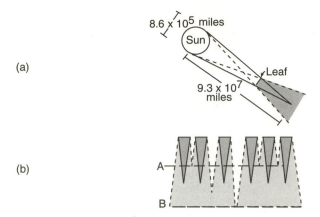

(a)

(b)

Figure 12.7. The interception of sunlight by leaves; (a) the basic geometry for a single leaf; (b) the shielding effect of leaf layers. Redrawn from *The Adaptive Geometry of Trees*, by H. S. Horn (1971), Princeton University Press: Princeton.

sunlight in order for the maximum benefits of photosynthesis to be aquired. A detailed discussion of this and many other aspects of sunlight and shade in trees may be found in the book by Horn. Consider the simplified geometry of the sun/leaf configuration in figure 12.7a, b. The following comments and questions can be made:

(i) The sun is completely blocked off by an opaque circular leaf of diameter d for a distance of αd; what is α (to the nearest integer)? What does this result imply about the optimal spacing between leaf layers?

(ii) There is in fact a brightly lit "aureole" around the sun (often noticed in cloudy weather, but very pronounced on clear days—*but don't look directly at it!*). Because of this, the distance at which the shadow of a circle vanishes on a clear day is about 50 to 70 times its diameter when the sun is at the zenith (directly overhead). On a uniformly cloudy day, the shadow of a circle (if at all visible) vanishes at a distance about equal to its diameter.

What, if anything, do these facts tell us about (i) the mean spacing between leaf layers, and (ii) the mean size of leaves (relative to the spacing between leaf layers) in sunny and cloudy climates? The basic criterion to consider is that no leaf should be subjected to total shade during the day. From the geometry of similar triangles discussed in chapter 4, the constant $\alpha = 108$. This means that a leaf layer blocks the sun completely in some leaves or parts of leaves in the next layer down, up to 108 leaf diameters below the top layer. Farther away than this, leaves are in the penumbra but not the umbra. For many such leaf layers, the overall effect is that of a light filter, reducing the light intensity by the proportion of light flux intercepted by the leaves (this will be quantified below). Furthermore, the average length of the umbra of a leaf is much greater in sunny climates than in cloudy ones (even including the effect of the aureole), but in both cases it is proportional to the leaf diameter d. If no leaves are to be subjected to total shade, the minimum distance between leaf layers should be large in sunny climates and could be much smaller in cloudy climates (though this is not a necessary condition). Correspondingly, we should expect that, for a given distance between leaf layers, the size of a leaf is much smaller in a sunny climate than in a cloudy one.

These are of course generalizations, and there are some qualifications to be made. It doesn't take a Sherlock Holmes (or even the usual rocket scientist) to notice that leaves are generally not circular in shape; what we are concerned with here is the largest circular area of the leaf that completely blocks off light, so we are dealing with an effective diameter—the diameter of the largest circle that can be fully inscribed within the perimeter of the leaf. Thus a thin, elongated pine leaf will generally have a smaller effective diameter than a round or elliptical leaf of equal area. Also, leaves are not entirely opaque, but typically they absorb about 95 percent of the light incident upon them.

And what is this I mentioned about leaves acting like a light filter? Well, in a sense it is rather obvious that they do, but we can flesh this out a bit with some elementary mathematical considerations. Again let us take circular leaves of radius r (in the generalized sense mentioned in the preceding paragraph) and "scatter" them randomly in n layers with an average density (number) of p leaves per unit area of ground. This means that this density p is uniformly distributed vertically among n layers, and we furthermore assume for simplicity that the leaves are spaced within each layer so that none overlap (this is not completely justifiable, of course, but it does provide us with an upper bound on the area intercepted by the totality of leaves).

We define the projection of leaves as the proportion of ground covered by the leaves' shadows when the leaves are illuminated by sunlight on a cloudless day; it is thus the proportion of the light flux that they intercept. The average projection of each of the n layers is therefore $p\pi r^2/n$. If the layers are far enough apart (in terms of leaf diameter in the sense described above), then a proportion $1 - p\pi r^2/n$ of the incident light gets through the first layer, and of that, the same proportion $1 - p\pi r^2/n$ is transmitted, and so on down to the nth layer. The proportion of incident light penetrating all n layers (multilayers) is thus

$$\left(1 - \frac{p\pi r^2}{n}\right)^n.$$

At this point there are two limiting cases or extremes of interest. The first is the case of the *monolayer* for which $n = 1$ (this is not uncommon, e.g., some types of palm tree); the second is the limiting case $n \to \infty$, for which, of course,

$$\lim_{n \to \infty} \left(1 - \frac{\rho \pi r^2}{n} \right)^n = e^{-\rho \pi r^2}.$$

Thus in assuming that the leaves are distributed evenly among the n layers (infinity is a whole *lot* of leaves, to be sure) and that the layers are far enough apart to be independent of each other, we have arrived at a well-known result for the exponential decay of the incident light as it passes through the "filters." This type of result also arises if we examine the opaqueness of a wood. Note that for an obviously finite number of leaves there is an upper bound to n; one cannot have fewer than one leaf per layer!

AEOLIAN TONES

When wind blows across a stretched wire (i.e., perpendicular to it), notes or humming sounds may be produced. If u is the speed of the wind and d is the diameter of the obstructing wire, then the frequency v of air vibrations produced is given approximately by the formula

$$v = \frac{0.185u}{d},$$

where the units of length and time are, respectively, centimeters and seconds.

The stretched wire itself may take no part in the vibration. However, if the frequency v is close to the fundamental resonant frequency of the wire (or a harmonic thereof), the wire *will* vibrate in a direction perpendicular to that of the wind, often vigorously, and the sound intensity will increase. The sound is due to the phenomenon of vortex shedding, whereby vortices develop at regular intervals, alternatively above and below the obstructing cylindrical wire. The phenomenon is an example of the instability of vortex sheets. The *Strouhal* number (0.185 above) is of course a dimensionless quantity, the ratio vd/u being dimensionless. The humming of telegraph and telephone wires is loudest when the wires are tightly stretched (as is often the case in cold weather). The eddies produced in the wind vary with the tension in the wire, and these variations in tension are transmitted to the posts, which in turn act like sounding boards to increase the volume of the sound.

Whether or not vortex shedding (essential to the production of Aeolian sound) is likely in any given situation depends on a dimensionless quantity called the Reynolds number Re, defined by

$$\text{Re} = \frac{ul}{v},$$

where u and l are typical velocity and length scales for the fluid motion, and now v is the (kinematic) viscosity of the air. Experimental and theoretical fluid dynamical studies lead to the conclusion that vortices will be shed when the Reynolds number lies in the range $100 \lesssim Re \lesssim 1000$.

What about sound produced by needles, leaves, and trees? Can this mechanism apply to such situations? For pine needles, let us take as typical values those provided by Suzuki and use $u \approx 3$ m/s, $l \approx 1.5$ mm and $v \approx 0.15$ cm^2/s. Thus, being careful to use consistent units we find that $Re \approx 300$, well within the range for exciting Aeolian tones.

In addition to pine needles, leafless twigs and small branches must by the same principle all produce Aeolian tones, so trees have voices characteristic of their species. The relatively large twigs and branches of the oak tree may be expected to produce many low tones compared with innumerable fine needles of the pine tree, which will produce a range of higher-pitched notes. How do such notes blend together? How are the loudness and pitch resulting from the symphony of combined notes related to the individual properties of each "musician"?

Good questions. Let us play around with some simple ideas first and then try to draw some plausible conclusions. Consider the result of adding two sinusoidal wave trains of equal amplitude a, angular frequencies ω_m, ω_n, and phases δ_m, δ_n respectively. Then

$$a\{\cos(\omega_m t - \delta_m) + \cos(\omega_n t - \delta_n)\}$$

$$= 2a \cos \frac{1}{2}[(\omega_m + \omega_n)t - \delta_m - \delta_n] \times \cos \frac{1}{2}[(\omega_m - \omega_n)t - \delta_m + \delta_n].$$

As is well known, if $|\omega_m - \omega_n|$ is small enough, then this describes the phenomenon of "beats," where an oscillation with frequency and phase exactly equal to the means for the two original oscillations (the first cosine on the right) is modulated by a slowly varying "carrier wave" (the second cosine on the right). In general, however, we can see that the higher frequency part (if $\omega_m > 0$ and $\omega_n > 0$) has frequency depending on the sum of ω_m and ω_n, whereas the lower frequency part depends on the difference. In short, the combination "note" will have a pitch intermediate between its components. The same result holds true for any number of constituent sounds, it seems; their resultant has a "quasi-average" pitch, though this seems much more difficult to prove.

An alternative way of summarizing this result for two components, due to Lord Rayleigh (in his *Theory of Sound*, vol. 2), is as follows (for simplicity let $\delta_m = \delta_n = \delta$). By writing $\cos(\omega_n t - \delta)$ as $\cos(\omega_m t - [\omega_m - \omega_n]t - \delta)$, it follows that

$$a\{\cos(\omega_m t - \delta_m) + \cos(\omega_n t - \delta_n)\} = r\cos(\omega_m t - \theta),$$

where

$$r^2 = 2a^2\{1 + \cos(\omega_m - \omega_n)t\}$$

and

$$\tan\theta = \frac{\sin\delta + \sin[(\omega_m - \omega_n)t + \delta]}{\cos\delta + \cos[(\omega_m - \omega_n)t + \delta]}.$$

This formulation also shows that the combination is quasi-harmonic in r and θ, each of which has angular frequency $\omega_m - \omega_n$. If both the amplitudes and initial phases of the two notes were different, a more complicated version of the above arguments would yield essentially the same conclusions, and in general many such sounds would be expected to merge into a "quasi-note" whose pitch is the approximate average of those of the individual components. Thus, as Humphreys points out in his book *Physics of the Air*, it may be concluded that "the whisper of a tree ... has substantially the same pitch as that of its individual twigs or needles; just as the hum of a swarm of bees is pitched to that of the average bee." Again, following Humphreys (who in turn based his analysis on that of Lord Rayleigh) we now examine, in an approximate manner, the *intensity* of the blended notes.

The basic idea is to consider n individual sounds of unit amplitude and the same pitch or frequency but arbitrary phase (at least for the moment). This is believed to be a reasonable approximation to the aeolian blend of a tree. If all the notes had the same phase at a given instant, the combined amplitude would be equal to n and the combined intensity would be n^2. If instead exactly half of the notes had a given phase (if n is even!) and the other half had exactly the opposite phase (180° difference), then the combined intensity would be zero. What we do now for simplicity is to consider that the n sounds have phases which are either "$+$" or the exact opposite ("$-$"), with equal probability. By confining the phases to these options we can gain insight into the more realistic case by means of some simple mathematics. The probability that all n sounds (or oscillations) have the same phase (say $+$) is $(1/2)^n$, and the intensity expectation is therefore $n^2(1/2)^n$. If, however, one of the oscillations has the ($-$) phase, then it will cancel out one of the remaining $n - 1$ ($+$) oscillations, leaving $n - 2$. Now there are n ways to choose this "nonconformist" sound, and so the expectation for the intensity in this case is $n(n - 2)^2(1/2)^n$. Continuing in this way, we see that for *two* oscillations of ($-$) phase, two ($+$) sounds will be canceled, leaving $n - 4$ of them, and this can occur in $n(n-1)/2!$ ways. The total expectation is therefore given by the expression

$$I = \left(\frac{1}{2}\right)^n \left\{ 1 \cdot n^2 + n(n-2)^2 + \frac{n(n-1)}{2!}(n-4)^2 \right.$$

$$\left. + \frac{n(n-1)(n-2)}{3!}(n-6)^2 + \cdots \right\}.$$

Does this series converge (of course! $n < \infty$), and if so, to what? Lord Rayleigh found that it has a sum equal to n, and he did so in rather an ingenious manner. Consider the binomial expansion of $f(x) = (2\cosh x)^n$ in exponential form, namely

$$f(x) = (e^x + e^{-x})^n = e^{nx} + ne^{(n-2)x} + \frac{n(n-1)}{2!}e^{(n-4)x} + \cdots .$$

By expanding the exponentials into sums of algebraic terms, we find that

$$(e^x + e^{-x})^n = \left\{ 1 + nx + \frac{n^2 x^2}{2!} + \cdots + n\left[1 + (n-2)x + \frac{(n-2)^2 x^2}{2!} + \cdots \right] \right.$$

$$\left. + \frac{n(n-1)}{2!}\left[1 + (n-4)x + \frac{(n-4)^2 x^2}{2!} + \cdots \right] + \cdots \right\}.$$

Next, we consider the Maclaurin series for $f(x)$:

$$f(x) = f(0) + xf'(0) + \frac{x^2}{2!}f''(0) + \cdots$$

so that, since $f'(0) = 0$,

$$f(x) = 2^n\left(1 + \frac{nx^2}{2!} + O(x^3) \right),$$

where we do not need to expand beyond the terms in x^2 for the present purposes (application of the ratio test shows that the series converges for $|x| < 1$; note that n is fixed here). Equating the coefficients of x^2 in the binomial and Maclaurin series, we find that

$$\frac{1}{2!}\left\{ n^2 + n(n-2)^2 + \frac{n(n-1)(n-4)^2}{2!} + \cdots \right\} = \frac{2^n n}{2!},$$

whence $I = n$. This means that on the average, given n sounds of unit amplitude with random phases confined to two opposite phases, their resultant intensity is always n. After lengthy calculations Rayleigh was also able to show that the same result is valid if the phases are truly random (i.e., not confined to two values only). This result is only a mean intensity in a possible range from 0 to n^2, but as pointed out by Humphreys, if the changes are rapid then the fluctuations from the mean are correspondingly small.

What, then, has been demonstrated here? First, that the pitch of a composite note is approximately given by the average of those of its components, and second, that the mean intensity is approximately the sum of the individual intensities. In practical terms this means that the pitch of the aeolian "whisper" of a pine tree, for example, is about the same as that of its average "leaf" or needle; furthermore, while the sound of an individual needle may be inaudible, that of the tree itself may be heard some distance away. One tree has many twigs or needles, but what about a forest of many trees? By the same arguments we can see that the vastly greater number of twigs/needles and trees will merge into the famous "murmur of the forest."

This problem and Rayleigh's analysis of it have been discussed in a rather more sophisticated manner by Denny and Gaines, who refer to it as the "Cocktail Party Problem." The idea is that you are in a crowded room with perhaps fifty other people, all of whom are engaged in the kind of loud conversations that seem to be necessary at such gatherings. Can the noise level reach a point where it is literally deafening? In their chapter on the statistics of extremes, Denny and Gaines discuss both the probability density function and the cumulative probability distribution for the pressure amplitudes at the party. Using reasonable numerical estimates they show that you would have to attend the cocktail party continuously for more than four and a half years before it was "an even bet" that your ears would be damaged. Hopefully most people would have left by that time anyway.

Let us play with a couple more things while we're here. Suppose we have a collection of N oscillations of unit amplitude, integral angular frequency $\omega_n = n$, $n = 1, 2, 3, \ldots, N$, and zero phase (for simplicity). How do these combine? In other words, what is $\sum_{n=1}^{N} \cos n\theta$ (or $\sum_{n=1}^{N} \sin n\theta$)? We can answer this question relatively simply by using complex polar forms, and we get a bonus result in so doing. Consider the finite series

$$\sum_{n=1}^{N} e^{in\theta} = e^{i\theta} + e^{2i\theta} + e^{3i\theta} + \cdots + e^{iN\theta}.$$

Of course, this is a complex (but not complicated) geometric progression with sum

$$\sum_{n=1}^{N} e^{in\theta} = \frac{e^{i\theta}(1 - e^{iN\theta})}{1 - e^{i\theta}}.$$

After separating the real and imaginary parts of this equation (this involves spending some pleasant minutes doing both algebra and trigonometry), the real part yields the desired result, namely,

$$\sum_{n=1}^{N} \cos n\theta = \frac{\cos \frac{1}{2}(N + 1)\theta \sin \frac{1}{2}N\theta}{\sin \frac{1}{2}\theta}.$$

The bonus is of course that the *imaginary* part provides us with

$$\sum_{n=1}^{N} \sin n\theta = \frac{\sin \frac{1}{2}(N+1)\theta \sin \frac{1}{2}N\theta}{\sin \frac{1}{2}\theta}.$$

APPENDIX: THE STATICS AND BENDING OF A SIMPLE BEAM: BASIC EQUATIONS

Consider a horizontal beam in equilibrium under the combined effects of *shearing force* $V(x)$, *torque* (or *bending moment*) $M(x)$, and *distributed load* (e.g., weight) $w(x)$, and examine the equilibrium of a small element of the beam, of length δx (see figures 12.8 and 12.9a, b).

For *force equilibrium*,

$$w(x)\delta x + V + \delta V = V$$

so in the limit as $\delta x \to 0$ (assuming $V(x)$ is a differentiable function)

$$\frac{dV}{dx} = -w(x).$$

For *torque equilibrium* (about the center of mass at O),

$$M + V \times \frac{1}{2}\delta x + (V + \delta V) \times \frac{1}{2}\delta x = M + \delta M,$$

so again, in the limit as $\delta x \to 0$, provided $M(x)$ is a differentiable function, it follows that

$$\frac{dM}{dx} = V(x)$$

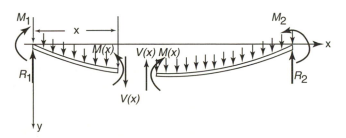

Figure 12.8. The shear forces and moments acting on a beam with weight distribution $w(x)$ under the influence of gravity. Redrawn from *Basic Equations of Engineering Science*, by W. F. Hughes and E. W. Gaylord (1964), McGraw-Hill: New York.

(a)

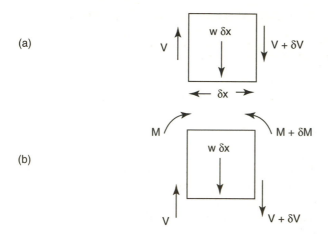

(b)

Figure 12.9. The basic (a) shear forces and (b) moments acting on a small element of the beam, used to derive the governing differential equations for the static equilibrium of the beam.

and obviously

$$\frac{d^2 M}{dx^2} = -w(x).$$

This means that if we know the distributed load $w(x)$ we can find both the shearing force and the bending moment. The constants of integration can be determined from the appropriate boundary conditions (at a *free* end of the beam, both V and M must vanish). (Note that in many texts and papers y is measured downward, as in the beam figure, but the moment M is measured in the opposite sense. This can lead to different signs in the above differential equations, and in the equations below for $\eta(x)$.)

Consider now the case of bending moments M distorting a small portion δx of the beam into a circular arc (see the greatly exaggerated figure 12.10). The upper and lower parts of the beam element are under compression and extension, respectively. In what follows the notation of Elmore and Heald is used. The neutral surface is indicated by the solid line, and has a radius of curvature R. A layer a distance y above the neutral surface therefore has a radius of curvature $R - y$ and has been compressed the fractional amount

$$\varepsilon_{xx} = \frac{(R - y)\delta\theta - R\delta\theta}{R\delta\theta} = -\frac{y}{R},$$

where $\delta\theta$ is the angle subtended by the element δx at the center of curvature C. The tension required to sustain this strain is therefore

$$f_{xx} = -E\frac{y}{R}.$$

If the area of an infinitesimal strip across the beam subject to this tensile stress is dS, the tensile force $f_{xx}\,dS$ acting at a distance y from the neutral surface corresponds to a torque or bending moment $yf_{xx}\,dS$ about a transverse axis (out of the plane) lying in the neutral surface. The cumulative elastic bending moment is

$$M_{total} = \int yf_{xx}\,dS = -\frac{E}{R}\int y^2\,dS = -\frac{E}{R}I,$$

where

$$I = \int y^2\,dS$$

is the moment of inertia of the section of the beam about a transverse axis through the neutral surface, that is, about the z-axis, if that is the axis through the centroid of the beam (taken as the origin) and lying in the neutral surface. The total bending moment opposes the applied bending moment M responsible for the curvature R, so that in general

$$M(x) = \frac{EI}{R(x)}.$$

Let $\eta(x)$ be the displacement of the neutral surface from its equilibrium position. The radius of curvature of the plane curve $\eta = \eta(x)$ is

$$R = \frac{[1 + (d\eta/dx)^2]^{3/2}}{d^2\eta/dx^2} \approx \left(\frac{d^2\eta}{dx^2}\right)^{-1}$$

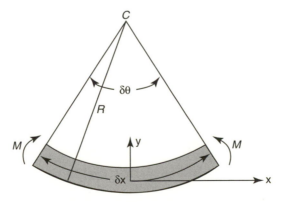

Figure 12.10. A highly exaggerated drawing of a small portion δx of the beam bent under the action of the moments M (only the upper half of δx is shown).

when $|d\eta/dx|$ is small. Under these circumstances we have, to a good approximation,

$$M(x) = EI \frac{d^2\eta}{dx^2}.$$

Note that if the beam is concave up with respect to the variable y then M is necessarily positive. Utilizing the above for V', M', and M'' it follows that

$$V(x) = \frac{dM}{dx} = EI \frac{d^3\eta}{dx^3}$$

and

$$-w(x) = \frac{d^2M}{dx^2} = EI \frac{d^4\eta}{dx^4},$$

subject to the above comments about consistent sign changes in these equations.

Note also that for small angular deviations θ from beam equilibrium,

$$\theta \approx \frac{d\eta}{dx}$$

and so

$$\frac{d\theta}{dx} \approx \frac{d^2\eta}{dx^2}.$$

This term provides the basis for the differential equation used above in the analysis concerning the heights of trees.

CHAPTER THIRTEEN

Bird Flight

> Does the hawk take flight by your wisdom and
> spread his wings toward the south? Does the eagle
> soar at your command and build his nest on high?
> —Job 39:26–27

In this chapter we discuss some principles relating to flight in general and bird flight in particular. Some of these principles pertain to the type of dimensional arguments discussed in chapter 3. The first topic to reencounter is that of *wing loading*. The power necessary for sustained flight for birds and airplanes is proportional to the wing loading, which is the weight of the bird or airplane divided by the area of the wings. For geometrically similar objects, weight increases as the cube of the length of the bird (or plane), and wing area as the square of the length, so wing loading is proportional to the length of the flying object. Thus if a bird is scaled up proportionally by a factor of 3, say, it will weigh 27 times as much, but only have 9 times the wing area, so each square inch of wing area will have to "support" 3 times as much weight. Also, once you're "up" you have to stay up, so an airborne object must fly fast enough to maintain "lift" on the wings. The minimum necessary speed is proportional to the square root of the wing loading. This means that the minimum speed is proportional to the square root of the length of the object. Our scaled-up bird must fly $\sqrt{3}$ or about 1.7 times as fast as its little cousin.

Let us consider some more examples. Recall in connection with the sparrow/ostrich discussion in chapter 3 that heavy birds have to fly fast or not at all. Now it is necessary to make a slight qualification: obviously ostriches are not just large sparrows, and neither are eagles. Larger flying birds have disproportionately larger wings to keep the wing loading down; for example, the pterodactyl-like creature *Quetzalcoatlus northropi* is believed to have had a wingspan of 36 feet and a weight of about 100 pounds. Today, the most efficient gliders or sailplanes have very long, narrow wings. Moreover, the smallest flying animals hover, the largest ones glide and soar (with occasional flapping of wings), and intermediate sizes may have "bounding" flight as well as more continuous flapping. Bounding flight consists of flying up in the air, folding the wings and sailing in a parabolic arc, and then flying again by flapping (notice how swallows in particular do this). McMahon and Bonner note that

- The frequency of wing beats is (approximately) inversely proportional to the wing length.
- The drag force on a bird or plane consists of two components: the *parasitic* drag (skin friction and pressure drag) and the *induced* drag (associated with lift).

The parasitic drag is proportional to the surface area of the bird and the square of the speed, so the power required to overcome the parasitic drag is proportional to the cube of the flying speed. Here is a basic *aerodynamic* principle: a flying object gets its lift by creating downward momentum within a deflected column of air under its "span," almost a "wake" of downward moving air. As will be demonstrated below, when the rate of change of this downward momentum with time equals the weight of the bird, it has lift! The power needed to generate this equal and opposite lifting momentum is called the induced power, and it is inversely proportional to the flying speed. The induced power required can be minimized if the shapes of the leading and trailing wing edges are elliptical—the British Spitfire was a very maneuverable plane built on this principle. At low speeds, the induced power requirement dominates, whereas at high speeds most of the power generated is required to overcome parasitic drag (as bicyclists usually know from experience). For low speeds, a stronger downwash is needed to generate the same lift. In general, the largest birds expend energy at rates close to the upper limits of their abilities to sustain flight in still air, so they search for free rides! This brings us (soon) to the topic of soaring flight. Let us first make some general comments.

FLAPPING FLIGHT

Birds of intermediate size, including even swans and geese, can cruise at relatively high speeds, but they also require high take-off speeds because of their high wing loadings. They can reach such speeds by running in an ungainly fashion along the water using their big feet as paddles (there is even a lizard known colloquially as the "Jesus Christ" lizard; it can run for many yards across the water, but it doesn't get airborne!). The large birds of prey such as hawks and eagles have an even better way—they jump off tall trees to get up flying speed! But how do they get up the tree? Well of course they land on one. . . .

SOARING FLIGHT

The above-mentioned free rides are most commonly available on the wind-ward sides of cliffs, hills, or mountains, where the incoming air is forced to rise; from thermal updrafts or "bubbles" of air above differentially warmed ground; by making use of the changes in wind speed with altitude over oceans, or in lee waves, the standing waves in the lee of hills or mountains. It is the use of "thermals" we will primarily consider here (see also chapter 6). As the sun warms the ground in the morning, the intensity of vertical updrafts (thermals) slowly increases, reaching a peak in mid-afternoon. The speed of the smallest vertical updraft necessary for soaring is proportional to the minimum airspeed for the bird in question, and therefore also to the square root of the bird's wing loading. Since wing loading is proportional to the size of the bird, conditions for thermal soaring are reached for small birds earlier in the morning. Kites soar earlier in the day than vultures. So the small birds get the first crack at the juicy insects that have accumulated during the night!

So how does all this fun come about? As indicated already, birds are carried aloft in great bubbles of buoyant air that boil upward from the warm earth during the day. These thermals are in reality invisible thermal shells, closed *vortex rings* (see below for a discussion of vorticity) of warm circulating air that pump a continuous flow of colder air upward through the center and downward on the outside. This is called *static soaring*, but in order to explore this, we need to distinguish it from another soaring mechanism frequently used by birds vacationing at the seaside. As long ago as 1883 (in the journal *Nature*) Lord Rayleigh showed that for soaring flight to be possible, either the air must have a horizontal velocity nonuniform in space or time, or it must have a local upward velocity. The first of these is satisfied at sea, so the albatross gracefully soars (sometimes over thousands of miles) in the wind shear layers over the open ocean. This is called *dynamic soaring*: the bird extracts energy from the velocity gradient (or shear) during its flight. It does this by gliding downwind, converting potential energy to kinetic energy, and picking up air speed as well as ground speed. Just above the waves it turns into the wind, aquiring initial lift from an abrupt increase in wing "angle of attack." As the bird rises it encounters higher wind speeds, and with its airspeed thereby maintained it is able to rise to the altitude from which it first descended with little or no expenditure of its own "internal" energy. It can do this for hours at a time.

We now return to static soaring. A British chemist, E. H. Hankin, stationed in India from 1910 to 1913, took a great interest in the variety and abundance of Indian vultures and made many observations of them. He noticed that by far the best soaring conditions occur when the air at the ground is almost completely calm. This disposed of the theory that a continuous vertical column of heated air was responsible for the birds' flight patterns. The "bubble" theory based on the generation of thermals has proved a much more likely explanation; although the bird sinks relative to the bubble, it can maintain an equilibrium position if the updraft exactly balances the sinking speed. A common phenomenon in East Africa is the alignment of thermals into "streets" that are marked by parallel lines of cumulus clouds. Sometimes the thermals can be so close that a bird (or glider pilot) can fly from one to another without losing height for as far as 80 km. Vultures readily fly along such streets. The European white stork is the *Guinness Book of Records* cross-country flier as far as distance is concerned. It relies almost entirely on thermal soaring to make its annual migration between northern Europe and the southern half of Africa (think of all those frequent-flier miles). In fact, these storks rely on coordinated social behavior to increase their chance of locating thermals.

FORMATION FLIGHT OF BIRDS

Formation flight of birds improves aerodynamic efficiency. Frequently I see (and hear) a flock of Canadian geese flying toward the west on spring evenings; and usually they fly in formation, approximating a V, most of the time with uneven sides, so they appear as a "flying check mark." In theory, according to the article by Lissaman and Shollenberger, a V formation of 25 birds has approximately 71 percent more range than a lone bird, this formation being optimal. The advantage is more pronounced when there is a tail wind. The V formation is required to distribute the drag saving equally and the lead bird does not necessarily have the most strenuous position. There are certain stability mechanisms that may make it easy and natural for the bird to sense optimal formations. It will be noticed from figure 13.1 that the upwash just beyond the wing tip creates a favorable lift for the birds on either side: less lift power is needed (so induced power is reduced), and this is true not only for line-abreast formations, but for echelons or Vs also. For line-abreast formations, the bird in the center has maximum drag savings, but in a V formation the drag is much more evenly distributed. The optimal shape is not an exact V: it is more swept at the tip than at the apex.

How do birds know about aerodynamic efficiency and drag savings? Perhaps in this context the main sensory mechanisms for a bird are the "bending moments" it feels in its various "joints" associated with muscle-powered flight. Furthermore, if a bird flies ahead of the V line, it requires more power to keep up with the formation, so at the same power level it will fall back until it is in line. Aft of the V line requires less power. Malingerers

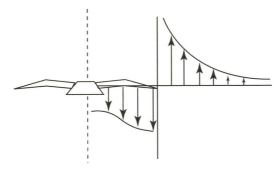

Figure 13.1. The backwash profile behind the wing of a bird in steady flight. From "Formation Flight of Birds," by P.B.S. Lissaman and C. A. Shollenberger (1970), *Science* 168: 1003–5. Reproduced by permission of Peter Lissaman, Caltech, and the American Association for the Advancement of Science.

who do "try it on" may not be tolerated by the others who would have to work harder on their account. It is possible that weaker or injured birds may use this sheltered position, however. Aerodynamic studies show that it is not necessary for the V to be symmetric: there may be 15 birds on one arm and 10 on the other. Small changes in lateral spacing can still give an even work loading for each bird. The most important requirement is that the front bird should have others on both sides. The optimal formation flight is at a lower speed than that of a single bird of the same species, which in part explains the increased range for the flock over the lone bird.

Now let us proceed with some mathematical details of bird (or glider) flight. The variable dependence of drag force on speed has been discussed at several junctures in this book (particularly in chapters 2 and 3). In the present context—flight—wind tunnel experiments on different types of bodies have shown that the drag force D resisting its motion through a medium (air or water, say) depends on the fluid density ρ and the shape, size, and speed u of the body. Specifically,

$$D = \frac{1}{2}\rho u^2 A C_D, \tag{1}$$

where A is a representative surface area (which may vary depending on context) and C_D is called the drag coefficient. It depends on both the shape of the body and the Reynolds number $\mathrm{Re} = ul/v$, l being a characteristic length scale for the body and v the kinematic viscosity of the fluid. The lift L on an airfoil (or hydrofoil) is similarly defined in terms of a lift coefficient C_L, being

$$L = \frac{1}{2}\rho u^2 A C_L, \tag{2}$$

where the area A may not be the same in each formula. The lift coefficient depends on both the Reynolds number and the angle of attack (the inclination of the wing to the flow of fluid). There is an equation relating these two coefficients in terms of the aspect ratio R, which is the wing span (tip-to-tip) divided by the average width of the wing perpendicular to the span:

$$C_D = C_{D0} + \frac{kC_l^2}{\pi R},\tag{3}$$

where C_{D0} is the drag coefficient at the angle of attack for which there is no lift (this depends on the shape of the wing section), and $k \approx 1$ is a constant. For further discussion of drag coefficients, see Tucker (2000).

Consider now a bird of mass m flying downward at a constant airspeed u at a glide angle θ (the angle relative to the horizontal at which the bird is flying). The bird is in equilibrium under the action of three forces: lift, drag, and gravity, where resolving perpendicular and parallel to the glide angle (see figure 13.2) we have

$$L = mg \cos \theta \text{ and } D = mg \sin \theta\tag{4}$$

The bird sinks at a speed $u \sin \theta$, and for small gliding angles it can be shown that

$$u_s = u \sin \theta = \left(\frac{\rho A C_{D0}}{2mg}\right)u^3 + \left(\frac{2kmg}{\pi \rho R A}\right)u^{-1}.\tag{5}$$

As the form of this equation implies, there is an optimum value of u for which the right-hand side is least, corresponding to a minimum sinking speed. Writing (5) more succinctly as

$$u_s = \alpha u^3 + \beta u^{-1},$$

it follows that

$$\frac{du_s}{du} = 0 \text{ when } u = \left(\frac{\beta}{3\alpha}\right)^{1/4} \text{ and } \frac{d^2u_s}{du^2} > 0,$$

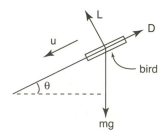

Figure 13.2. A figure depicting the glide angle θ, the glide speed u, and the directions of lift L, drag D, and weight ($W = mg$) forces acting on the gliding bird.

so that the sinking speed is least for a given airspeed when

$$A = \frac{2mg}{\rho u} \sqrt{\frac{k}{3\pi R C_{D0}}} \tag{6}$$

(correcting a minor error in the book by Alexander). This means that optimal wing area should be large at low speeds and vice versa. The wing loading (proportional to m/A) is low at low speeds and high at high speeds.

Many birds of prey spiral around as they descend relative to the air. For a circular (actually helical) path of radius r there is an associated centripetal force mu^2/r required to prevent the bird from flying straight. If the wing line is tilted inward by an angle ϕ to provide this force (figure 13.3), it follows, resolving forces horizontally and vertically, that

$$\frac{mu^2}{r} = L \sin \phi \tag{7}$$

and

$$mg = L \cos \phi, \tag{8}$$

from which

$$L = m\sqrt{g^2 + \frac{u^4}{r^2}}. \tag{9}$$

Thus if the speed is increased or the radius decreased (or both), the lift must necessarily be increased. The banking angle (a new term for investors) follows from (7) and (8) as

$$\phi = \arctan \frac{u^2}{gr}. \tag{10}$$

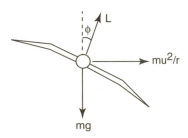

Figure 13.3. The geometry for a bird circling in a circle of radius r under the action of the lift L, weight ($W = mg$), and centripetal (mu^2/r) forces.

For a bird gliding in a small thermal, a smaller radius and correspondingly more lift are required than in a large one. Summaries of the gliding behavior of fulmars, buzzards, and albatrosses can be found in the book by Alexander (also see Pennycuick). For the fulmar, the minimum gliding angle is about 7°, with a minimum sinking speed of about 1.2 m/s. The soaring characteristics of buzzards are as follows: a minimum gliding angle of about 2.5°, with a minimum sinking speed of about 0.6 m/s, remarkably similar to those for gliders! Walkden has estimated that an albatross gliding with an airspeed of u m/s can only glide upward without losing airspeed if the wind speed increases by at least $13/u$ m/s for every meter increase in height. Thus the required wind shear (or gradient) decreases with gliding speed. Since their airspeeds may reach 28 m/s, the minimum wind gradient must be about 0.5 s^{-1}; such are only found near the surface of the sea, which may explain why the birds rarely soar high above the sea. At the top of a climb they may glide down again in another direction, including against the wind or downwind. Their gliding behavior can be characterized as "high airspeeds in weak wind gradients." By contrast, buzzards can only soar in weak thermals if they have low airspeeds. Vultures soar in a similar manner to buzzards (though they are considerably larger birds). The following wing loading table for vultures and albatrosses is an approximation based on data provided by Greenewalt (and reproduced by Alexander).

	Weight (gm)	Wing span (cm)	Wing area (cm²)	Wing loading (gm/cm²)
Vulture	7,300	260	10,500	0.7
Albatross	8,500	340	6,200	1.4

There are some interesting features associated with this table. Notice that, although the albatross has much longer wings than the vulture, they must be much narrower in view of their smaller area. Indeed, the wing loading for the vulture is half that of the albatross. As we have seen, a high value is beneficial for the albatross, and a low value is for the vulture. The aspect ratio for these two birds are, respectively, 18 and 6; there is a limiting useful value to this ratio, because long wings have large bending moments at their bases (see chapter 12), and would require a strong (heavy) skeleton to support them. The vulture makes do with a relatively low aspect ratio in order to get a large enough wing area without having too long a wing.

Finally, we do a little calculation about hummingbirds. While they do not really look or sound like little helicopters, the analogy is quite a good one as far as their relative power output is concerned. They operate on a similar principle: air is accelerated downward, creating the force necessary to keep the object more or less stationary, hovering in the air. For a helicopter of mass m the minimum power P to do this in air of density ρ is

$$P = \sqrt{\frac{m^3 g^3}{2A\rho}}, \tag{11}$$

where now A is the area of the circle swept out by the rotating blades. Unlike these, the hummingbird wing can change direction, turning upside down as it swings back, each wing swinging forward and back by about 120°. Following Alexander, we put the appropriate details for these little creatures into equation (11). For a small hummingbird such as the eastern rubythroat, weighing about 3 gm, with wingspan about 4 cm and each wing swinging through about 120°, the equivalent area swept out is about 2/3 the area of a circle of radius 4 cm—about 34 cm². The density of air is about 1.3×10^{-3} gm/cm³. Hence $P \approx 5.4 \times 10^5$, or about 0.05 W. At this rate eight hundred of these little birds could just about maintain your dim 40 W lamp!

We can arrive at an interesting result about hovering based on pure scaling arguments, by the way (see chapter 3). As we have seen, the hovering bird produces a downward jet of air, and the momentum (mass times speed) of this jet per unit time is the "lift" generated, which, if the bird is to be successful, must equal its weight! Suppose that the jet speed is v, its density is ρ, and its cross-sectional area is A (we do not need to specify units for this argument); then the mass of air projected downward per unit time is given by $Av\rho$, so the downward momentum per unit time must equal the weight of the bird,

$$Av^2\rho = mg,$$

where m is the mass of the bird. Dimensionally, this is proportional to the cube of the size L (however that is defined; it must be consistent for each bird, however, e.g., beak to tip, wingspan, where we are assuming geometrically similar birds for the purposes of this simple argument), so we have that

$$m \sim L^3 \text{ and } A \sim L^2,$$

which is generally true for birds, but not small insects. Thus

$$v \sim \sqrt{L}.$$

Now the power output P used in generating the jet must be equal to the kinetic energy per unit time in the jet, whence

$$P \sim Av\rho \times v^2 \sim L^2 \times L^{3/2} \sim L^{3.5}.$$

For a variety of reasons (see below) the power available to the bird increases as L^2, which means that there must be an upper limit to the size (and hence weight) of birds that are capable of hovering in the absence of wind, convection, or other external aids (parachutes, etc.), because the power required increases faster with size than the power available. According to Maynard Smith this upper limit is found to occur for birds about 35–40 lb in weight.

LIFT AND BERNOULLI

Thus far, the theory of lift has not been addressed mathematically, and this has been deliberately left to the end of the chapter. It has been pointed out in several places (see the references) that there is some confusion in pre-college textbooks about lift, flight, and the relevance of Bernoulli's theorem to both, so before proceeding with a more mathematical discussion of the basis for flight in general, it would be well to address this issue and seek some clarification. Smith points out that Bernoulli's theorem is really an example of the law of energy conservation. It is concerned only with internal relationships inside the fluid, so that dynamic lift, as it is called, must be examined as an encounter between air and an object (e.g., an airfoil), and this will invariably involve the application of Newton's third law describing action and reaction.

What is Bernoulli's theorem? A common form is written

$$\frac{1}{2}v^2 + \frac{p}{\rho} + gz = \text{constant},$$

where v is fluid velocity, p and ρ are pressure and density, respectively, g is gravitational acceleration, and z is height above some reference plane. The first term on the left represents kinetic energy of the fluid per unit mass, while the remaining two correspond to potential energy (arising from pressure and gravity). In many applications the gravitational term is negligible and can be neglected; also the density can often be considered constant. Thus in a system where the energy is constant, this theorem describes the interchange between the pressure and velocity, but it should not be invoked to explain a net force (e.g., lift) occurring in that system.

A common "explanation" for lift on an airfoil is that with more curvature at the top, the air must somehow travel faster over the top in order to meet the slower-moving air traveling around the bottom (why does it care?). This results in lower pressure on the upper side (à la Bernoulli) and consequent lift. But planes can fly with perfectly symmetric airfoils, and even those with asymmetric profiles have been known to fly upside down! And by Newton's third law there must be a corresponding reaction on the airstream.

Weltner writes that "Textbooks stating that the higher streaming velocity is the *reason* for the low pressure are *wrong*. It is the other way round. The low pressure is the reason for the higher velocity of the streaming air." Put another way, the pressure differences are evidence of lift, not its cause. The cause is the reaction; the wing produces a downward force on the air which is accelerated (and therefore deflected) downward. This is particularly noticeable (as Smith points out) when helicopters hover over bodies of water: the spray and waves are created by the downdraft of air. Air is pushed down continuously to create lift.

The lifting force of (on) a given airfoil depends on the angle of attack α (its orientation to the oncoming airstream) and the airstream velocity. The dependence on α is "concave down"; it increases from its value at $\alpha = 0°$ to a maximum, but decreases thereafter. It is important to note that there can be lift even when $\alpha = 0°$ because the airstream behind the wing follows the trailing edge, which usually has a downward direction. As α increases, so does the lift until an angle is reached at which turbulence behind the wing interferes with lift, resulting ultimately in a stall if α continues to increase. The angle at which this occurs depends on the airstream speed, but it will usually be considerably less than 45°.

We need to define an important term in connection with the problem of lift: the *streamline*. A streamline is a curve in the fluid, the direction of which coincides at each point with the direction of the velocity of the fluid; that is, at all locations it is tangent to the velocity vector (they are made visible in wind tunnels by using smoke filaments). In a steady (or time independent) fluid flow the fluid elements themselves move on streamlines. Around an airfoil the streamlines are curved, which signifies that there are associated radial accelerations and pressure gradients in the airstream, the latter being given at a point by the centripetal formula

$$\frac{dp}{dR} = \frac{\rho v^2}{R},$$

where R is the radius of curvature of the streamline at that point. The pressure is reduced as the airstream follows the upper curved surface because the cross sections of the streamline "tubes" increase, thus gaining a higher speed. It will be noted below that viscosity and vorticity have an important role to play in this regard, but for now, following Weltner, we examine the air flow in a near-rectangular curved region of height a and width L defined by two planes shaped like airfoils to deflect the air downward by a small angle α (see figure 13.4). The mass flow is given by

$$\frac{dm}{dt} = aLv\rho.$$

To the left, the airspeed is v and entirely horizontal; to the right, it has a downward component $v \sin \alpha$, which gives rise to a vertical reaction force in the upward direction of

$$F = \frac{dm}{dt} v \sin \alpha = aL\rho v^2 \sin \alpha.$$

The *circulation* Γ of the air is defined as the line integral around any simple closed contour in the fluid; in theory for such calculations we can choose any contour enclosing the wing, but here we choose the curved region $ABCD$ so that

$$\Gamma = \oint_{ABCD} v \cdot ds.$$

The contributions from paths 2 and 4 are antisymmetric and therefore cancel. For path 1, $\alpha = 0$ so there is no contribution, so the entire contribution to Γ comes from path 3 and is

$$v \sin \alpha \frac{a}{\cos \alpha} \approx av \sin \alpha$$

for small values of α. Thus we can write the reaction force (lift) as

$$F = \rho v L \Gamma.$$

This is called the *Kutta-Joukowski formula* (which is derived much more rigorously in texts on fluid dynamics); it relates the lifting force to the circulation around the wing.

The terms vorticity and viscosity were mentioned earlier, both being extremely important in the analysis of flight. Vorticity (a vector quantity: it is $\nabla \times v$) is a measure of the local rotation or curvature of the velocity in a fluid. A vortex line is a line defined at every point by the direction of the vorticity vector; vortex lines always form closed loops (or end at a surface). Vorticity can only be generated at a solid surface bordering or immersed in a fluid. Thus, for example, viscosity on the surface of a wing causes vorticity

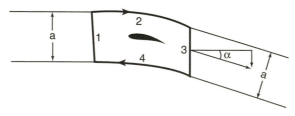

Figure 13.4. A simplified geometrical model used to establish the expression for lift of an airfoil. From "A Comparison of Explanations of the Aerodynamic Lifting Force" by K. Weltner (1987), *American Jounral of Physics* 55: 50–54. Reproduced by permission of Klaus Weltner and the *American Journal of Physics*.

to be generated, and this accelerates (decelerates) the flow over the upper (lower) surface, which, as we have seen, is evidence of the lift generated by the pressure difference between these two surfaces. Familiar examples of fluid flow with vorticity are dust devils, tornados (one hopes this is not too familiar an example), and whirlpools. The first two (especially the latter, which is a much stronger version of the former) are essentially vortex lines stretching from the ground to the clouds. A smoke ring is an example of a closed line vortex, and as such it is called a vortex ring. It is by means of such rings occurring at the top of a rising bubble of air—a thermal—that many medium-sized and large birds are able to soar (noted above). Expensive equipment is not necessary to demonstrate vortex rings if you don't smoke, but it helps if you are a tea or coffee drinker. Just draw a spoon across the surface of the drink (it doesn't matter which way the spoon is facing) and watch the small dimple on either side travel behind it and "peel off."

The complete closed vortex system of a wing in steady flow consists of *four* vortices. The starting vortex is directly analogous to the teaspoon vortices mentioned above and is generated by the air moving at different speeds above and below the wing (recall the shear instability discussed in the section on billow clouds in chapters 6 and 8). This is left far behind as the aircraft accelerates away. However, physics (specifically, conservation of angular momentum) demands the existence of another vortex with circulation equal and opposite to the starting vortex. As pointed out by Wegener, when you fly from New York to San Francisco, the starting vortex is deposited in New York and its "mate" appears on the runway in San Francisco. In a loose sense, what started out as two equal and opposite vortices at the front and back ends of the wing at the start of the flight have now become separated by 3000 miles! The two remaining vortices are generated at the wing tips and are sometimes referred to as trailing vortices; they may be visible as vapor trails (not to be confused with contrails resulting from the jet exhaust) formed by condensation of water vapor in the low pressure centers of the vortex tubes.

As has been noted on several occasions throughout this book, an important dimensionless number associated with all phenomena taking place in a viscous fluid is the Reynolds number defined by

$$\mathrm{Re} = \frac{ul}{\nu},$$

where u and l are typical velocity and length scales and ν is the (kinematic) viscosity. It bears repeating that Re describes the relative importance of the inertial forces (due to the momentum of the fluid) to the viscous forces. If Re is large (say $\geq 10^3$), then viscous forces are less important than for lower values. According to Rayner, $\mathrm{Re} \gtrsim 10^4$ in most aerodynamic situations, and values between 10^3 and 10^5 are typical of swimming fish or flying birds.

For small insects and water dwellers Re is low, because their environments are dominated by viscosity at these scales. Since no fluid flow can contain momentum in the absence of external forces unless there are closed vortices present, and there are no external forces acting on the wake of a fish or bird, it follows that closed vortices must be present in their wakes, and in the case of birds these are induced by flapping flight. Elements of vorticity are shed by each successive downstroke and spaced out along the wake; this is quite a technical subject and the reader is referred to the papers by Rayner for further details.

Finally, since we have previously mentioned playing with spoons in cups of tea, let's have some fun with water flowing from a tap by harking back to Bernoulli's theorem. Consider a steady streamlined flow of water from a tap, the mouth of which has cross-sectional area A_0 and is at level $z = 0$. If the pressure at different points in the flow is constant (a reasonable assumption), then applying Bernoulli's theorem to water that exits the tap with velocity v_0 we have that

$$\frac{1}{2}\rho v_0^2 = \frac{1}{2}\rho v^2 - gz,$$

from which (just as for a particle projected downward with velocity v_0)

$$v = (v_0^2 + 2gz)^{1/2}.$$

The equation of continuity in this context relates the cross-sectional area of the flow to its velocity,

$$A_0 v_0 = A v.$$

Assuming that the cross section at each level is circular, it follows that the diameter of the stream at each level z is given by

$$d = d_0\left(1 + \frac{2gz}{v_0^2}\right)^{-1/4}.$$

Clearly, for freely falling water, d is a monotonically decreasing function of distance from the tap, as it should be. The diameter is a rapidly decreasing function of depth when v_0 is small, at least initially. Solving for v_0 we have

$$v_0 = \frac{2gz}{([d_0/d]^4 - 1)}, \quad d < d_0.$$

So the next time you want a drink of water take a glass and a small ruler, and try and find v_0 (and consult the article by Tan).

CHAPTER FOURTEEN

How Did the Leopard Get Its Spots?

> By his stripes we are healed . . . ; Can . . . the leopard
> [change] his spots?
>> —Isaiah 53:5a (RSV); Jeremiah 13:23a

(. . . or legs, or tail, or toes; or the feathers, or wings in a bird. . . .) Before we can even attempt an answer to this question (or these questions), it is necessary to try to understand something about the phenomenon of diffusion. Denny and Gaines have done an excellent job explaining the physical and mathematical principles behind diffusion, and in considerable detail, but since you are not reading their book at this moment, for completeness I will attempt to explain the ideas more succinctly but somewhat heuristically here.

RANDOM WALKS AND DIFFUSION

Diffusion arises because of random molecular motion; generally molecules move around at great speed and continually change their directions of travel as a result of mutual collisions. If you put a spoonful of sugar in a cup of tea, for example, but do not stir it in, given enough time the sugar molecules will eventually become uniformly distributed in the tea as a result of these random molecular motions. However (and here's the rub), you will have to wait a really long time because the process of diffusion is extremely slow and inefficient for a container the size of a cup, as will be demonstrated below. This is why we stir our sugared hot drinks.

The motion of each molecule is analogous to a *random walk* in three dimensions. What is a random walk? In heuristic terms we can think of it using the analogy of a drunkard's staggering gait; to simplify things for the moment, we examine the problem in one dimension, along the x-axis, for example. The following table is almost self-explanatory, being based on a one-dimensional random walk of a particle (or inebriated individual) initially at zero, but able to move one unit in the positive or negative x-direction at each timestep. (Budding statisticians please note: this is actually a binomial distribution that becomes normal as $n \to \infty$, with standard deviation $\sqrt{n}$.)

Position	−4	−3	−2	−1	0	1	2	3	4
Probability of being here after 1 jump				$\frac{1}{2}$		$\frac{1}{2}$			
... after 2 jumps			$\frac{1}{4}$		$\frac{1}{2}$		$\frac{1}{4}$		
... after 3 jumps		$\frac{1}{8}$		$\frac{3}{8}$		$\frac{3}{8}$		$\frac{1}{8}$	
... after 4 jumps	$\frac{1}{16}$		$\frac{1}{4}$		$\frac{3}{8}$		$\frac{1}{4}$		$\frac{1}{16}$

You begin to get the picture, I am sure. Now we "up the dimension," so to speak, and consider the drunkard able to lurch to the right ($s < 0$) or to the left ($s > 0$) with each step forward (this forward distance is denoted by r). This is a model of small, unconnected random movements from an initial point $(0,0)$. Note that in this case we have taken $x \geq 0$, though there is no restriction in general for random walks (except when a barrier of some kind is present). Let us suppose that the "lurches" to the left or the right all have the same magnitude ($\pm s$), and that at the nth step the lurch is denoted by s_n. It is important to note that each value of s_n is independent of every other value: they have no memory of the past. After n steps the net sideways position (or distance from the x-axis) of the man is $Y_n = \sum_{i=1}^{n} s_i$, so the average value of Y_n (taken over many such random walks) is zero. However,

$$Y_n^2 = \left(\sum_{i=1}^{n} s_i \right)^2 = s_1^2 + s_2^2 + \cdots + s_n^2 + 2(s_1 s_2 + \cdots + s_2 s_3$$

$$+ \cdots + \text{all other products of different displacements}),$$

and again, on the average over many such random walks, the sum of the off-diagonal products is zero. Then the average of Y_n^2 is ns^2. But in n lurches, the drunkard has moved directly forward a distance of $x = nr$ units, so $n = x/r$. Thus the average of Y_n^2 at x, $y^2(x)$ say, can be written in terms of $x, s,$ and r,

$$y^2 = \left(\frac{s^2}{r} \right) x,$$

so $y \propto \sqrt{x}$. Similar considerations and results arise in three dimensions. This concept suitably generalized accounts for the approximate width of a plume of pollutant downstream from its point of discharge at $(0,0)$, although it would be the width based on a time-exposed photograph, not at an instant of time. On a much smaller scale, and of more interest to us here, the random walk is a very good approximation to what happens in molecular diffusion, when heat or some chemical spreads through a system. More often than not, however, *it is diffusion as a function of time that is of interest.* Thus (replacing y by d and x by t) sugar in tea or an enzyme in a biological fluid will diffuse a certain distance L in a time T by collisions of period r over small distances s, where $L \propto \sqrt{T}$. In fact, dimensional arguments based on the equation of diffusion show that $L \approx \sqrt{DT}$, where D is the *coefficient of diffusion*. This relationship is illustrated below in terms of the approximate time taken for oxygen or sucrose to diffuse across different distances L in a system, assuming no change in shape of the system. From the formula for L we see that

$$T \approx \frac{L^2}{D}.$$

What value should be assigned to D? For oxygen in water at 25°C it transpires that $D \approx 2.4 \times 10^{-5} \mathrm{cm^2/sec}$, whereas for sucrose in water at 20°C, $D \approx 4.6 \times 10^{-6} \mathrm{cm^2/sec}$. We therefore take $D \approx 10^{-5} \mathrm{cm^2/sec}$ as typical for our purposes. Values of L and the corresponding diffusion times T are calculated and recorded in the table below. Note that T is proportional to L^2, so that, for example, increasing L by a factor of ten increases T by a factor of one hundred.

L	T
1 μm	10^{-3} sec
10 μm	10^{-1} sec
100 μm	10 sec
1 mm	10^3 sec $\approx$ 15 min
1 cm	10^5 sec $\approx$ 1 day
1 m	10^9 sec $\approx$ 30 years

It would therefore take about three months for that sugar to disperse throughout the cup of tea unless I stirred it. Now we are going to put some mathematical meat on these bones of diffusion theory by deriving in a nonrigorous manner the governing equation for diffusion in one spatial dimension. We follow the excellent description by J. Maynard Smith (in his book *Mathematical Ideas in Biology*).

Consider the concentration x of some substance in a tube (e.g., sugar solution) plotted against distance s along the tube. If the tube has uniform cross-sectional area a, in a small portion of the tube with length δs, there is an amount $ax\delta s$ of the substance present (approximately; it is assumed that δs is small enough that x is approximately constant therein). As seen from figure 14.1, the rate at which the amount of substance in this small element δs is increasing is the difference in the rates of diffusion *into* the element (A) and *out of* the element (B), across the two faces, that is, $A - B$. Mathematically, this is expressed as

$$\frac{\partial}{\partial t}(ax\delta s) = a\delta s \frac{\partial x}{\partial t} = A - B.$$

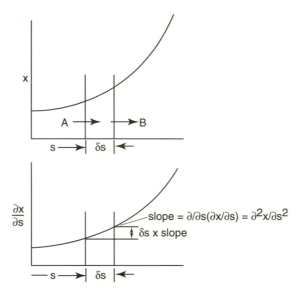

Figure 14.1. A simple geometric derivation of the diffusion equation in one spatial dimension. From *Mathematical Ideas in Biology*, by J. Maynard Smith (1968). Reprinted by permission of Cambridge University Press.

Since the rate at which a substance diffuses across a surface is proportional to the area of the surface and to the concentration gradient at right angles to the surface, we may write

$$A = -aD\left(\frac{\partial x}{\partial s}\right) \quad \text{at } s, \quad B = -aD\left(\frac{\partial x}{\partial s}\right) \quad \text{at } s + \delta s.$$

What is D, and why is there a negative sign in these expressions? The quantity D, already encountered in fact, is a constant of proportionality that depends on various features of the system: the temperature and properties of the diffusing substance. It is called the diffusion coefficient, and is usually treated as constant in simple analyses. The negative sign arises because if $\partial x/\partial s > 0$, diffusion would occur from right to left (from the region of higher concentration to that of lower concentration), and so A and B would be negative. Therefore

$$A - B = -aD\left[\left(\frac{\partial x}{\partial s}\right)_s - \left(\frac{\partial x}{\partial s}\right)_{s+\delta s}\right]$$

$$\approx -aD \times \delta s \times \left(-\frac{\partial^2 x}{\partial s^2}\right).$$

Thus combining the above equations we have the famous diffusion (or heat) equation

$$\frac{\partial x}{\partial t} = D\frac{\partial^2 x}{\partial s^2}.$$

What does this tell us physically? Well, if $x(s, t)$ is concave upward (downward) in s at a point, then x is increasing (decreasing) in time at that point, and vice versa. We will discuss various analytical solutions of this equation toward the end of the chapter; for now we concern ourselves with the steady state (equilibrium) solution. Under such circumstances, $\partial x/\partial t = 0$. Suppose that the tube has length l, and that the imposed boundary conditions are $x(0) = X_0$, $x(l) = X_1$. The equation

$$\frac{d^2 x}{ds^2} = 0$$

has the solution

$$x = X_0 - \frac{(X_0 - X_1)}{l}s.$$

Note that the concentration gradient $\partial x/\partial s$ along the tube is inversely proportional to l, the length of the tube. In order to find how long it will take for a particular concentration to be reached at a particular point within the tube, we must revert to the full time-dependent diffusion equation and perform some elementary dimensional analysis. Now consider again

$$\frac{\partial x}{\partial t} = D\frac{\partial^2 x}{\partial s^2}$$

and suppose that (as in the list above) typical length- and time-scales in the system are denoted by L and T, respectively. For any given concentration x_* it is clear that, dimensionally,

$$\frac{x_*}{T} \sim D\frac{x_*}{L^2},$$

that is,

$$T \sim \frac{L^2}{D} \quad \text{or} \quad T \propto L^2.$$

This fact has major ramifications in the realm of biology, as we have already seen. Note furthermore that D must have dimensions of (length)^2/(time), so all is consistent with the dimensional arguments and numerical values used above.

Now: what has all this got to do with butterflies, leopards, tigers, and even bones, feathers, and teeth? The British mathematician Alan Turing wrote what turned out to be a seminal paper in 1952, in which he showed that spatial patterns could arise spontaneously (under the right circumstances). (Turing is perhaps better known for his work on decoding the Enigma machine during World War II and his subsequent researches in the theory of computation.) As noted in connection with the sugar-in-tea problem, diffusion is generally a *smoothing* influence, ultimately destroying differences in chemical concentrations, so it is rather counterintuitive to find, as Turing proved, that diffusion can actually *destabilize* a chemical system and produce patterns in place of a uniform homogeneous equilibrium.

But it is not just a matter of chemistry: the coat patterns of animals, for example, are produced by hairs that are white, black, brown, or yellowish in color. Their color is determined by the pigment-producing cells (melanocytes) in the skin below. Thus skin coloration is caused by the pigment *melanin*, but (i) what makes the skin produce melanin or not, and (ii) what mechanism gives to melanin production in patterns of spots, stripes, and so on? The mathematician James Murray proposed that early in embryonic development a "pre-pattern" of chemical "morphogens" is laid down, to be implemented later by the melanocytes as the patterns that are observed. This may help to explain the apparent paradox that leopards have spots, tigers have stripes, but they are comparable in size when full grown (to within a factor of about two or three by weight; clearly the leopard is built to run very fast and is therefore less stocky and massive than the tiger, though it is about the same tip-to-tail length); resolution of the paradox would depend on the sizes of each embryo when the pre-pattern is laid down.

Mathematical models developed along these lines by Murray, Maini, and others describe the diffusion of chemicals and their reactions in terms of reaction-diffusion equations. These models are by no means considered definitive at this stage of their development, but they are becoming increasingly sophisticated and beneficial to the development of the field of biological pattern formation. As in any model, there are several underlying assumptions that need to be stated at the outset. These are

Assumption 1. Certain chemicals stimulate cells to produce melanin; high concentrations of it produce coloration but low concentrations do not.

Assumption 2. Two chemicals are produced in the skin. One stimulates the production of melanin and the other inhibits it.

Assumption 3. Production of the "stimulator" initiates production of the "inhibitor."

Assumption 4. The inhibitor diffuses *faster* than the stimulator.

These assumptions will be translated into equations below, but taken together they mean that, as a concentration of stimulator is produced, the inhibitor that is thereby released can move out faster and "encircle" the more slowly diffusing activator, thus inhibiting it (preventing further expansion). The result is a circular spot. To help visualize this, Murray gives an evocative analogy. Fire breaks out (stimulator) in a dry forest. Firefighters (inhibitor) are stationed throughout the forest, and with helicopters can travel more quickly than the fire (and dump water on the fire or spray fire-resistant chemicals onto the trees). Because of the intensity of the fire it cannot be contained at its core, so they outrun the fire and spray the trees. The fire's progress (hopefully) is stopped when it reaches these regions. Viewed from the air, a blackened spot marks the burned-out fire, and it is surrounded by a green ring of sprayed trees. This can be generalized to the more realistic case of fires breaking out all over the forest to give a pattern of circular spots.

Murray found the patterns produced by the mathematical models also depended on the shape and size of the skin region. Basically, as the geometric domain increases in size, more complex patterns are possible. For very small regions there may be no pattern. For larger regions there may be stripes, small spots, or large spots; in other words, patterns reminiscent of the stripes of a leopard's tail or those of a zebra, the spots of a cheetah, the larger "cells" of a giraffe. For even larger regions it was noticed that there may be no pattern once again. One implication of all this is that small animals with short gestation periods may have less complex coat patterns than larger animals, because the latter can support more complex modal solutions, and there is some evidence for this. Furthermore, for very large animals as the modes increase the features may start to merge (as in giraffe patterns, for example), squeezing out the boundary lines (for further descriptive details see Ball's epic *The Self-Made Tapestry*). The shape as well as the size of the domain is important; Murray found that if the skin area was sufficiently narrow, only stripes would form; there were no spots. This implies that spotted animals can have striped tails but no striped animals have spotted tails, which is exactly what is observed (note that leopards and cheetahs are "spotted" with striped tails; see plate 23).

Murray was able to generate many types of pattern combinations by varying different parameters in his models, and interestingly, the equations can support solutions similar to those governing the vibrations of appropriately shaped membranes. Murray showed that very similar patterns (spots, stripes, etc.) could be produced in acoustically excited membranes (some simple mathematical models of which are summarized in the appendix to this chapter). This is only an analogy—no one is suggesting that the vibrational patterns are produced by activation/inhibition mechanisms (they are not)—but analogies can be very helpful both visually and mathematically, as we have seen on several occasions throughout this book.

There are alternatives to the reaction-diffusion embryological pattern formation models. In the models discussed so far, it is assumed that the cell density is uniform and the chemical concentrations determine the spatial patterns. In cell movement models it is hypothesized that the observed patterns occur in cell density, and cells in high-density regions then differentiate (but not à la calculus). These models can also explain many observed coat patterns, but the most promising approach of all incorporates both types of model. A valuable nontechnical account of the subject can be found in the article by Maini; in it he also discusses skeletal patterns in the vertebrate limb, feather germ (or site) formation, and tooth development in the alligator. All these examples illustrate what has come to be called self-organization; the mechanisms can produce spatial patterns of varying complexity that are consistent with those that are observed. However, as Maini points out, in coat marking models the patterns are laid down simultaneously, whereas in limb development skeletal elements are laid down sequentially; the feather germ and tooth development illustrate respectively sequential regular and irregular pattern formation.

SEASHELLS

Seashells can exhibit a wonderful array of patterns: stripes, V-shapes, swirls, and various combinations of these. While a gut reaction would be to regard these as analogous to the coat markings mentioned above, they are more akin to limb development in that they represent, as Ball points out, a historical record of processes taking place continually as the shell grows. The following comments are adapted from an article entitled "Space-time on a Seashell" by Brian Hayes. As with all the references in this book, if you are interested in the topic, go and read the original article!

Seashells have a variety of shapes that can (generally) be generated from a single basic form: the logarithmic or equiangular spiral (not qualitatively unlike the seed patterns on the head of a sunflower or daisy discussed in chapter 10). The surface patterns, on the other hand, require more detailed mathematical analysis to be able to reproduce the observed features. Such analysis includes, in general, coupled partial differential equations describing the creation, diffusion, and decay of chemical growth or inhibitory factors that can promote or inhibit pigmentation. They are the reaction-diffusion equations discussed above; a simplified pair will be discussed shortly. As far as the mollusk shell is concerned, note that the pattern is a two-dimensional record of a one-dimensional process—the cells at the edge or margin of the shell are either "on" or "off" in terms of defining pigmentation, and a shell unrolled into a rectangle would record its pattern as a "space-time diagram." Thus a series of vertical stripes (perpendicular to a horizontal growing edge) implies a static distribution of pigment-secreting cells at the margin. If such a group of cells is permanently "on," then there develops a dark vertical stripe: adjacent cells that are "off" correspond to what appears to be a white vertical stripe. Horizontal stripes, on the other hand, growing parallel to the edge, result from a temporal rather than a spatial oscillation (all the secreting cells switch on and off in synchrony). What about oblique stripes? Here is the really fascinating thing: an oblique stripe records the passage of a traveling wave of excitation in the row of pigment-producing cells. The slope of the line indicates the speed of the traveling wave: the more nearly vertical it is, the slower the wave. All sorts of fascinating interactions between waves can occur.

MECHANISMS OF ACTIVATION AND INHIBITION

Coupled reaction-diffusion models often involve the existence of an "activator" factor A that promotes not only pigment production but its own synthesis—the more that is present in a cell, the more the cell makes—a positive feedback loop. The obvious resulting instability is "inhibited" by an inhibitor factor I, whose synthesis also depends on the activator concentration. The resulting patterns depend crucially on the rates of synthesis, diffusion, and decay of the two factors, and sometimes small changes in these rates can initiate large changes in the resulting patterns.

Here's one possible scenario: suppose that **I** diffuses much more rapidly along the margin than does **A**. If a small group of cells has a slightly enhanced concentration of **A** compared with neighboring groups, then this disparity will increase still further (instability: see chapter 8) as it enhances its own production. **I** will also be produced in these cells, but it will diffuse away faster than **A** and spread its effect over a wider area than **A**, and as a result production of **A** from cells in these regions, where **A** is already low, will be further suppressed. In this way, random spatial variations in the concentration of **A** and the resultant pigment production are amplified, and remain stable over long time intervals. In a one-dimensional model this mechanism produces (vertical) stripes; in two dimensions this produces spots.

It is also possible to produce horizontal stripes using ideas of this type. If production of **I** lags behind that of **A**, the delay allows **A** to overshoot its equilibrium level. When **I** eventually catches up, it also overshoots and heavily suppresses the production of **A**. This causes the level of **I** to fall and so **A** begins to recover. This cycle can repeat indefinitely with a period determined by the rates of synthesis and decay. If **A** diffuses faster than **I**, obliques line can be formed... details of that can be found in Hayes's article. This mechanism can give rise to a Λ-shaped pair of lines radiating from the site of activation of **A**. Two traveling waves can be annihilated at a **V**-shaped vertex. For this and further details read Hayes's article and the references contained therein.

REACTION-DIFFUSION EQUATIONS: A LINEAR MODEL

Consider the reaction and diffusion of two chemicals (an "activator" and an "inhibitor") in one spatial dimension, s. Furthermore, let their respective concentrations be X and Y, and assume (as is frequently the case) that there are values (X_e, Y_e) for which they are in chemical equilibrium ($\partial X_e/\partial t = 0 = \partial Y_e/\partial t$); this pair of values need not be unique. Now let there be a departure from this equilibrium such that

$$X = X_e + x, \quad Y = Y_e + y.$$

In view of the fact that when $x = y = 0$ (equilibrium), $\partial x/\partial t = \partial y/\partial t = 0$, we can state that for *small* displacements from equilibrium (neglecting diffusion for the moment)

$$\frac{\partial x}{\partial t} = ax + by, \quad \frac{\partial y}{\partial t} = cx + dy,$$

where a, b, c, and d are constants, not in general all of the same sign. Now including the effects of diffusion, with constant diffusion coefficients D_1 and D_2, respectively, the system of coupled partial differential equations becomes

$$\frac{\partial x}{\partial t} = ax + by + D_1 \frac{\partial^2 x}{\partial s^2}, \quad \frac{\partial y}{\partial t} = cx + dy + D_2 \frac{\partial^2 y}{\partial s^2}.$$

It was noted earlier that equations of this type were first derived and discussed by Alan Turing (of Enigma code-breaking fame). Although analytical solutions are available, we will follow J. Maynard Smith and discuss a simplified version in a qualitative manner.

The basic idea is to see how the solutions to the above coupled system evolve in space and time after the equilibrium state has been perturbed by the small deviations (x, y). For most members of the parameter set $\{a, b, c, d, D_1, D_2\}$ the equilibrium is restored; it is therefore *stable*, that is, $\lim_{t\to\infty} x(s, t) = 0 = \lim_{t\to\infty} y(s, t)$. However, there are some values of the parameters for which the equilibrium is *unstable*, and nonhomogeneity develops in space or time or both. So-called "standing waves" of morphogen concentration can develop. To illustrate this qualitatively and graphically consider the following simplifying assumptions:

(1) $a > 0$ and $c > 0$. This means that if the concentration of X rises above its equilibrium level, both X and Y increase with time (or, more accurately, the rate of synthesis of both X and Y rises).

(2) $b < 0$. This means that if the concentration of Y rises, it leads to the destruction of X (since it contributes a negative term to $\partial x / \partial t$). $d = 0$ without much loss of generality.

(3) $D_2 > D_1$; Y diffuses faster than X.

The development of a standing wave is illustrated in figure 14.2. Suppose that the homogeneous equilibrium is perturbed by a small local rise in the concentration of X (represented by the solid line; see 14.2a). This leads (b) to a further increase in both X and Y, but by assumption Y has diffused out farther. The arrow marks a point where $y > 0$ and $x \approx 0$, so that there is now a net destruction of X (see 14.2c). However, this will in turn lead to destruction of Y, so for each chemical a trough will develop on either side of the initial peak. By similar arguments, it can be seen that these troughs will give rise to further peaks, and so on, until a standing wave has developed. The "wavelength" of this pattern will depend on the particular set $\{a, b, c, d, D_1, D_2\}$.

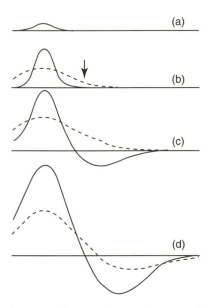

Figure 14.2. The development of a standing wave pattern in morphogenesis. From *Mathematical Ideas in Biology*, by J. Maynard Smith (1968). Reprinted by permission of Cambridge University Press.

BUTTERFLIES (AND MOTHS)

Philip Ball writes in *The Self-Made Tapestry*, "Consider . . . the butterfly, whose wings are a kaleidoscope of color. Not only is the range of hues fantastically rich, but the patterns seem to have a precision that goes beyond the zebra's stripes." Some of the colors of the butterfly wing are produced by pigments, but some are the result of *iridescence* arising from the interference of light, discussed in chapter 5. The wing pattern is laid down during the pupation stage, not unlike the situation for animals where the coat pre-patterns are determined during the embryonic stage. Again, Murray and others have developed mathematical models that can account for many of the features observed in the wings of butterflies and moths. One approach is that a chemical signal (morphogen) that "switches on" a particular gene in the wing cells is released from a source (or sources) located somewhere on the wing; some models invoke the existence of morphogen "sinks" as well as sources. The morphogens diffuse throughout the wing cells and "throw" biochemical switches when they exceed some critical threshold concentration. Nijhout has invoked an activator/inhibitor scheme to account for the rich and complex patterns, and a very readable account of this and many other examples of self-organization may be found in the excellent book by Ball, to which the reader is encouraged to go. What we will do below is examine a relatively simple single-morphogen model based on the description in the book by Murray; the model describes the gene-activation mechanism in terms of the diffusion and threshold morphogen levels and provides a length-scale for the pattern size. It turns out that the wing pattern depends crucially on the geometry and the scale of the wing. Since there are approximately 10^6 different types of butterflies and moths, we shall consider a generic "model."

Mathematically, we consider a one-dimensional problem for diffusing morphogen and gene activation system. A given amount of morphogen is released at $x = 0$. As it diffuses, cells react in response to the local morphogen level $S(x, t)$, and a gene G is activated by S to produce a product g, which in turn determines the pattern via pigment-generating cells. The governing equation is of the linear reaction-diffusion type (see Murray's account of this problem):

$$\frac{\partial S}{\partial t} = \frac{\partial^2 S}{\partial x^2} - \gamma S, \qquad \gamma > 0, x \geq 0, t > 0, \tag{1}$$

where the second term on the right-hand side represents the natural decay or depletion of the morphogen in time; initially (at $t = 0$) it is produced at a "point source" so that

$$S(x, 0) = \delta(x),$$

$\delta(x)$ being the Dirac delta function, and its concentration diminishes away from the source, so that

$$\lim_{x \to \infty} S(x, t) = 0.$$

Before proceeding further, the astute reader will have noticed that there is no diffusion coefficient D in equation (1) (more accurately, D has been set to unity). It is not necessary to do this, of course, but to render the equations nearly identical to those used by Murray, the x variable in equation (1) is defined in terms of the "old" spatial variable s above by $x = sD^{-1/2}$.

It is shown in standard texts on partial differential equations that this problem has the (Green's function) solution

$$S(x, t) = (4\pi t)^{-1/2} \exp\left(-\left(\gamma t + \frac{x^2}{4t}\right)\right), \qquad t > 0. \qquad (2)$$

It is a useful exercise in partial differentiation to show that equation (2) does indeed satisfy the differential equation (1). For a given value of x, $\max\{S\} = S_{\max}$ occurs at time t_m, defined by

$$S_t(t_m) = \frac{\partial S}{\partial t}(t_m) = 0$$

(of course, $S_{tt}(t_m) < 0$). Since $S_t(t) = 0$ when

$$t^{-1} = 2\left(\frac{x^2}{4t^2} - \gamma\right),$$

it follows that (on choosing the positive root)

$$t = t_m = \frac{-1 + \sqrt{1 + 4\gamma x^2}}{4\gamma} = \frac{-1 + z}{4\gamma}, \qquad (3)$$

so that

$$S_{\max}(x) = S(x, t_m) = \left(\frac{\pi}{\gamma}[z - 1]\right)^{-1/2} \exp\left(\frac{1 - z}{4} - \frac{x^2 \gamma}{z - 1}\right).$$

This can be simplified using the definition of z, so that

$$S_{\max}(x) = \left\{\frac{\gamma}{\pi(z - 1)}\right\}^{1/2} e^{-z/2} \qquad (4)$$

According to one model, the gene product g can be shown to satisfy the following nonlinear ordinary differential equation:

$$\frac{dg}{dt} = \gamma f(g; S) = \gamma\left[k_1 S + \frac{k_2 g^2}{1 + g^2} - k_3 g\right], \qquad k_i > 0, \qquad i = 1, 2, 3 \qquad (5)$$

A typical graph for $f(g; S)$ vs. g for given values of S is shown in figure 14.3; note that the curve for $S = S_c$ is the separatrix (or threshold concentration) between three equilibrium states (two stable, one unstable, $S < S_c$) and one stable one ($S > S_c$). Thus as S increases to $S_{max} = S_c$, this effects a "switch" mechanism from $g = 0$ to $g = g_3$. If S_c is known, the corresponding value of $z = z_c$ can be calculated from (4) and hence

$$x_c = \sqrt{\frac{z_c^2 - 1}{4\gamma}}. \tag{6}$$

This gives the distance of the threshold value S_c from the source of morphogen, and hence *the domain of a specific pigmentation*. In two spatial dimensions with radial diffusion, for example, this type of result would provide the radius of a circular spot (an "eyespot") induced by the source of morphogen at its center. The gene that determines the location of such eyespots was identified in 1996 by scientists in Wisconsin; apparently it is "switched on" during the late stages of larval growth, while the butterfly is still "cocooned." Interestingly, this same gene (called *Distal-less*) plays a different role in arthropods (e.g., beetles), determining where the legs grow.

Mention has been made of some of the many beautiful patterns that can be seen in the animal and insect kingdoms; it is perhaps fitting to end this subsection with a comment about *man*. I was flipping through the pages of *Reader's Digest* (January 2002 issue) in an auto shop while waiting for my car, and found a cartoon of a man, presumably at a zoo, looking at a zebra and a leopard in adjacent enclosures. He was wearing brightly colored shorts, and those two animals were returning his gaze. The cartoon caption read: "Man: the only animal to purposely wear plaid."

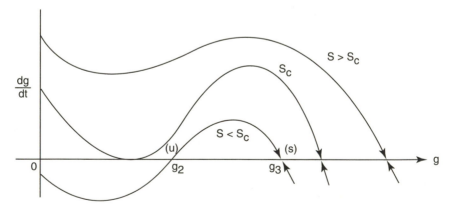

Figure 14.3. A biochemical switch mechanism invoked in a simple mathematical model of spot formation in a butterfly wing. From *Mathematical Biology*, by James D. Murray (1990), reprinted by permission of Springer-Verlag and James D. Murray.

We now examine another application of linear diffusion theory, but in an entirely different context.

THE SIZE OF PLANKTON BLOOMS

Consider a long narrow body of water, $0 \leq x \leq L$, limited on the long sides by two natural boundaries. This will define our idealized plankton bloom. Outside $[0, L]$ the water is environmentally unsuitable for plankton, for one reason or another that does not concern us here. Neglecting transverse diffusion, we consider the concentration of plankton $c(x, t)$ to be a continuous function of both its arguments. Adding a linear reproduction term gives us the *growth* analog of the decay-diffusion equation used above in connection with butterfly eyespots, namely

$$\frac{\partial c}{\partial t} = D \frac{\partial^2 c}{\partial x^2} + Kc, \qquad 0 \leq x \leq L, \qquad t > 0, \tag{1}$$

where $K > 0$ is the plankton reproduction rate. Following Kierstead and Slobodkin, we impose so-called rigid boundary conditions (meaning here that the concentration c drops to zero at the ends of the strip), that is,

$$c(0, t) = c(L, t) = 0. \tag{2}$$

Since we are interested in how a thin initial patch of plankton evolves in time and space, we also impose the initial condition

$$c(x, 0) = f(x), \quad x \in [0, L]. \tag{3}$$

We will use a device (an old trick that works at least twice) to reduce the original equation to the simpler diffusion equation, and then invoke the standard solution to that, found by separation of variables. Thus if

$$c(x, t) = y(x, t)e^{Kt} \tag{4}$$

equation (1) becomes

$$\frac{\partial y}{\partial t} = D \frac{\partial^2 y}{\partial x^2}, \tag{5}$$

where $y(x, t)$ instead of $c(x, t)$ now satisfies equations (2) and (3). The solution to this equation, subject to the boundary and initial conditions is

$$y(x, t) = \sum_{n=1}^{\infty} b_n e^{-(n\pi/L)^2 Dt} \sin \frac{n\pi x}{L}, \tag{6}$$

where

$$b_n = \frac{2}{L} \int_0^L f(x) \sin \frac{n\pi x}{L} \, dx, \qquad n = 1, 2, 3, \ldots. \tag{7}$$

In terms of the original dependent variable, the concentration of plankton bloom is therefore

$$c(x,t) = \sum_{n=1}^{\infty} b_n e^{[K-(n\pi/L)^2 D]t} \sin\frac{n\pi x}{L}. \tag{8}$$

For a given reproductive rate K there will be a smallest value of n (call it N) such that

$$K < \left(\frac{N\pi}{L}\right)^2 D.$$

Since $n \geq 1$, a necessary condition for the maintenance or growth of the population is the condition for linear instability (that word again; see chapter 8) is that

$$K \geq \frac{D\pi^2}{L^2} \tag{9}$$

(otherwise *all* the terms, and hence c, would decay exponentially in time). Equivalently, this implies that there is a critical size (or length here) of plankton bloom, given by

$$L_c = \pi\sqrt{\frac{D}{K}},$$

such that if $L > L_c$, the total plankton *population*

$$N(t) = A \int_0^L c(x,t)\,dx$$

will increase or *bloom*, and it will decrease if $L < L_c$ (remaining constant in time if $L = L_c$). Here A is the cross-sectional area of the thin plankton strip. It is interesting to note that if the same type of analysis is carried out for a circular patch of radius r, the corresponding critical radius is

$$R_c = 2.4048\sqrt{\frac{D}{K}},$$

the numerical factor being the location of the first zero of the Bessel function of the first kind and order zero, $J_0(x)$.

And now for something completely different. Some of the nature and consequences of diffusion in specific biological contexts have been discussed so far in this chapter, but the effects and implications are far-reaching (quite literally!). Diffusion of one type or another is quite ubiquitous, and so this chapter closes with two rather "earthy" examples illustrating the applications of the diffusion equation to the flow of heat in both the surface layers and deeper reaches of planet earth. In particular the latter is primarily of *historical* interest, because the theory of plate tectonics has rendered the discussion obselete. The first topic is a nice application of a useful solution-building technique that has been mentioned earlier—the method of separation of variables. The second application by contrast uses a complementary approach—integral superposition of solutions—to achieve its ends. Like its predecessor, it is an interesting example of the interplay between the mathematics and its interpretation, which is crucial in any type of mathematical modeling. In this case the results are woefully inadequate in explaining the age of the earth, but bear in mind that it is the self-correcting process in science that makes it so powerful. In this case the mathematics is telling us that we have not included something important; at the time the problem was originally formulated in mathematical terms, that "something" (radioactivity, convection in the earth, and plate tectonics) was undiscovered, so in a sense, the model made a prediction that there was more to come (without knowing exactly what!). The chapter ends with an appendix on the vibrational patterns in rectangular and circular membranes, which have provided a fascinating analogy with animal coat patterns derived by solving reaction-diffusion equations as discussed in the main text of the chapter.

ONE-DIMENSIONAL HEAT FLOW IN A SEMI-INFINITE SOLID WITH APPLICATION TO DIURNAL AND ANNUAL VARIATIONS IN THE TEMPERATURE BELOW THE SURFACE OF THE EARTH

Consider the following mathematical problem describing the flow of heat into the earth from the surface; such types of problems are posed in many books on applications of partial differential equations (e.g., that by Pinsky):

$$\frac{\partial u}{\partial t} = \alpha^2 \frac{\partial^2 u}{\partial x^2} \qquad 0 \le x \le \infty,$$

where x increases downward from the surface, and

$$|u(x,t)| < \infty,$$

with the initial condition

$$u(0,t) = T_0 \cos \omega t, \qquad -\infty < t < \infty.$$

In this problem, $u(x,t)$ is the temperature variation about some mean value $\bar{\theta}$ (assumed constant); the actual temperature is

$$\theta(x,t) = \bar{\theta} + u(x,t).$$

Other assumptions include constant diffusivity α^2 (quite acceptable if the nature of the earth does not change significantly as we proceed downward); also we assume here a flat earth. This is an excellent assumption provided

$$x \ll R,$$

where R is the radius of the earth. We shall see that this is well satisfied. We proceed by using the technique of separation of variables. Let

$$u(x,t) = X(x)T(t)$$

to obtain

$$T' + \lambda \alpha^2 T = 0,$$

$$X'' + \lambda X = 0,$$

where $-\lambda$ is the separation constant. Clearly, to within a multiplicative constant,

$$T = e^{-\lambda \alpha^2 t},$$

and since the solution must be bounded at all times, this forces λ to be an imaginary number,

$$\lambda = i\beta,$$

where β is real. The equation for $X(x)$ is of second-order, constant coefficient type, so we seek solutions of the form

$$X \sim e^{\gamma x},$$

which yields

$$\gamma^2 + i\beta = 0.$$

Taking $\beta > 0$ without loss of generality we note that

$$\gamma^2 = -i\beta = \beta e^{-i\pi/2},$$

whence

$$\gamma = \pm\sqrt{\beta}e^{-i\pi/4} = \pm\sqrt{\frac{\beta}{2}}(1-i).$$

Since $|X(x)|$ must also be bounded (we are assuming no heat sources in the interior, infinite or otherwise!), the negative root must be taken, and so

$$\gamma = \sqrt{\frac{\beta}{2}}(-1+i).$$

Therefore a set of solutions to this problem (subject to invoking the initial condition) is

$$u(x,t) = Ae^{-\sqrt{\beta/2}\,x}e^{i(\sqrt{\beta/2}\,x - \beta\alpha^2 t)},$$

where A is chosen to be a real constant. The above expression for $u(x,t)$ is of course a complex function of x and t, so it is customary to take the *real part* of the expression, since we are dealing with a real temperature variation. Thus,

$$\mathrm{Re}[u(x,t)] = u_r(x,t) = Ae^{-\sqrt{\beta/2}\,x}\cos\left(\sqrt{\frac{\beta}{2}}\,x - \beta\alpha^2 t\right).$$

Therefore

$$u_r(0,t) = A\cos(-\beta\alpha^2 t) = A\cos(\beta\alpha^2 t) = T_0\cos\omega t,$$

from which it follows that

$$A = T_0, \quad \text{and} \quad \beta = \omega\alpha^{-2}.$$

Finally, the result we wish for is

$$u_r(x,t) = T_0 e^{-\sqrt{(\omega/2\alpha^2)}\,x}\cos\left(\sqrt{\frac{\omega}{2\alpha^2}}\,x - \omega t\right)$$

This result can be used to study seasonal or daily variations of temperature within the earth. Let us examine some consequences of this expression. For $x = 0$, the maximum of u_r occurs for

$$t = 0, \pm\frac{2\pi}{\omega}, \pm\frac{4\pi}{\omega}, \dots.$$

For

$$x_m = \sqrt{\frac{2\alpha^2}{\omega}}\,\pi$$

(in particular) the temperature is a *minimum* for the same times. In other words, when it is *summer* on the earth's surface, it is *winter* at this depth. A reasonably typical value of α^2 is

$$\alpha^2 \approx 2 \times 10^{-3}\ \mathrm{cm^2/s}.$$

For annual variations, the number of seconds in a year is required; this is

$$\widetilde{T} \approx 3.15 \times 10^7\ \mathrm{s},$$

and of course,

$$\omega = 2\pi/\widetilde{T}$$

if ω is the angular frequency of temperature variation. On substituting these values into the above expression for x_m, we find that

$$x_m \approx 4.5 \text{ m}.$$

A corresponding calculation for daily variations with $\tilde{T} \approx 8.64 \times 10^4$ s yields

$$x_m \approx 23 \text{ cm}.$$

Some other comments are in order. Let

$$u_1 = \max u(x_1, t) = T_0 \exp\left(-\sqrt{\frac{\omega}{2\alpha^2}} x_1\right)$$

at some depth $x = x_1$, and let u_2 correspond similarly to the maximum at depth x_2. These can be measured (in principle) if the depths are not too great. Thus

$$\frac{u_1}{u_2} = \exp\left(-\sqrt{\frac{\omega}{2\alpha^2}}(x_1 - x_2)\right).$$

This means that the *diffusivity of the earth, α^2, can be determined* (everything else is known).

What about "skin depth"? By this term I merely mean to ask how far down we have to go before seasonal or daily temperature variations are "small." And what do I mean by small? These are good questions. Let me take 5% of the surface variation as the critical value—thus now I ask the question, how far down is the temperature variation 5% of the corresponding variation at the surface? We get a quick and dirty estimate by noting that, since $e^{-3} \approx 0.05$, this will occur when

$$\sqrt{\frac{\omega}{2\alpha^2}} x = 3, \quad \text{or} \quad x = 3\alpha\sqrt{\frac{\tilde{T}}{\pi}} \approx 0.076\sqrt{\tilde{T}}.$$

For the annual variation this is about 4.3 m; for the daily variation it is about 22 cm. This means, in the light of the earlier calculations, that summer up top means a mild winter below, and vice versa! Note that the condition

$$x \ll R$$

is well satisfied in all cases discussed above ($R \approx 4000$ *miles*), so the assumption of a flat earth is well justified here.

As with all good science and applied mathematics (and I *hope* this is in that category), assumptions made usually have domains of validity, within which they are appropriate, and outside of which they are not. These domains will vary with context, of course. A flat earth will be a good approximation if we are interested in air-sea interactions over Chesapeake Bay; it will not be over the whole Atlantic or Pacific Ocean.

HEAT FLOW IN A SEMI-INFINITE SOLID AND
THE "AGE" OF THE EARTH

Once again we study the diffusion of heat in the earth, so the governing equation is our old friend

$$\frac{\partial u}{\partial t} = \alpha^2 \frac{\partial^2 u}{\partial x^2}, \qquad 0 \le x < \infty,$$

and

$$u(0, t) = 0, \qquad 0 < t < \infty,$$

$$u(x, 0) = f(x), \qquad 0 < x < \infty,$$

$$|u(x, t)| < \infty, \qquad 0 < x, \qquad t < \infty.$$

Now we use the Fourier sine transform because of the boundary condition at $x = 0$. Also define $\alpha^2 = \kappa > 0$. The formal solution of this problem is

$$u(x, t) = \frac{2}{\pi} \int_0^\infty \left[\int_0^\infty u(\xi, t) \sin \omega\xi \, d\xi \right] \sin \omega x \, d\omega.$$

Thus, upon rearranging the integrals (legally, I might add!) we find

$$u(x, t) = \frac{2}{\pi} \int_0^\infty d\xi \int_0^\infty f(\xi) e^{-\kappa\omega^2 t} \sin \omega\xi \sin \omega x \, d\omega$$

or

$$u(x, t) = \frac{1}{\pi} \int_0^\infty d\xi \int_0^\infty f(\xi) e^{-\kappa\omega^2 t} [\cos \omega(\xi - x) - \cos \omega(\xi + x)] \, d\omega.$$

Now, since

$$\int_0^\infty e^{-az^2} \cos bz \, dz = \frac{1}{2}\sqrt{\frac{\pi}{a}} e^{-b^2/4a},$$

it follows after a little rearrangement that

$$u(x, t) = \frac{1}{2\sqrt{\pi\kappa t}} \int_0^\infty f(\xi) \left[e^{-(\xi-x)^2/4\kappa t} - e^{-(\xi+x)^2/4\kappa t} \right] d\xi.$$

(Had the boundary condition at $x = 0$ been a condition on the flux, κu_x for example, then the Fourier cosine transform would have been used.)

In order to apply this to the "age" of the earth, let $f(x) = c$, a constant. If we assume for simplicity that the earth was originally at this uniform temperature c throughout, that κ is constant, and that the surface has always been at the temperature zero, then we can use the above result to estimate how long it has taken the earth to cool thus far. This of course makes the big (and unwarranted) assumption that there are no internal sources of heat in the earth (such as radioactivity in rocks and convective motions). Such internal sources were unknown at the time that Lord Kelvin performed his calculations (late 19th century). The above equation now becomes

$$u(x,t) = \frac{c}{2\sqrt{\pi \kappa t}} \int_0^\infty \left[e^{-(\xi-x)^2/4\kappa t} - e^{-(\xi+x)^2/4\kappa t} \right] d\xi,$$

which may be rewritten as

$$u(x,t) = \frac{c}{2\sqrt{\pi \kappa t}} \left[\int_{-x}^\infty e^{-\beta^2/4\kappa t} \, d\beta - \int_x^\infty e^{-\beta^2/4\kappa t} \, d\beta \right]$$

with appropriate changes of variables; because the integrand is in each case an even function of its argument, $u(x,t)$ may be simplified further to

$$u(x,t) = \frac{c}{\sqrt{\pi \kappa t}} \int_0^x e^{-\beta^2/4\kappa t} \, d\beta.$$

Let

$$\eta^2 = \frac{\beta^2}{4\kappa t},$$

from which we may write

$$u(x,t) = \frac{2c}{\sqrt{\pi}} \int_0^{x/2\sqrt{\kappa t}} e^{-\eta^2} \, d\eta = c \, \mathrm{erf}\left(\frac{x}{2\sqrt{\kappa t}} \right).$$

This is a tabulated function called the *error function*. If we descend into the earth (assuming we have nothing better with which to occupy our time, of course) it transpires that after we reach the regions where the diurnal and annual temperature variations cease to be appreciable, the temperature begins to increase. This is in general a function of position, but a crude average is about 1°F for every 50 feet of descent for depths up to about a mile. This would of course be a consequence of heat flowing outward from the hotter central regions of the earth. Again (as in the problem of diurnal and annual variations) we ignore the convexity of the earth's surface. From the immediately preceding equation

$$\frac{\partial u}{\partial x} = \frac{c}{\sqrt{\pi \kappa t}} e^{-x^2/4\kappa t}$$

Kelvin found that κ was about 400 ft²/yr; this is about five times larger than the value we used before (2×10^{-3} cm²/sec) but the value used by Kelvin may be an average over the much larger depths he was concerned with, and the accuracy of the value is open to question given the equipment available at that time. Nevertheless, we will use his value here to illustrate the argument. If the earth was originally at the temperature of molten rock, about 7000°F, then inserting the temperature gradient of 1°F per 50 feet, we have that at $x = 0$,

$$\frac{\sqrt{t}}{50} = \frac{7000}{\sqrt{400\pi}}$$

or

$$t = \frac{7000^2 \times 50^2}{400\pi} \approx 10^8 \text{ years.}$$

We obtain a value about five times larger than this if we use the more recent estimate for κ. Notice that we have cheated a little here: $f(x) = c$ does not possess a Fourier sine transform (see Farlow)! We can get around this problem in several ways by restricting $f(x)$ appropriately; suffice it to say that this analysis nevertheless gives us the flavor of the problem posed by Kelvin.

A final point: the temperature gradient estimated above may not be constant as we descend into the deep dark earth, so that is yet another questionable assumption we have made! Is there no end to such limitations? Probably not, and that is what makes mathematical modeling as much an art as a science! A final point can be made concerning apparently strange units: the units used above for the value of κ are ft^2/year, but remember that just as any *speed* can be expressed in, for example, furlongs per fortnight, so too any diffusivity can be expressed in units of (dimension)^2/time, for example, acres/millennium. Use whatever units are appropriate for the context of the problem!

APPENDIX: THE NORMAL MODES OF RECTANGULAR AND CIRCULAR MEMBRANES

Suppose that in a plane Cartesian coordinate system the spatial coordinates are denoted by x and y. If $u(x, y, t)$ is the small amplitude perturbation in the surface of an otherwise flat rectangular elastic membrane rigidly clamped at the boundaries $x = 0$, $x = a$, $y = 0$, $y = b$. The governing wave equation is

$$\frac{\partial^2 u}{\partial t^2} = c^2 \left(\frac{\partial^2 u}{\partial x^2} + \frac{\partial^2 u}{\partial y^2} \right),$$

where c is the speed of the waves. The boundary conditions are

$$u(0, y, t) = 0, \qquad u(x, 0, t) = 0,$$
$$u(a, y, t) = 0, \qquad u(x, b, t) = 0.$$

As is shown in elementary texts on partial differential equations, this equation and many like it can be solved by the method of separation of variables. Briefly, solutions are sought in the form of a product,

$$u(x, y, t) = f(x)g(y)h(t),$$

and upon substitution of this product into the governing partial differential equation, three ordinary differential equations are forthcoming (one for each of the dependent variables f, g, and h). These are

$$\frac{d^2f}{dx^2} = \mu f, \quad \frac{d^2g}{dy^2} = (\mu - \lambda)g, \quad \frac{d^2h}{dt^2} = -\lambda c^2 h.$$

The constants μ and λ are called separation constants, and there are a countable infinity of each. As such they must be supplemented by appropriate subscripts. The first two equations (in the spatial variables) are supplemented by the now decoupled boundary conditions, which are

$$f(0) = 0, \qquad f(a) = 0,$$
$$g(0) = 0, \qquad g(b) = 0,$$

These two equations have respective solutions depending on the positive integers m and n

$$f_n(x) = \sin \frac{n\pi x}{a}; \quad \text{let } \mu_n = \left(\frac{n\pi}{a}\right)^2, \quad n = 1, 2, 3, \ldots,$$

and

$$g_{nm}(y) = \sin \frac{m\pi y}{b}; \quad \text{let } \lambda_{nm} = \mu_n + \left(\frac{m\pi}{b}\right)^2$$

$$= \left(\frac{n\pi}{a}\right)^2 + \left(\frac{m\pi}{b}\right)^2, \quad m = 1, 2, 3, \ldots.$$

Thus the formal solution of our two-dimensional eigenvalue problem depends in particular on the spatial product terms

$$\phi(x, y) = f(x)g(y) = \sin\left(\frac{n\pi x}{a}\right) \sin\left(\frac{m\pi y}{b}\right),$$

$$n = 1, 2, 3, \ldots, m = 1, 2, 3, \ldots$$

The solution of the differential equation for the temporal part of the solution is a linear combination of the terms

$$\sin(c\sqrt{\lambda_{nm}}\,t) \quad \text{and} \quad \cos(c\sqrt{\lambda_{nm}}\,t).$$

Each of the products

$$u_{nm} = \sin\left(\frac{n\pi x}{a}\right) \sin\left(\frac{m\pi y}{b}\right) \sin(c\sqrt{\lambda_{nm}}t) \qquad \text{and}$$

$$u_{nm} = \sin\left(\frac{n\pi x}{a}\right) \sin\left(\frac{m\pi y}{b}\right) \cos(c\sqrt{\lambda_{nm}}t)$$

represents a standing wave with amplitude varying periodically in time. By the principle of linear superposition the general solution of the full wave equation is a double infinite sum over the integers m and n:

$$u(x, y, t) = \sum_{m=1}^{\infty} \sum_{n=1}^{\infty} A_{nm} \sin\left(\frac{n\pi x}{a}\right) \sin\left(\frac{m\pi y}{b}\right) \cos(c\sqrt{\lambda_{nm}}t)$$

$$+ \sum_{m=1}^{\infty} \sum_{n=1}^{\infty} B_{nm} \sin\left(\frac{n\pi x}{a}\right) \sin\left(\frac{m\pi y}{b}\right) \sin(c\sqrt{\lambda_{nm}}t).$$

The coefficients A_{nm} and B_{nm} (which do not concern us here) are determined by the initial conditions $u(x, y, 0)$ and $\partial u(x, y, 0)/\partial t$, that is, the displacement and velocity of the membrane at time $t = 0$.

Returning to the modes u_{nm} above, we note that there will be *zero displacement* in the (x, y)-plane along the lines for which

$$\sin\left(\frac{n\pi x}{a}\right) = 0 \qquad \text{and} \qquad \sin\left(\frac{m\pi y}{b}\right) = 0,$$

that is, where

$$\frac{nx}{a} = 0, 1, 2, 3, \dots, n \qquad \text{or} \qquad x = 0, \frac{a}{n}, \frac{2a}{n}, \frac{3a}{n}, \dots, a,$$

and

$$\frac{my}{b} = 0, 1, 2, 3, \dots, n \qquad \text{or} \qquad y = 0, \frac{b}{m}, \frac{2b}{m}, \frac{3b}{m}, \dots, b.$$

These are respectively vertical and horizontal *nodal lines*, and they form lines of demarcation between rectangular regions or "cells" that have positive and negative displacements perpendicular to the plane of the membrane. The modal diagram for the (k, j)th mode has k segments of length a/k in the x-direction and j segments of length b/j in the y-direction, constituting a total of $k \times j$ cells, in each of which the displacement is 180° out of phase with its neighbor. Thus, for example, the $(3, 2)$ modal diagram has a total of six cells, three in the x-direction and two in the y-direction. Some of these diagrams are illustrated in figure 14.4(a).

The corresponding problem for a circular membrane is governed by the equation

$$\frac{\partial^2 u}{\partial t^2} = c^2 \left(\frac{\partial^2 u}{\partial r^2} + \frac{1}{r}\frac{\partial u}{\partial r} + \frac{\partial^2 u}{\partial \theta^2} \right), \qquad 0 \leq r \leq a.$$

The so-called normal mode solutions for this equation can be shown by the method of separation of variables to be of the form

$$u(r, \theta, t) = u_0 (J_m(nr)\cos m\theta) e^{-i\omega t},$$

where the θ-axis has been oriented in such a way as to eliminate the corresponding term in $\sin m\theta$. This can be done without loss of generality and simplifies the ensuing mathematics somewhat. The order m of the Bessel function can be $0, 1, 2, \ldots$, $\omega = nc$, and u_0 is a constant. Since at the boundary the displacement u is zero, it follows that $u(a, \theta, t) = 0$ or

$$J_m(na) = 0.$$

This equation has an infinite number of solutions for each m-value, determined by the zeros of the Bessel function. If x_{mk} denotes the kth zero of $J_m(x)$, $k = 1, 2, 3, \ldots$, then

$$n = \frac{x_{mk}}{a} \equiv n_{mk}.$$

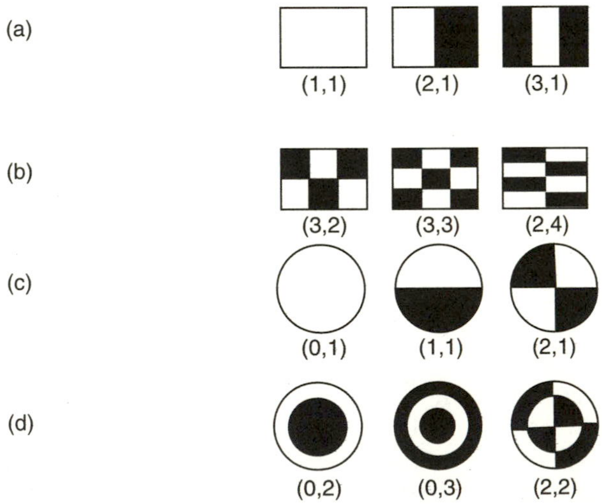

(a)

(1,1)   (2,1)   (3,1)

(b)

(3,2)   (3,3)   (2,4)

(c)

(0,1)   (1,1)   (2,1)

(d)

(0,2)   (0,3)   (2,2)

Figure 14.4. Some standing wave patterns exhibited in both rectangular and circular membranes with mode numbers specifying the number of cells (a) in the (x, y) directions and (b) in the (θ, r) directions. From *The Science of Soap Films and Soap Bubbles*, by C. Isenberg (1992), reprinted by permission of Dover Publications.

The normal mode frequencies are determined by the number pair (m, k) because $\omega = n_{mk}c$. The nodal lines are where the displacement vanishes, that is, where

$$J_m(nr) \cos m\theta = 0.$$

This means that there are two classes of nodal lines: those for which $J_m(nr) = 0$ and those for which $\cos m\theta = 0, m \neq 0$. The first of these corresponds to values of r for which $nr = x_{mk}$, that is, a set of circles (because r is constant for a given value of n), and the second condition implies that

$$m\theta = \left(s + \frac{1}{2}\right)\pi, \quad \text{or} \quad \theta = \left(s + \frac{1}{2}\right)\frac{\pi}{m}, \quad s = 0, 1, 2, \ldots, (2m - 1).$$

These are radial lines; adjacent lines are separated by an angle π/m. Some low-value pairs are illustrated in figure 14.4b.

Fractals: An Appetite Whetter . . .

Fractal geometry will make you see everything differently. There is danger in reading further. You risk the loss of your childhood vision of clouds, forests, galaxies, leaves, feathers, flowers, rocks, mountains, torrents of water, carpets, bricks, and much else besides. Never again will your interpretation of these things be quite the same.

The observation by Mandelbrot . . . of the existence of a "Geometry of Nature" has led us to think in a new scientific way about the edges of clouds, the profiles of the tops of forests on the horizon, and the intricate moving arrangement of the feathers on the wings of a bird as it flies. Geometry is concerned with making our spatial intuitions objective. Classical geometry provides a first approximation to the structure of physical objects; it is the language which we use to communicate the designs of technological products, and, very approximately, the forms of natural creations. Fractal geometry is an extension of classical geometry. It can be used to make precise models of physical structures from ferns to galaxies. Fractal geometry is a new language. Once you speak it, you can describe the shape of a cloud as precisely as an architect can describe a house.

—Michael Barnsley (in *Fractals Everywhere*)

Introducing chance as small disturbances while constructing fractals, we can make fractals serve as a model for natural objects like trees, plants, sponges and corals.

—Hans Lauwerier (in *Fractals*)

In order to appreciate the underlying features of fractals it is necessary to examine a fundamental procedure common to many topics within the mathematical and computational sciences, namely, *iteration*. Some humorist has said that under the dictionary definition of iteration it says "See iteration"; that is in fact a good pragmatic and self-referential definition! Sometimes we may be concerned with iteration of functions in the complex plane, rather than restricting ourselves to the non-negative real line (which is, of course, contained in the complex plane), but in the spirit of this book we briefly discuss the iteration of geometric patterns at smaller and smaller scales to mimic one of the shapes we observe in nature—coastlines.

The person most often associated with the discovery of fractals is the mathematician Benoit B. Mandelbrot (the Mandelbrot set obviously being named after its discoverer; for details of this amazing set, consult the references). The underlying mathematics, however (geometric measure theory), had been developed long before the "computer revolution" made possible the visualization of such complicated mathematical objects. In the 1960s Mandelbrot pointed out some interesting but very surprising results in a paper entitled "How Long Is the Coastline of Britain?" published posthumously by the English meteorologist Lewis Fry Richardson. That author had noticed that the measured length of the west coast of Britain depended heavily on the scale of the map used to make those measurements: a map with scale 1:10,000,000 (1 cm being equivalent to 100 km) has less detail than a map with scale 1:100,000 (1 cm equivalent to 1 km). The more detailed map, with more "nooks and crannies," gives a larger value for the coastline! Alternatively, one can imagine measuring a given map with smaller and smaller measuring units, or even walking around the coastline with smaller and smaller graduations on our meter rule! Of course this presumes that at such small scales we can meaningfully define the coastline, but of course this process cannot be continued indefinitely due to the atomic structure of matter, unlike the "continuum" mathematical models to which we have referred, and in which there is no smallest scale.

Richardson also investigated the behavior with scale for other geographical regions: the Australian coast, the South African coast, the German Land Frontier (1900), and the Portugese Land Frontier. For the west coast of Britain in particular, he found the following relationship between the total length s in km and the numerical value a of the measuring unit (in km, so a is dimensionless):

$$s = s_1 a^{-0.22},$$

where s_1 is the length when $a = 1$. Clearly, as a is reduced, s increases! If the measuring unit were one meter instead of one km, the value of s would increase by a factor of about 4.6 according to this model! Clearly, the concept of length in this context is a rather unstable one; is there a better way of describing the coastline? Can we measure the "crinkliness" or "roughness" or "degree of meander" or some other such quantity? Mandelbrot showed that the answer to this question is *yes*, and the answer is intimately connected with a generalization of our familiar concept of "dimension." This is the so-called topological dimension, expressed in the natural numbers 0, 1, 2, 3, ... (there is no reason to stop at three, by the way). It turns out that the concept of fractal dimension used by Mandelbrot (the Hausdorff-Besiscovich dimension), being a ratio of logarithms, is not generally an integer.

Consider the measurement of a continuous curve by a "measuring rod" of length a. Suppose that it fits N times along the length of the curve, so that the measured length $L = Na$. Obviously, then, $N = L/a$ is a function of a, that is, $N = N(a)$. Thus if $a = 1, N(1) = L$. Similarly, if $a = \frac{1}{2}, N(\frac{1}{2}) = 2L$, $N(\frac{1}{3}) = 3L$, and so on. For fractal curves, $N = La^{-D}$, where $D > 1$ in general, and it is called the *fractal dimension*. This means that making the scale three times as large (or a one-third of the size) as previously may lead to the measuring rod fitting around the curve *more* than three times the previous amount. This is because, if $N(1) = L$ as before,

$$N\left(\frac{1}{3}\right) = L\left(\frac{1}{3}\right)^{-D} = L(3^D) > 3$$

if $D > 1$. In what follows we shall use unit length, $L = 1$, when $a = 1$, without loss of generality.

Given that

$$N = a^{-D}$$

it follows that

$$D = \frac{\log N}{\log(1/a)}.$$

More precisely, Mandelbrot used the definition of the fractal dimension as

$$D = \lim_{a \to 0} \frac{\log N}{\log(1/a)}.$$

If this has the same value at each step, then the former definition is perfectly general. Let us apply the definition to what has come to be called the Koch snowflake curve. The basic iteration step is to take each line segment or side of an equilateral triangle, remove the middle third, and replace it by two sides of an equilateral triangle (each side of which is equal in length to the middle third, so now it is a "Star of David"). Each time this procedure is carried out the previous line segment is increased in length by a factor

$\frac{4}{3}$. Thus $a = \frac{1}{3}$ and $N = 4$ (there now being four smaller line segments in place of the original one), so

$$D = \frac{\log 4}{\log 3} \approx 1.26186.$$

The limiting snowflake curve (as $a \to 0$) thus "intrudes" a little into the second dimension; this intrusion is indicated by the degree of "meander" as expressed by the fractal dimension D. This curve is *everywhere continuous but nowhere differentiable!* Such curves, continuous but without tangents, were first defined over a century ago by the German mathematician Karl Weierstrass (1815–1897); their existence horrified many of his peers. Physicists, however, were more welcoming: Ludwig Boltzmann (1844–1906) wrote to Felix Klein (1849–1925) in January 1898 with the comment that such functions might well have been invented by physicists because there are problems in statistical mechanics "that absolutely necessitate the use of nondifferentiable functions." He had in mind, no doubt, *Brownian motion* (this is the constant and highly erratic movement of tiny particles [e.g., pollen] suspended in a liquid or a gas, or on the surface of a liquid).

Consider next the box fractal: this is a square in which the basic iteration is to divide it into 9 identical smaller squares and remove the middle squares from each side leaving 5 of the original 9. It is readily seen that $a = \frac{1}{3}$ and $N = 5$, so

$$D = \frac{\log 5}{\log 3} \approx 1.46497.$$

The Sierpinski triangle is any triangle for which the basic iteration is to join the midpoints of the sides with line segments and remove the middle triangle. Now $a = \frac{1}{2}$ and $N = 3$, so

$$D = \frac{\log 3}{\log 2} \approx 1.58496.$$

Thus it seems that these fractals penetrate increasingly more into the second dimension. We will mention two more at this juncture: the Menger sponge and Cantor dust. For the former, we do in three dimensions what was done in two for the box fractal. Divide a cube into 27 identical cubes and "push out" the middle ones in each face (and the central one). Now it follows that $a = \frac{1}{3}$ and $N = 20$ (seven smaller cubes having been removed in the basic iterative step), so in the limit of the requisite infinite number of iterations

$$D = \frac{\log 20}{\log 3} \approx 2.72683$$

(intruding well into the third dimension). Cantor dust: what is it? Take any line segment and remove the middle third; this is the basic iteration. In the limit of an infinity of such iterations for which $a = \frac{1}{3}$ and $N = 2$, it follows that

$$D = \frac{\log 2}{\log 3} \approx 0.63093,$$

which is obviously less than one. Quite amazing.

Fractals are being used in many different subject areas from graphic design to medicine; for a fascinating collection of applications to the latter the book edited by Nonnenmacher et al. is recommended. There are many excellent theoretical and "applied" descriptions of fractals (including the Mandelbrot set), fractal dimension, and the related subject of chaos (see the bibliography for details). The book by Gleick—*Chaos*—is an excellent introduction to these topics, and *Fractals Everywhere* by Barnsley is a valuable resource for mathematicians interested in the subject of fractals. For applications to geophysics, see the book by Turcotte.

However, it is worth noting that another perspective on the study of many natural pheonomena—Adrian Bejan's *constructal theory*—and summarized in his comprehensive book *Shape and Structure, from Engineering to Nature* (published in 2000), represents a different but extremely important perspective on "patterns and nature." Bejan's view is that fractal geometry is descriptive but not predictive (or explanatory, it might be added). Indeed, Kadanoff has asked the question, Fractals: where's the physics? (see also Avnir et al.). References to these publications and many others can be found in Bejan's book; it addresses many of the topics presented in this book in a complementary fashion, and is well worth reading.

BIBLIOGRAPHY

Includes both sources used and suggestions for further reading; the relevant chapters in this book are in parentheses; G refers to general interest reading.

Acheson, D. J. (1990). *Elementary Fluid Dynamics*. Clarendon Press: Oxford. (7–9)

Adam, J. A. (1989). "A Nonlinear Eigenvalue Problem in Astrophysical Magnetohydrodynamics: Some Properties of the Spectrum." *Journal of Mathematical Physics* 30: 744–756. (7,8)

Adam, J. A., and Maggelakis, S. (1990). "Diffusion Regulated Growth Characteristics of a Prevascular Carcinoma." *Bulletin of Mathematical Biology* 52: 549–582. (14)

Adam, J. A. (1995). "Educated Guesses." *Quantum* September/October: 20–24. (2)

Adam, J. A., and Bellomo, N. (1997). *A Survey of Models for Tumor-Immune System Dynamics*. Birkhäuser: Boston. (1,14)

Adam, J. A. (2002a). "The Mathematical Physics of Rainbows and Glories." *Physics Reports* 356: 229–365. (5)

Adam, J. A. (2002b). "Like a Bridge over Colored Water: A Mathematical Review of *The Rainbow Bridge: Rainbows in Art, Myth and Science*." *Notices of the American Mathematical Society* 49: 1360–1371. (5)

Adler, F. (1998). *Modeling the Dynamics of Life*. Brooks/Cole: Pacific Grove. (11)

Alexander, R. M. (1984). "Walking and Running." *American Scientist* 72: 348–354. (3)

Anderson, R. S. (1996). "The Attraction of Sand Dunes." *Nature* 379: 24–25. (6)

Ahrens, D. (2000). *Meteorology Today*. Brookes Cole: Pacific Grove. (G,5,6)

Ahrens, D. (2001). *Essentials of Meteorology*. Brookes Cole: Pacific Grove. (G,5,6)

Alexander, R. M. (1968). *Animal Mechanics*. University of Washington Press: Seattle. (3,13)

Allen, O. E. (1983). *Atmosphere*. Time Life: Alexandria. (G,5,6)

Anderson, B. R. (1975). *Weather in the West*. American West Publishing Company: Palo Alto. (G,5,6)

Arfken, G. (1985). *Mathematical Methods for Physicists*. 3rd edition. Academic Press: Orlando. (G,12)

Asimov, I. (1983). *The Measure of the Universe*. Harper and Row: New York. (G,2)

Austin, J. D., and Dunning, F. B. (1988). "Mathematics of the Rainbow." *The Mathematics Teacher* September: 484–488. (5)

Avnir, D., Biham, O., Lidar, D., and Malcai, O. (1998). "Is the Geometry of Nature Fractal?" *Science* 279: 39–40. (Preface, Appendix)

Bachman, C. H. (1988). "Ubiquitous Geometry. Some Examples Showing the Significance of Size and Shape in the Works of Man and Nature." *The Physics Teacher* 26: 341–370. (3)

Baldock, G. R., and Bridgeman, T. (1981). *The Mathematical Theory of Wave Motion.* Ellis Horwood: Chichester. (7–9)

Bagnold, R. A. (1965). *The Physics of Blown Sand and Desert Dunes.* Methuen: London. (6)

Ball, P. (1999a). *The Self-Made Tapestry.* Oxford University Press: Oxford. (G,6,8,10–12,14)

Ball, P. (1999b). *H_2O: A Biography of Water.* Phoenix: London. (5)

Banks, R. B. (1998). *Towing Icebergs, Falling Dominoes, and Other Adventures in Applied Mathematics.* Princeton University Press: Princeton. (G)

Banks, R. B. (1999). *Slicing Pizzas, Racing Turtles, and Further Adventures in Applied Mathematics.* Princeton University Press: Princeton. (G,5)

Barber, N. F. (1969). *Water Waves.* Wykeham Press: London. (7,8)

Barnes, G. (1989). "Physics and Size in Biological Systems." *The Physics Teacher* April: 234–253. (3)

Barnes, G. (1990). "Food, Eating and Mathematical Scaling." *The Physics Teacher* December: 614–615. (3)

Barnsley, M. F. (1988). *Fractals Everywhere.* Academic Press: San Diego. (Appendix)

Bascom, W. (1980). *Waves and Beaches.* Anchor Press: New York. (7,9)

Batschelet, E. (1975). *Introduction to Mathematics for Life Scientists.* Springer-Verlag: New York. (10,11)

Beckmann, P. (1971). *The History of π.* St. Martin's Press: New York. (11)

Bejan, A. (2000). *Shape and Structure, from Engineering to Nature.* Cambridge University Press: Cambridge. (Appendix)

Belloc, H. (1959). "The Water Beetle," in *Cautionary Verses.* Knopf: New York. (7)

Bender, C. (1978). *An Introduction to Mathematical Modeling.* John Wiley: New York. (3)

Bentley, W. A., and Humphreys, W. J. (1962). *Snow Crystals.* Dover: New York. (5)

Bhatnagar, P. L. (1979). *Nonlinear Waves in One-Dimensional Dispersive Systems.* Clarendon Press: Oxford. (7–9)

Biddle, W. (1989). "Skeleton Alleged in the Stealth Bomber's Closet." *Science* 244: 650–651. (8)

Blasiak, P. (1995). "Laws of Physics, Little People, and Gluttony." *The Physics Teacher* 33: 122–123. (3)

Bohren, C. F., and Huffman, D. R. (1983). *Absorption and Scattering of Light by Small Particles.* Wiley: New York. (4,5)

Bohren, C. F., and Fraser, A. B. (1985). "Colors of the Sky." *The Physics Teacher* May: 267–272. (4)

Bohren, C. F., and Fraser, A. B. (1986). "At What Altitude Does the Horizon Cease to Be Visible?" *American Journal of Physics* 54: 222–227. (4)

Bohren, C. (1987). *Clouds in a Glass of Beer.* Wiley: New York. (G,4,5)

Bohren, C. (1991). *What Light Through Yonder Window Breaks?* Wiley: New York. (G,4,5)

Botley, C. M. (1946). "Halos and Coronae." *Weather* July: 85–88. (5)

Boyer, C. B. (1987). *The Rainbow, from Myth to Mathematics*. Princeton University Press: Princeton. (5)

Boys, C. V. (1962). *Soap Bubbles*. Thomas Crowell: New York. (11)

Briggs, J., and Peat, F. D. (1989). *Turbulent Mirror*. Harper and Row: New York. (G)

Briggs, J. (1992). *Fractals—The Patterns of Chaos*. Simon and Schuster: New York. (G)

Bryant, H. C., and Jarmie, N. (1974). Reprinted in *Light from the Sky* (1980; Readings from *Scientific American*), pp. 66–74. Freeman: San Francisco. (5)

Brookhart, C. (1998). *Go Figure!* Contemporary Books: Chicago. (G,2)

Brousseau, A. (1968). "On the Trail of the California Pine." *Fibonacci Quarterly* 6: 69–76. (10)

Brousseau, A. (1969). "Fibonacci Statistics in Conifers." *Fibonacci Quarterly* 7: 525–532. (10)

Brown, J. H., and West, G. B., eds. (2000). *Scaling in Biology*. Oxford University Press: Oxford. (3)

Buchanan, M. (1998). "Pure Genius." *New Scientist* 19/26 December: 56–57. (11)

Buckley, P. A., and Buckley, F. G. (1977). "Hexagonal Packing of Royal Tern Nests." *The Auk* 94: 36–43. (11)

Burton, R. F. (1998). *Biology by Numbers*. Cambridge University Press: Cambridge. (2,3)

Calder, W. A. III, (1996). *Size, Function and Life History*. Dover: New York. (3)

Callander, R. A. (1978). "River Meandering." *Annual Review of Fluid Mechanics* 10: 129–158. (12)

Canny, M. J. (1998). "Transporting Water in Plants." *American Scientist* 86: 152–159. (12)

Cardon, B. L. (1977). "An Unusual Lunar Halo." *American Journal of Physics* 45: 331–335. (5)

Cassie, B. (National Audobon Society). (1999). *Trees*. Knopf: New York. (G,12)

Chandrasekhar, S. (1981). *Hydrodynamic and Hydromagnetic Stability*. Dover: New York. (8)

Cipra, B. A. (1995). "Touring Turing." *SIAM News* May/June: 24. (14)

Coffey, M. T. (1980). "An Observation of Billow Clouds." *Weather* 35: 261. (5,8)

Cohen, J., and Stewart, I. (1993). "Let T Equal Tiger. . . " *New Scientist* 6 November: 40–44. (14)

Colinvaux, P. (1990). *Why Big Fierce Animals Are Rare*. Penguin: London. (G,3)

Colson, D. (1954). "Wave-Cloud Formation at Denver." *Weatherwise* 7: 34–35. (5,8)

COMAP (1991). *For All Practical Purposes*. 2nd edition. Freeman: New York. (3,10)

Cook, T. A. (1979). *The Curves of Life*. Dover: New York. (G,10)

Coveney, P., and Highfield, R. (1995). *Frontiers of Complexity*. Faber and Faber: London. (G,14)

Coxeter, H.S.M. (1969). *Introduction to Geometry*. John Wiley: New York. (10,11)

Crawford, F. S. (1970). "Douglas Fir Echo Chamber." *American Journal of Physics* 38: 1477. (4,12)

Crawford, F. S. (1971). "Culvert Whistlers." *American Journal of Physics* 39: 610–615. (4,12)

Criswell, D. R., Lindsay, J. F., and Reasoner, D. L. (1975). "Seismic and Acoustic Emissions of a Booming Dune." *Journal of Geophysical Research* 80: 4963–4974. (6)

Cromer, A. H. (1977). *Physics for the Life Sciences*. McGraw-Hill: New York. (3,12)

Davidson, N. (1993). *Sky Phenomena*. Lindisfarne: New York. (G)

Davis, P. J. (1961). *The Lore of Large Numbers*. Mathematical Association of America: Washington, D.C. (2)

Denny, M., and Gaines, S. (2000). *Chance in Biology*. Princeton University Press: Princeton. (12,14)

Devaney, R. L. (1990). *Chaos, Fractals, and Dynamics*. Addison-Wesley: New York. (G)

Devlin, K. (1994). "Why Runners Go Round the Bend." Chapter 21 in his book *All the Math That's Fit to Print*. Mathematical Association of America: Washington, D.C. (12)

Devlin, K. (1998a). *The Language of Mathematics*. Freeman: New York. (G,1)

Devlin, K. (1998b). *Life by the Numbers*. Wiley: New York. (G,1,3,10,11,14)

Dickinson, T. (1988). *Exploring the Sky by Day*. Camden House: Camden East, Ontario. (4–6)

Ditchburn, R. W. (1991). *Light*. Dover: New York. (G,4,5)

Dixon, R. (1981). "The Mathematical Daisy." *New Scientist* 17 December: 792–795. (10)

Dixon, R. (1989). "Spiral Phyllotaxis." *Computers and Mathematics with Applications* 17: 535–538. (10)

Dormer, K. J. (1980). *Fundamental Tissue Geometry for Biologists*. Cambridge University Press: Cambridge. (11,12)

Drazin, P., and Reid, W. (1981). *Hydrodynamic Stability*. Cambridge University Press: Cambridge. (8)

Dutton, J. A. (1986). *The Ceaseless Wind*. Dover: New York. (6–8)

Edelstein-Keshet, L. (1988). *Mathematical Models in Biology*. Random House: New York. (14)

Ehrlich, R. (1993). *The Cosmological Milkshake*. Rutgers University Press: New Brunswick. (G)

Ehrlich, R. (1997). *Why Toast Lands Jelly-Side Down*. Princeton University Press: Princeton. (G)

Elmore, W. C., and Heald, M. A. (1969). *Physics of Waves*. Dover: New York. (7–9)

Epstein, L. C. (1994). *Thinking Physics*. Insight Press: San Francisco. (G)

Etkina, E., Holton, B., and Horton, G. (1998). "Planar Motion, Complex Numbers, and Falling Leaves: An Intriguing Minilab." *The Physics Teacher* 36: 135–138. (12)

Farlow, S. J. (1993). *Partial Differential Equations for Scientists and Engineers*. Dover: New York. (14)

Ferguson, R. I. (1973). "Regular Meander Path Models." *Water Resources Research* 9: 1079–1086. (12)

Ferguson, R. I. (1976). "Disturbed Periodic Model for River Meanders." *Earth Surface Processes* 1: 337–347. (12)

Field, M., and Golubitsky, M. (1995). *Symmetry in Chaos*. Oxford University Press: Oxford. (G)

Fowler, A. C. (1997). *Mathematical Methods in the Applied Sciences*. Cambridge University Press: Cambridge. (G,11)

Fraser, A. B. (1972). "Inhomogeneities in the Color and Intensity of the Rainbow." *Journal of the Atmospheric Sciences* 29: 211–212. (5)

Fraser, A. B. (1975). "Theological Optics." *Applied Optics* 14: A92–A93. (5)

Fraser, A. B., and Mach, W. H. (1976). "Mirages." Reprinted in *Light from the Sky* (1980; Readings from *Scientific American*), pp. 29–37. Freeman: San Francisco. (5)

Fraser, A. B. (1983). "Why Can the Supernumerary Bows Be Seen in a Rain Shower?" *Journal of the Optical Society of America* 73: 1626–1628. (5)

Freier, G. D. (1992). *Weather Proverbs*. Fisher Books: Tucson. (G,5,6)

Gallant, R. A. (1987). *Rainbows, Mirages, and Sundogs*. Macmillan: New York. (5)

Gamow, R. I. (1979). "Spirals in Nature." *The Physics Teacher* January: 14–22. (G,10)

Garland, T. H. (1987). *Fascinating Fibonaccis*. Dale Seymour Publications: Palo Alto. (10)

Gay, D. (1998). *Geometry by Discovery*. Wiley: New York. (G,11)

Gedzelman, S. D. (1980). *The Science and Wonders of the Atmosphere*. Wiley: New York. (G,4–6)

Gleick, J. (1987). *Chaos*. Viking: New York. (G)

Gleick, J., and Porter, E. (2001). *Nature's Chaos*. Little, Brown: Boston. (G)

Goldstein, H. (1980). *Classical Mechanics*. Addison-Wesley: Reading, Mass. (5)

Goodwin, B. (1994). *How the Leopard Changed Its Spots*. Phoenix: London. (G,14)

Gordon, J. E. (1979). *Structures, or Why Things Don't Fall Down*. Penguin: London. (G,12)

Gray, J. (1959). *How Animals Move*. Pelican: Edinburgh. (3,13)

Greenewalt, C. H. (1990). *Hummingbirds*. Dover: New York. (5,13)

Greenler, R. G., and Mallmann, A. J. (1972). "Circumscribed Halos." *Science* 176: 128–131. (5)

Greenler, R. G., Drinkwine, M., Mallmann, A. J., and Blumenthal, G. (1972). "The Origin of Sun Pillars." *American Scientist* 60: 292–302. (5)

Greenler, R. G. (1980). *Rainbows, Halos and Glories*. Cambridge University Press: Cambridge. (5)

Griffin, O. M. (1982). "Vortex Streets and Patterns." *Mechanical Engineering* March: 56–61. (6–8)

Haldane, J.B.S. (1956). "On Being the Right Size." In Newman, J. R., *The World of Mathematics*. (Volume 2). Simon and Schuster: New York. (G,3)

Hales, T. (2000). "Cannonballs and Honeycombs." *Notices of the American Mathematical Society* 47: 440–449. (11)

Halliday, D. (1990). "Ballpark Estimates." *Quantum* May: 30–31. (2)

Hanna, S. R. (1969). "The Formation of Longitudinal Sand Dunes by Large Helical Eddies in the Atmosphere." *Journal of Applied Meteorology* 8: 874–883. (6)

Harte, J. (1985). *Consider a Spherical Cow: A Course in Environmental Problem Solving.* Kauffman: Los Altos. (G,2)

Hastings, J. D. (1971). "Sand Streets." *Meteorological Magazine* 100: 155–159. (6)

Hayes, B. (1995). "Space-Time on a Seashell." *American Scientist* 83: 214–218. (14)

Hejnowicz, Z. (1973). "Morphogenetic Waves in Cambia of Trees." *Plant Science Letters* 1: 359–366. (7,12)

Hejnowicz, Z. (1974). "Pulsations of Domain Length as Support for the Hypothesis of Morphogenetic Waves in the Cambium." *Acta Societatis Botanicorum Poloniae* XLIII: 263–271. (7,12)

Hewes, L. I. (1948). "A Theory of Surface Cracks in Mud and Lava and Resulting Geometrical Relations." *American Journal of Science* 246: 138–149. (11)

Hidy, G. M. (1971). *The Waves.* Van Nostrand Reinhold: New York. (G,7,8)

Higdon, J.J.L., and Corrsin, S. (1978). "Induced Drag of a Bird Flock." *American Naturalist* 112: 727–744. (13)

Higgins, P. M. (1998). *Mathematics for the Curious.* Oxford University Press: Oxford. (G,10)

Hildebrandt, S., and Tromba, A. (1996). *The Parsimonious Universe.* Springer: New York. (G,11)

Hill, A. V. (1950). "The Dynamics of Animals and Their Muscular Dynamics." *Science Progress* XXXVIII: 209–230. (3)

Hodges, L. (1987). "Tides, Eclipses, and the Densities of the Sun and the Moon." *The Physics Teacher* October: 427. (9)

Hodges, L. (1991). "Gravitational Field Strength inside the Earth." *American Journal of Physics* 59: 954–956. (9)

Hofer, T., Maini, P. K., Kondo, S., and Asai, R. (1996). "Letter to Nature." *Nature* 380: 678. (14)

Hokkanen, J.E.I. (1986). "Notes Concerning Elastic Similarity." *Journal of Theoretical Biology* 120: 499–501. (3)

Hopkins, G. M. (1995). *Hopkins: Poems and Prose.* Everyman's Library Pocket Poets. Knopf: New York. (1)

Horn, H. S. (1971). *The Adaptive Geometry of Trees.* Princeton University Press: Princeton. (12)

Houghton, J. T. (1977). *Physics of Atmospheres.* Cambridge University Press: Cambridge. (6,7)

Hoyle, F. (1957). *The Black Cloud.* Harper: New York. (2)

Humphreys, W. J. (1964). *Physics of the Air.* Dover: New York. (4,5)

Huntley, H. E. (1970). *The Divine Proportion: A Study in Mathematical Beauty.* Dover: New York. (3)

Isenberg, C. (1992). *The Science of Soap Films and Soap Bubbles.* Dover: New York. (5,11,14)

Janke, S. (1992). *Somewhere Within the Rainbow.* Module 724. COMAP, Inc.: Lexington. (5)

Jaquin, F., Steele, K., and Hafmeister, D. (1982). "The Apparent Ellipticity of the Setting Sun." *The Physics Teacher* September: 404–405. (4)

Jargodzki, C., and Potter, F. (2001). *Mad about Physics*. Wiley: New York. (G,4,12)

Jean, R. V., and Johnson, M. (1976). "An Adventure into Applied Mathematics with Fibonacci Numbers." *School Science and Mathematics* 89: 487–498. (12)

Jean, R. V. (1994). *Phyllotaxis*. Cambridge University Press: Cambridge. (10)

Judson, H. F. (1980). *The Search for Solutions*. Holt, Rinehart, Winston: New York. (G,11)

Kadanoff, L. P. (1986). "Fractals: Where's the Physics?" *Physics Today*, February: 6–7. (Preface, Appendix)

Kappraff, J. (1991). *Connections*. McGraw-Hill: New York. (G,10)

Kerr, D. E. (ed.). (1965). *Propagation of Short Radio Waves*. Dover: New York. (5,7)

Khinchin, A. Y. (1992). *Continued Fractions*. Dover: New York. (10)

Khular, E., Thyagarajan, K., and Ghatak, A. K. (1977). "A Note on Mirage Formation." *American Journal of Physics* 45: 90–92. (5)

Kiersted, H., and Slobodkin, L. B. (1953). "The Size of Water Masses Containing Plankton Blooms." *Journal of Marine Research* 12: 141–147. (14)

Kinsey, L. C., and Moore, T. E. (2001). *Symmetry, Shape, and Space*. Key College Publishing: Emoryville. (G,10,11)

Kinsman, B. (1965). *Wind Waves: Their Generation and Propagation on the Ocean Surface*. Prentice-Hall: Englewood Cliffs. (7–9)

Kline, M. (1985). *Mathematics and the Search for Knowledge*. Oxford University Press: Oxford. (G,1)

Kolman, B. (1997). *Introductory Linear Algebra with Applications*. Prentice-Hall: Englewood Cliffs. (3)

Kondo, S., and Asai, R. (1995). "A Reaction-Diffusion Wave on the Skin of the Marine Angelfish *Pomacanthus*." *Nature* 376: 765–768. (14)

Können, G. P., and de Boer, J. H. (1979). "Polarized Rainbow." *Applied Optics* 18: 1961–1965. (5)

Können, G. P. (1985). *Polarized Light in Nature*. Cambridge University Press: Cambridge. (4,5)

Laithwaite, E. (1994). *An Inventor in the Garden of Eden*. Cambridge University Press: Cambridge. (G,3,12)

Lamb, H. (1945). *Hydrodynamics*. Dover: New York. (7–9)

Landau, L. D., and Lifshitz, E. M. (1959). *Fluid Dynamics*. Pergamon: London. (3,7,13)

Langbein, W. B., and Leopold, L. B. (1966). *River Meanders—Theory of Minimum Variance*. Geological Survey Professional Paper 422-H. U.S. Government Printing Office: Washington, D.C. (12)

Lanzara, E., Mantegna, R. N., Spagnolo, B., and Zangara, R. (1977). "Experimental Study of a Nonlinear System in the Presence of Noise: The Stochastic Resonance." *American Journal of Physics* 65: 341–349. (7)

Lauwerier, H. (1991). *Fractals*. Princeton University Press: Princeton. (Appendix)

Lavers, C. (2001). *Why Elephants Have Big Ears*. Phoenix: New York. (G,3)

Lee, R. L. (1991). "What Are All the Colors of the Rainbow?" *Applied Optics* 30: 3401–3407. (5)

Lee, R. L., Jr., and Fraser, A. B. (2001). *The Rainbow Bridge: Rainbows in Art, Myth, and Science*. Pennsylvania State University Press: University Park. (5)

Lemons, D. S., and Lipscombe, T. C. (2002). "The Shape of a Randomly Lying Cord." *American Journal of Physics* 70: 570–574. (12)

Leopold, L. B., and Langbein, W. B. (1966). "River Meanders." *Scientific American* 214: 60–70. (12)

Leopold, L. B. (1971). "Trees and Streams: The Efficiency of Branching Patterns." *Journal of Theoretical Biology* 31: 339–354. (12)

Leopold, L. B. (1974). *Water*. Freeman: San Francisco. (12)

Leopold, L. B. (1997). *Water, Rivers, and Creeks*. University Science Books: Sausalito. (12)

Levine, Z. H. (1993). "How to Measure the Radius of the Earth on Your Beach Vacation." *The Physics Teacher* 31: 440–441. (4)

Lin, C. C., and Segel, L. A. (1988). *Mathematics Applied to Deterministic Problems in the Natural Sciences*. Society for Industrial and Applied Mathematics: Philadelphia. (G,7–9)

Lin, H. (1978). "Newtonian Mechanics and the Human Body: Some Estimates of Performance." *American Journal of Physics* 46: 15–18. (3)

Lin, H. (1982). "Fundamentals of Zoological Scaling." *American Journal of Physics* 50: 72–81, 567. (3)

Lindsay, J. F., Criswell, D. R., Criswell, T. L., and Criswell, B. S. (1976). "Sound-producing Dune and Beach Sands." *Geological Society of America Bulletin* 87: 463–473. (6)

Lissaman, P.B.S., and Shollenberger, C. A. (1970). "Formation Flight of Birds." *Science* 168: 1003–1005. (13)

Livio, M. (2002). *The Golden Ratio*. Broadway Books: New York. (10)

Longwell, C. R., and Flint, R. F. (1962). *Introduction to Physical Geology*. Wiley: New York. (6)

Lovett, D. (1994). *Demonstrating Science with Soap Films*. Institute of Physics (UK): Bristol. (11)

Ludlam, F. H. (1967). "Characteristics of Billow Clouds and Their Relation to Clear-Air Turbulence." *Quarterly Journal of the Royal Meteorological Society* 93: 419–435. (6,8)

Ludlam, D. M. (National Audubon Society). (1998). *Guide to North American Weather*. Knopf: New York. (G,4–6)

Lynch, D. K. (1978). "Atmospheric Halos." Reprinted in *Light from the Sky* (1980; Readings from *Scientific American*), pp. 38–46. Freeman: San Francisco. (5)

Lynch, D. K. (1982). "Tidal Bores." *Scientific American* 247: 146–156. (9)

Lynch, D. K. (1987). "Optics of Sunbeams." *Journal of the Optical Society of America* A4: 609–611. (4)

Lynch, D. K., and Livingston, W. (1995). *Color and Light in Nature*. Cambridge University Press: Cambridge. (G,4–5)

MacDonald, N. (1983). *Trees and Networks in Biological Models*. Wiley: New York. (12)

MacMillan, W. D. (1958). *Statics and the Dynamics of a Particle*. Dover: New York. (12)

Maini, P. K. (1997). "Bones, Feathers, Teeth, and Coat Markings: A Unified Model." *Science Progress* 80: 217–229. (14)

Malkus, W.V.R. (1955). "Rainbows and Cloudbows." *Weather* 10: 331–335. (5)

Mandelbrot, B. B. (1967). "How Long Is the Coastline of Britain? Statistical Self-Similarity and Fractional Dimension." *Science* 55: 636–638. (Appendix)

Mandelbrot, B. B. (1983). *The Fractal Geometry of Nature*. Freeman: New York. (G)

Marchand, P. J. (1995). "Waves in the Forest." *Natural History* 2/95: 26–32. (7,12)

Markowsky, G. (1992). "Misconceptions about the Golden Ratio." *The College Mathematics Journal* 23: 2–19. (10)

Maynard Smith, J. (1968). *Mathematical Ideas in Biology*. Cambridge University Press: Cambridge. (3,14)

Mayo, N. (1994). "A Hurricane for Physics Students." *The Physics Teacher* 32: 148–154. (6)

Mayo, N. (1997). "Ocean Waves—Their Energy and Power." *The Physics Teacher* 35: 352–356. (7,8)

McMahon, T. (1973). "Size and Shape in Biology." *Science* 179: 1201–1204. (3,12)

McMahon, T. A. (1975). "The Mechanical Design of Trees." *Scientific American* 233: 92–102. (12)

McMahon, T. A., and Bonner, J. T. (1983). *On Size and Life*. Scientific American Library: New York. (3,12)

Meinel, A., and Meinel, M. (1983). *Sunsets, Twilights, and Evening Skies*. Cambridge University Press: Cambridge. (4)

Meinhardt, H., and Klingler, M. (1987). "A Model for Pattern Formation on the Shells of Molluscs." *Journal of Theoretical Biology* 126: 63–89. (14)

Meinhardt, H. (1995a). *The Algorithmic Beauty of Seashells*. Springer: New York. (14)

Meinhardt, H. (1995b). "Dynamics of Stripe Formation." *Nature* 376: 722–723. (14)

Memory, J. D., and Jenkins, A. W., Jr. (1977). "Estimating Orders of Magnitude." *The Physics Teacher* 15: 43–44. (2)

Menninger, E. A. (1995). *Fantastic Trees*. Timber Press: Portland. (G,12)

Minnaert, M.G.J. (1954). *Light and Colour in the Open Air*. Dover: New York. (G,4–5)

Minnaert, M.G.J. (1993). *Light and Colour in the Outdoors*. Springer: New York. (G,4–5)

Mitchison, G. J. (1977). "Phyllotaxis and the Fibonacci Series." *Science* 196: 270–275. (10)

Morrison, P., and Morrison, P. (1994). *Powers of Ten*. Scientific American Library: New York. (G,2)

Munson, B. R., Young, D. F., and Okiishi, T. H. (1998). *Fundamentals of Fluid Mechanics*. John Wiley: New York. (2)

Murchie, G. (1954). *The Song of the Sky*. Houghton Mifflin: Boston. (G,5,6)

Murphy, P. (1993). *By Nature's Design*. Chronicle Books: San Francisco. (G,1)

Murphy, P., and Doherty, P. (1996). *The Color of Nature*. Chronicle Books: San Francisco. (G,1)

Murray, J. D. (1988). "How the Leopard Gets Its Spots." *Scientific American* 258: 80–87. (14)

Murray, J. D. (1990). *Mathematical Biology*. Springer: New York. (14)

National Audubon Society (1995a). *Clouds and Storms*. Knopf: New York. (G,6)

National Audubon Society (1995b). *Sun and Moon*. Knopf: New York. (G,4,5)

National Audubon Society (1998). *Familiar Trees of North America (East)*. Knopf: New York. (G,12)

Naylor, M. (2002). "Golden, $\sqrt{2}$, and π Flowers: A Spiral Story." *Mathematics Magazine* 75: 163–172. (10)

Nijhout, H. F. (1978). "Wing Pattern Formation in Lepidoptera: A Model." *Journal of Experimental Zoology* 206: 119–136. (14)

Nijhout, H. F. (1981). "The Color Patterns of Butterflies and Moths." *Scientific American* 245: 140–151. (14)

Niklas, K. J. (1992). *Plant Biomechanics*. University of Chicago Press: Chicago. (G,12)

Niklas, K. J. (1994). *Plant Allometry*. University of Chicago Press: Chicago. (G,12)

Niklas, K. J. (1996). "How to Build a Tree." *Natural History* 2/96: 48–57. (12)

Niven, I. (1961). *Numbers, Rational and Irrational*. Random House: New York. (10)

Nonnenmacher, T. F., Losa, G. A., and Weibel, E. R. (eds.) (1994). *Fractals in Biology and Medicine*. Birkhauser Verlag: Basel. (G)

Nori, F., Sholtz, P., and Bretz, M. (1997). "Booming Sand." *Scientific American* September: 84–89. (6)

Nussenzveig, H. M. (1977). "The Theory of the Rainbow." Reprinted in *Light from the Sky* (1980; Readings from *Scientific American*), pp. 54–65. Freeman: San Francisco. (5)

Nussenzveig, H. M. (1979). "Complex Angular Momentum Theory of the Rainbow and the Glory." *Journal of the Optical Society of America* 69: 1068–1079. (5)

Nussenzveig, H. M. (1992). *Diffraction Effects in Semiclassical Scattering*. Cambridge University Press: Cambridge. (5)

Olds, C. D. (1963). *Continued Fractions*. Random House: New York. (10)

Officer, C. B. (1974). *Introduction to Theoretical Geophysics*. Springer: New York. (5)

Oprea, J. (2000). *The Mathematics of Soap Films: Explorations with Maple®*. American Mathematical Society: Providence, R.I. (11)

Packard, E. (1994). *Imagining the Universe*. Perigee: New York. (G,2)

Painter, K. J., Maini, P. K., and Othmer, H. G. (1999). "Stripe Formation in Juvenile *Pomacanthus* Explained by a Generalized Turing Mechanism with Chemotaxis." *Proceedings of the National Academy of Sciences* 96: 5549–5554. (14)

Pascuzzi, E. (1998). "The Glorious Glory." *The Physics Teacher* 36: 164–166. (5)

Paterson, A. R. (1983). *A First Course in Fluid Dynamics*. Cambridge University Press: Cambridge. (6–9)

Pedgely, D. E. (1986). "The Tertiary Rainbow." *Weather* 41: 401. (5)

Pennycuick, C. J. (1992). *Newton Rules Biology*. Oxford University Press: Oxford. (3)

Peterson, M. A. (2002). "Galileo's Discovery of Scaling Laws." *American Journal of Physics* 70: 575–581. (3)

Petrides, G. A., and Wehr, J. (1998). *The Peterson Field Guide to Eastern Trees.* Houghton Mifflin: Boston. (G,12)

Petroski, H. (1992). *To Engineer Is Human.* Vintage: New York. (G)

Pickersgill, A. O., and Hunt, G. E. (1979). "The Formation of Martian Lee Waves Generated by a Crater." *Journal of Geophysical Research* 84: 8317–8331. (6–8)

Pickersgill, A. O., and Hunt, G. E. (1981). "An Examination of the Formation of Linear Lee Waves Generated by Giant Martian Volcanoes." *Journal of the Atmospheric Sciences* 38: 40–51. (6–8)

Pinsky, M. A. (1984). *Introduction to Partial Differential Equations with Applications.* McGraw-Hill: New York. (G,14)

Pirraglia, J. A. (1976). "Martian Atmospheric Lee Waves." *Icarus* 27: 517–530. (6–8)

Polya, G. (1977). *Mathematical Methods in Science.* Mathematical Association of America: Washington, D.C. (G)

Prandtl, L. (1952). *Essentials of Fluid Dynamics.* Hafner: New York. (G,6–8)

Prescott, J. (1961). *Applied Elasticity.* Dover: New York. (12)

Prusinkiewicz, P., and Lindenmayer, A. (1990). *The Algorithmic Beauty of Plants.* Springer: New York. (10,14)

Purcell, E. M. (1977). "Life at Low Reynolds' Number." *American Journal of Physics* 45: 3–11. (3,7)

Rayleigh, J.W.S. (1880). "On the Resultant of a Large Number of Vibrations of the Same Pitch and of Arbitrary Phase." *Philosophical Magazine and Journal of Science,* 10: 73–78. (12)

Rayleigh, J.W.S. (1945). *The Theory of Sound.* (Volumes 1 and 2). Dover: New York. (4,12)

Rayner, J.M.V. (1979a). "A Vortex Theory of Animal Flight. Part I: The Vortex Wake of a Hovering Animal." *Journal of Fluid Mechanics* 91: 697–730. (13)

Rayner, J.M.V. (1979b). "A Vortex Theory of Animal Flight. Part II: The Forward Flight of Birds." *Journal of Fluid Mechanics* 91: 731–763. (13)

Reade, J. B. (1997). "Icebows." *Mathematical Gazette,* 81, No. 490, March: 3–6. (5)

Reade, J. B. (June 2003). "On the Scientific Explanation of Parhelia." *Mathematical Gazette.* (5)

Research Reporter (1971). "The Six-Sided Snowflake." *Chemistry* 44: 19–20. (5)

Rinard, P. M. (1972). "Rayleigh, Echoes, Chirps, and Culverts." *American Journal of Physics* 40: 923–924. (4,12)

Rivier, N., Occelli, R., Pantaloni, J., and Lissowdki, A. (1984). "Structure of Benard Convection Cells, Phyllotaxis and Crystallography in Cylindrical Symmetry." *Journal de Physique* 45: 49–63. (10)

Roberts, W. J. (1991). "Honeycomb Geometry: Applied Mathematics in Nature." *Applications of Secondary School Mathematics* (Readings from the *Mathematics Teacher*; J. D. Austin, ed.). National Council of Teachers of Mathematics: Reston. (11)

Rosen, J. (1998). *Symmetry Discovered.* Dover: New York. (G,10,11)

Rossotti, H. (1983). *Colour*. Princeton University Press: Princeton. (G,4)

Sassen, K., Arnott, W. P., Barnett, J. M., and Aulenbach, S. (1998). "Can Cirrus Clouds Produce Glories?" *Applied Optics* 37: 1427–1433. (5,6)

Schaefer, V. J., and Day, J. A. (1981). *Field Guide to the Atmosphere*. Houghton Mifflin: Boston. (G,4–6)

Scheidegger, A. E., and Langbein, W. B. (1966). *Probability Concepts in Geomorphology*. Geological Survey Professional Paper 500-C. U.S. Government Printing Office: Washington, D.C. (12)

Schelling, H. von (1951). "Most Frequent Particle Paths in a Plane." *Transactions of the American Geophysical Union* 32: 222–226. (12)

Schiffer, M. M., and Bowden, L. (1984). *The Role of Mathematics in Science*. Mathematical Association of America: Washington, D.C. (G)

Schmidt-Nielson, K. (1972). "Locomotion: Energy Cost of Swimming, Flying and Running." *Science* 177: 222–228. (3,13)

Schroeder, M. (1991). *Fractals, Chaos, Power Laws*. Freeman: New York. (G)

Scorer, R. S. (1978). *Environmental Aerodynamics*. Ellis Horwood: Chichester. (6–8,13)

Scorer, R. S., and Verkaik, A. (1989). *Spacious Skies*. David and Charles: London. (G,5,6)

Scorer, R. S. (1997). *Dynamics of Meteorology and Climate*. Wiley-Praxis: Chichester. (7)

Segel, L. A. (1987). *Mathematics Applied to Continuum Mechanics*. Dover: New York. (G,7–9,14)

Sharp, W. E. (1971). "An Analysis of the Laws of Stream Order for Fibonacci Drainage Patterns." *Water Resources Research* 7: 1548–1557. (12)

Shea, N. M. (1987). "Estimating the Power in the Tides." *The Physics Teacher* October: 426. (9)

Shea, N. M., and Loomis, J. (1997). "Estimating the Effective Depth of the Atmosphere Using the *Old Farmer's Almanac*." *The Physics Teacher* 35: 90. (4)

Shen, L. C., and Kong, J. A. (1995). *Applied Electromagnetism*. Prindle, Weber, Schmidt: Boston. (G,4)

Smith, N. F. (1972). "Bernoulli and Newton in Fluid Mechanics." *The Physics Teacher* 10: 451–455. (13)

Smith, S. S. (1981). *A Search for Structure*. MIT Press: Cambridge. (G,11)

Sodha, M. S., Aggarwal, A. K., and Kaw, P. K. (1967). "Image Formation by an Optically Stratified Medium: Optics of Mirage and Looming." *British Journal of Applied Physics* 18: 503–511. (5)

Stevens, P. S. (1974). *Patterns in Nature*. Atlantic-Little, Brown: Boston. (G,1,10–12)

Stewart, I., and Golubitsky, M. (1992). *Fearful Symmetry*. Blackwell: Oxford. (G,10,11)

Stewart, I. (1995a). "Daisy, Daisy, Give Me Your Answer, Do." *Scientific American* January: 96–99. (10)

Stewart, I. (1995b). *Nature's Numbers*. Basic Books: New York. (G,10,14)

Stewart, I. (1998). *Life's Other Secret*. John Wiley: New York. (10, 14)

Stewart, I. (2000). "Spiral Slime." *Scientific American* November: 116–118. (14)

Stewart, I. (2001). *What Shape Is a Snowflake?* Freeman: New York. (G,6,10,11,14)

Suzuki, S. (1958). "Aeolian Tones in a Forest and Flowing Cloudlets over a Hill." *Weather* 13: 20–25. (12)

Tan, A. (1985). "The Shape of Streamlined Tap Water Flow." *The Physics Teacher* 23: 94. (13)

Tape, W. (1982). "Folds, Pleats and Halos." *American Scientist* 70: 467–474. (5)

Tape, W. (1985). "The Topology of Mirages." *Scientific American* 252: 120–129. (5)

Tape, W. (1994). *Atmospheric Halos*. American Geophysical Union: Washington D.C. (5)

Taubes, C. H. (2001). *Modeling Differential Equations in Biology*. Prentice-Hall: Englewood Cliffs. (14)

Taylor, G. I. (1923). "Stability of a Viscous Liquid Contained between Two Rotating Cylinders." *Philosophical Transactions of the Royal Society of London* A223, 289–343. (8)

Thompson, D. W. (1992). *On Growth and Form*. Dover: New York. (G,3,10,11,13)

Thompson, G. R., and Turk, J. (1999). *Earth Science and the Environment*. Saunders: Orlando. (G)

Thurman, H. V. (1994). *Introductory Oceanography*. Macmillan: New York. (3)

Trefil, J. S. (1975). *Introduction to the Physics of Fluids and Solids*. Pergamon: Oxford. (G)

Trefil, J. S. (1983). *The Unexpected Vista*. Scribner's: New York. (G,4,5)

Trefil, J. S. (1987). *Meditations at Sunset*. Collier: New York. (G,4,5,6)

Tricker, R.A.R. (1964). *Bores, Breakers, Waves, and Wakes*. Elsevier: New York. (7,9)

Tricker, R.A.R. (1970). *Introduction to Meteorological Optics*. Elsevier: New York. (5)

Tucker, V. A. (1969). "Wave-Making by Whirligig Beetles (Gyrinidae)." *Science* 166: 897–899. (7)

Tucker, V. A. (1971). "Waves and Water Beetles." *The Physics Teacher* 9: 10–14,19. (7)

Tucker, V. A. (2000a). "Gliding Flight: Drag and Torque of a Hawk and a Falcon with Straight and Turned Heads, and a Lower Value for the Parasite Drag Coefficient." *Journal of Experimental Biology* 203: 3733–3744. (10,13)

Tucker, V. A. (2000b). "The Deep Fovea, Sideways Vision, and Spiral Flight Paths in Raptors." *Journal of Experimental Biology* 203: 3745–3754. (10,13)

Tucker, V. A., Tucker, A. E., Akers, K., and Enderson, J. H. (2000c). "Curved Flight Paths and Sideways Vision in Peregrine Falcons (*Falco Peregrinus*)." *Journal of Experimental Biology* 203: 3755–3763. (10,13)

Turcotte, D. L., and Oxburgh, E. R. (1967). "Finite Amplitude Convective Cells and Continental Drift." *Journal of Fluid Mechanics* 28: 29–42. (8,14)

Turcotte, D. L. (1997). *Fractals and Chaos in Geology and Geophysics*. Cambridge University Press: Cambridge. (G,A)

Turing, A. (1952). "The Chemical Basis of Morphogenesis." *Philosophical Transactions of the Royal Society* B 237: 37–72. (14)

Tyson, N. de Grasse (1994). *Universe Down to Earth*. Columbia University Press: New York. (2)

van de Hulst, H. C. (1947). "A Theory of the Anti-Coronae." *Journal of the Optical Society of America* 37: 16–22. (5)

van de Hulst, H. C. (1981). *Light Scattering by Small Particles*. Dover: New York. (G,5)

Velarde, M. G., and Normand, C. (1980). "Convection." *Scientific American* 243: 92–108. (8)

Vergara, W. C. (1959). *Mathematics in Everyday Things*. Harper: New York. (G)

Vincent, J. (1996). "Tricks of Nature." *New Scientist* 17 August: 38–40. (12)

Vogel, H. (1979). "A Better Way to Construct a Sunflower Head." *Mathematical Biosciences* 44: 145–174. (10)

Vogel, S. (1993). "When Leaves Save the Tree." *Natural History* 9/93: 59–62. (12)

Vogel, S. (1994). "Dealing Honestly with Diffusion." *The American Biology Teacher* 56: 405–407. (14)

Vogel, S. (1998a). "Exposing Life's Limits with Dimensionless Numbers." *Physics Today* November: 22–27. (G,12)

Vogel, S. (1998b). *Cat's Paws and Catapults*. Norton: New York. (G)

von Baeyer, H. C. (1994). *The Fermi Solution*. Penguin: London. (G,2)

von Karman, T., and Biot, M. A. (1940). *Mathematical Methods in Engineering*. McGraw-Hill: New York. (12)

Walkden, S. L. (1925). "Experimental Study of the Soaring Albatrosses." *Nature, London* 116: 132–134. (13)

Walker, D. (1950). "A Rainbow and Supernumeraries with Graduated Separations." *Weather* 5: 324–325. (5)

Walker, J. D. (1976). "Multiple Rainbows from Single Drops of Water and Other Liquids." *American Journal of Physics* 44: 421–433. (5)

Walker, J. D. (1977). *The Flying Circus of Physics (with Answers)*. Wiley: New York. (G)

Walker, J. D. (1982). "Walking on the Shore, Watching the Waves and Thinking on How They Can Shape the Beach." *Scientific American* 247: 144–148. (7)

Walker, J. D. (1980). "Billows in the Ionosphere Are Tracked with Transistor Radios." *Scientific American* 243: 232–243 (especially p. 243 on gravity waves). (6–8)

Walker, J. D. (1982). "Looking into the Ways of Water Striders, the Insects that Walk (and Run) on Water." *Scientific American* 247: 188–196. (7)

Walker, J. D. (1984). "Edge Waves Form a Spokelike Pattern When Vibrations Are Set Up in a Liquid." *Scientific American* 251: 130–138. (7)

Waltham, C. (1998) "Flight without Bernoulli." *The Physics Teacher* 36: 457–462. (13)

Wegener, P. P. (1997). *What Makes Airplanes Fly?* Springer: New York. (13)

Weltner, K. (1987). "A Comparison of Explanations of the Aerodynamic Lifting Force." *American Journal of Physics* 55: 50–54. (13)

Went, F. W. (1968). "The Size of Man." *American Scientist* 56: 400–413. (3)

Weyl, H. (1989). *Symmetry*. Princeton University Press: Princeton. (G,10,11)

Whitaker, R. J. (1974). "Physics of the Rainbow." *The Physics Teacher* 12: 283–286. (5)

Williams, J. (1997). *The Weather Book*. Vintage: New York. (G,5,6)

Wilson, I. (1972). "Sand Waves." *New Scientist* 23 March: 634–637. (6)

Winfree, A. T. (1974). "Rotating Chemical Reactions." *Scientific American* 230: 82–95. (14)

Wolfram, S. (2002). *A New Kind of Science*. Wolfram Media, Inc.: Champaign. (Preface)

Wood, E. A. (1975). *Science from Your Airplane Window*. Dover: New York. (G,4,5)

Wurtele, M. G., Sharman, R. D., and Datta, A. (1996). "Atmospheric Lee Waves." *Annual Review of Fluid Mechanics* 28: 429–476. (6–8)

Zebrowski, E., Jr. (1999). *A History of the Circle*. Rutgers University Press: New Brunswick. (G,2)

INDEX